Lecture Notes in Computer Science   11297

Commenced Publication in 1973
Founding and Former Series Editors:
Gerhard Goos, Juris Hartmanis, and Jan van Leeuwen

More information about this series at http://www.springer.com/series/7409

Anirban Mondal · Himanshu Gupta
Jaideep Srivastava · P. Krishna Reddy
D. V. L. N. Somayajulu (Eds.)

Big Data Analytics

6th International Conference, BDA 2018
Warangal, India, December 18–21, 2018
Proceedings

Springer

Editors
Anirban Mondal
Ashoka University
Sonepat, India

Himanshu Gupta
IBM Research - India
New Delhi, India

Jaideep Srivastava
University of Minnesota
Minneapolis, MN, USA

P. Krishna Reddy
IIIT
Hyderabad, India

D. V. L. N. Somayajulu
National Institute of Technology
Warangal, India

ISSN 0302-9743   ISSN 1611-3349   (electronic)
Lecture Notes in Computer Science
ISBN 978-3-030-04779-5   ISBN 978-3-030-04780-1   (eBook)
https://doi.org/10.1007/978-3-030-04780-1

Library of Congress Control Number: 2018962144

LNCS Sublibrary: SL3 – Information Systems and Applications, incl. Internet/Web, and HCI

This Springer imprint is published by the registered company Springer Nature Switzerland AG
The registered company address is: Gewerbestrasse 11, 6330 Cham, Switzerland

Preface

Big data analytics has been gradually becoming an integral aspect of decision-support systems in several important and diverse domains such as finance, health care, e-governance, agriculture, education, telecommunications, astronomy, security and surveillance, media and entertainment, smart cities and transportation, and so on. Given the ever-increasing amounts of data and the inherent heterogeneity and complexity of the data in these domains, we need to build capabilities for designing the next-generation of big data processing systems in order to perform actionable analytics.

Incidentally, new kinds of applications involving artificial intelligence, machine learning, and reinforcement learning are gradually increasing in importance for gaining valuable insights for decision-making, while considering the contextual aspect of services deployment and delivery in various domains. These applications are essentially data-driven and require the efficient management of massive, heterogeneous, and disparate data repositories to enable knowledge extraction. This necessitates innovations in the underlying Cloud computing infrastructure as well as the design and modeling of data for supporting these complex cognitive technology-based applications.

The 6th International Conference on Big Data Analytics (BDA) was held during December 18–21, 2018, at the National Institute of Technology (NIT), Warangal, Telangana State, India. The program included peer-reviewed research contributions as well as invited contributions. A view of research activity across a wide gamut of big data analytics was provided by the sessions on related topics. This volume contains the contributions of keynote speakers, invited speakers, and tutorial speakers. This volume also contains the peer-reviewed research contributions in the areas of the vision and perspective of big data analytics, financial data analytics and data streams, Web and social media data, big data systems and frameworks, predictive analytics in health care and agricultural domains, as well as machine learning and pattern mining.

Within the big data analytics framework, the conference attracted submissions under a wide gamut of diverse topics, including Web and social media, machine learning, predictive analytics, modeling big data and data streams. Furthermore, the submissions covered big data applications and frameworks in a wide variety of domains including (but not limited to) finance, health care, and agriculture. The selected papers in these proceedings, along with keynotes, invited talks, and tutorials on a variety of relevant topics, are expected to further stimulate research and exposure to cutting-edge research in the area of big data analytics.

The conference received 93 submissions. The Program Committee (PC) consisted of researchers from both academia as well as industry from nine different countries, namely, Japan, USA, Australia, India, China, Estonia, France, Canada, and Sweden. Each submission was reviewed by three to four PC members. A PC discussion phase was also incorporated to facilitate the final decision-making regarding the papers.

Based on this review process, the PC selected 12 full papers and six short papers. The overall acceptance rate was 19.3%.

We would like to extend our sincere thanks to the members of the PC and external reviewers for their time, energy, and expertise in providing support to BDA 2018. Additionally, we would like to thank all the authors who considered BDA 2018 as the forum to publish their research contributions.

The sponsoring organizations, the Steering Committee, and the Organizing Committee deserve praise for the support they provided. A number of individuals contributed to the success of the conference. We thank Prof. Divyakant Agrawal, Dr. Mukesh Mohania, and Prof. Yoshiharu Ishikawa for their insightful keynote talks.

The conference received invaluable support from NIT, Warangal. In this context, we thank Prof. N. V. Ramana Rao, Director, NIT Warangal, India. Many thanks are also extended to the faculty members of the Department of Computer Science and Engineering, National Institute of Technology, Warangal, for their constant cooperation and support.

Anirban Mondal

Himanshu Gupta

Jaideep Srivastava

P. Krishna Reddy

D. V. L. N. Somayajulu

Organization

BDA 2018 was organized by Department of Computer Science and Engineering, National Institute of Technology (NIT), Warangal, Telangana State, India.

Chief Patrons

Ajay Prakash Sawhney (Secretary)	Meity, GOI, India
N. V. Ramana Rao (Director)	NIT Warangal, India

Honorary Chairs

S. K. Gupta	IIT Delhi, India
V. Rajanna (Vice President & Regional Head)	Tata Consultancy Services, Hyderabad, India

General Chair

D. V. L. N. Somayajulu	NIT Warangal, India

Steering Committee Chair

P. Krishna Reddy	IIIT Hyderabad, India

Steering Committee

S. K. Gupta	IIT Delhi, India
Srinath Srinivasa	IIIT Bangalore, India
Krithi Ramamritham	IIT Bombay, India
Sanjay Kumar Madria	Missouri University of Science and Technology, USA
Masaru Kitsuregawa	University of Tokyo, Japan
Raj K. Bhatnagar	University of Cincinnati, USA
Vasudha Bhatnagar	University of Delhi, India
Mukesh Mohania	IBM Research, Australia
H. V. Jagadish	University of Michigan, USA
Ramesh Kumar Agrawal	Jawaharlal Nehru University, India
Divyakant Agrawal	University of California at Santa Barbara, USA
Arun Agarwal	University of Hyderabad, India
Subhash Bhalla	University of Aizu, Japan
Jaideep Srivastava	University of Minnesota, USA

Anirban Mondal   Ashoka University, India
Sharma Chakravarthy   University of Texas at Arlington, USA

Program Committee Chairs

Anirban Mondal   Ashoka University, India
Himanshu Gupta   IBM Research, India
Jaideep Srivastava   University of Minnesota, USA

Organizing Chair

R. B. V. Subramanyam   NIT Warangal, India

Finance Chair

S. Ravi Chandra   NIT Warangal, India

Sponsorship Chair

D. V. L. N. Somayajulu   NIT Warangal, India

Publication Chair

P. Krishna Reddy   IIIT Hyderabad, India

Workshop Chairs

Sanjay Chaudhary   Ahmedabad University, India
Punam Bedi   University of Delhi, India
Subhash Bhalla   University of Aizu, Japan

Tutorial Chairs

Vikram Goyal   IIIT Delhi, India
Sanjay Kumar Madria   Missouri University of Science and Technology, USA

Publicity Chairs

Shelly Sachdeva   National Institute of Technology Delhi, India
Vasudha Bhatnagar   University of Delhi, India

Panel Chairs

Naresh Manwani   IIIT Hyderabad, India
Sharma Chakravarthy   University of Texas at Arlington, USA

Website Chair

T. Ramakrishnudu   NIT Warangal, India

Organizing Committee

B. B. Amberker   Department of Computer Science and Engineering,
                                  NIT Warangal, Telangana State, India
S. G. Sanjeevi   Department of Computer Science and Engineering,
                                  NIT Warangal, Telangana State, India
K. Ramesh   Department of Computer Science and Engineering,
                                  NIT Warangal, Telangana State, India
Ch. Sudhakar   Department of Computer Science and Engineering,
                                  NIT Warangal, Telangana State, India
S. Ravi Chandra   Department of Computer Science and Engineering,
                                  NIT Warangal, Telangana State, India
Raju Bhukya   Department of Computer Science and Engineering,
                                  NIT Warangal, Telangana State, India
R. Padmavathy   Department of Computer Science and Engineering,
                                  NIT Warangal, Telangana State, India
K. V. Kadambari   Department of Computer Science and Engineering,
                                  NIT Warangal, Telangana State, India
U. S. N. Raju   Department of Computer Science and Engineering,
                                  NIT Warangal, Telangana State, India
P. V. Subba Reddy   Department of Computer Science and Engineering,
                                  NIT Warangal, Telangana State, India
Rashmi Ranjan Rout   Department of Computer Science and Engineering,
                                  NIT Warangal, Telangana State, India

Program Committee

Alok Singh   University of Hyderabad, India
Arkady Zaslavsky   CSIRO, Australia
Arnab Basu   IIM Bangalore, India
Asoke Talukder   Precision Genomics, India
Atul Singh   Fidelity Investments, India
Bharat Bhargava   Purdue University, USA
Danish Contractor   IBM Research, India
Dhaval Patel   IBM Research, USA
Dhruba Bhattacharyya   Tezpur University, India
Girish Agrawal   Jindal Global University, India
Hoang Tam Vo   IBM Research, Australia
Ladjel Bellatreche   ENSMA, France
Lili Jiang   Umea University, Sweden
Lukas Pichl   International Christian University, Japan
Naresh Manwani   IIIT Hyderabad, India

Niloy Ganguly	IIT Kharagpur, India
Philippe Fournier-Viger	Harbin Institute of Technology, China
Pradeep Kumar	IIM Lucknow, India
Prasad Pathak	FLAME University, India
Prem Jayaraman	Swinburne University of Technology, Australia
R. K. Agrawal	JNU, India
Raja Sengupta	McGill University, Canada
Rajmohan C.	IBM Research, India
Rakesh Pimplikar	IBM Research, India
Samant Saurabh	Shiv Nadar University, India
Samiulla Shaikh	IBM Research, India
Santhanagopalan Rajagopalan	IIIT Bangalore, India
Shelly Sachdeva	NIT Delhi, India
Soumyava Das	Teradata-Aster, USA
Srinath Srinivasa	IIIT, Bangalore, India
Uday Kiran	University of Tokyo, Japan
Vadlamani Ravi	IDRBT Hyderabad, India
Vasudha Bhatnagar	University of Delhi, India
Vijil Chenthamarakshan	IBM Research, USA
Vikram Goyal	IIIT Delhi, India
Rema Ananthnarayanan	IBM Research, India
Sonia Khetarpaul	Shiv Nadar University, India
Nitin Gupta	IBM Research, India
Manju Bharadwaj	University of Delhi, India
Satish Narayana Srirama	University of Tartu, Estonia
Deepak Vijaykeerthy	IBM Research, India
Manu Awasthi	Ashoka University, India
K. Shashi Prabh	ABB Research, India
Shikha Mehta	Jaypee Institute of Information Technology, India
Subhash Bhalla	University of Aizu, Japan
Ravi Kothari	Ashoka University, India
Akhil Kumar	Penn State University, USA
Aswin Kannan	IBM Research, India
Harshit Kumar	IBM Research, India
Sanjay Madria	Missouri S&T, USA
Avinash Sharma	IIIT Hyderabad, India
Manu Sood	Himachal Pradesh University, India
Anand Gupta	NSIT Delhi, India

External Reviewers

Sharanjit Kaur	Acharya Narendra Dev College, University of Delhi, India
Alo Peets	University of Tartu, Estonia
Pelle Jakovits	University of Tartu, Estonia

Mohan Liyanage	University of Tartu, Estonia
Jakob Mass	University of Tartu, Estonia
Parul Agarwal	Jaypee Institute of Information Technology, Noida, India
Anuradha Gupta	Jaypee Institute of Information Technology, Noida, India

Sponsoring Institutions

Department of Computer Science and Engineering, NIT, Warangal;
International Institute of Information Technology, Bangalore;
University of Aizu, Japan;
Indraprastha Institute of Information Technology, Delhi;
School of Computer and Information Sciences, University of Hyderabad, Hyderabad;
Department of Computer Science and Engineering,
    Indian Institute of Technology, Delhi;
Department of Computer Science, University of Delhi, India;
International Institute of Information Technology, Hyderabad.

Contents

Big Data Analytics: Vision and Perspectives

Fault Tolerant Data Stream Processing in Cooperation with OLTP Engine . . .　3
　Yoshiharu Ishikawa, Kento Sugiura, and Daiki Takao

Blockchain-Powered Big Data Analytics Platform . . . .　15
　Hoang Tam Vo, Mukesh Mohania, Dinesh Verma, and Lenin Mehedy

Humble Data Management to Big Data Analytics/Science:
A Retrospective Stroll . . . .　33
　Sharma Chakravarthy, Abhishek Santra, and Kanthi Sannappa Komar

Fusion of Game Theory and Big Data for AI Applications . . . .　55
　Praveen Paruchuri and Sujit Gujar

Financial Data Analytics and Data Streams

Distributed Financial Calculation Framework on Cloud
Computing Environment . . . .　73
　Rao Casturi and Rajshekhar Sunderraman

Testing Concept Drift Detection Technique on Data Stream . . . .　89
　Narinder Singh Punn and Sonali Agarwal

Homogenous Ensemble of Time-Series Models for Indian Stock Market . . . .　100
　Sourabh Yadav and Nonita Sharma

Improving Time Series Forecasting Using Mathematical and Deep
Learning Models . . . .　115
　Mohit Gupta, Ayushi Asthana, Nishant Joshi, and Pulkit Mehndiratta

Emerging Technologies and Opportunities for Innovation in Financial
Data Analytics: A Perspective . . . .　126
　Anirban Mondal and Atul Singh

Web and Social Media Data

Design of the Cogno Web Observatory for Characterizing Online
Social Cognition . . . .　139
　Srinath Srinivasa and Raksha Pavagada Subbanarasimha

Automated Credibility Assessment of Web Page Based on Genre 155
   Shriyansh Agrawal, S. Lalit Mohan, and Y. Raghu Reddy

CbI: Improving Credibility of User-Generated Content on Facebook 170
   Sonu Gupta, Shelly Sachdeva, Prateek Dewan,
   and Ponnurangam Kumaraguru

A Parallel Approach to Detect Communities in Evolving Networks 188
   Keshab Nath and Swarup Roy

Modeling Sparse and Evolving Data 204
   Shivani Batra, Shelly Sachdeva, Aayushi Bansal, and Suyash Bansal

Big Data Systems and Frameworks

Polystore Data Management Systems for Managing Scientific Data-sets
in Big Data Archives. 217
   Rashmi Girirajkumar Patidar, Shashank Shrestha, and Subhash Bhalla

MPP SQL Query Optimization with RTCG 228
   K. T. Sridhar, M. A. Sakkeer, Shiju Andrews, and Jimson Johnson

Big Data Analytics Framework for Spatial Data 250
   Purnima Shah and Sanjay Chaudhary

An Ingestion Based Analytics Framework for Complex Event Processing
Engine in Internet of Things. 266
   Sanket Mishra, Mohit Jain, B. Siva Naga Sasank, and Chittaranjan Hota

An Energy-Efficient Greedy MapReduce Scheduler for Heterogeneous
Hadoop YARN Cluster 282
   Vaibhav Pandey and Poonam Saini

Predictive Analytics in Healthcare and Agricultural Domains

Analysis of Narcolepsy Based on Single-Channel EEG Signals. 295
   Jialin Wang, Yanchun Zhang, and Qinying Ma

Formal Methods, Artificial Intelligence, Big-Data Analytics,
and Knowledge Engineering in Medical Care to Reduce Disease
Burden and Health Disparities 307
   Sakthi Ganesh and Asoke K. Talukder

Adaboost.RT Based Soil N-P-K Prediction Model for Soil and Crop
Specific Data: A Predictive Modelling Approach. 322
   Rashmi Priya and Dharavath Ramesh

Machine Learning and Pattern Mining

Deep Neural Network Based Image Captioning.   335
  Anurag Tripathi, Siddharth Srivastava, and Ravi Kothari

Oversample Based Large Scale Support Vector Machine for Online
Class Imbalance Problem.   348
  D. Himaja, T. Maruthi Padmaja, and P. Radha Krishna

Using Crowd Sourced Data for Music Mood Classification   363
  Ashish Kumar Patel, Satyendra Singh Chouhan, and Rajdeep Niyogi

Applying Big Data Intelligence for Real Time Machine Fault Prediction   376
  Amrit Pal and Manish Kumar

PRISMO: Priority Based Spam Detection Using Multi Optimization   392
  Mohit Agrawal and R. Leela Velusamy

Malware Detection Using Machine Learning and Deep Learning.   402
  Hemant Rathore, Swati Agarwal, Sanjay K. Sahay, and Mohit Sewak

Spatial Co-location Pattern Mining   412
  Venkata M. V. Gunturi

Author Index   423

Big Data Analytics: Vision and Perspectives

Fault Tolerant Data Stream Processing in Cooperation with OLTP Engine

Yoshiharu Ishikawa$^{(\boxtimes)}$, Kento Sugiura, and Daiki Takao

Graduate School of Informatics, Nagoya University, Nagoya, Japan
ishikawa@i.nagoya-u.ac.jp,
{sugiura,takao}@db.is.i.nagoya-u.ac.jp

Abstract. In recent years, with the increase of big data and the spread of IoT technology and the continual evolution of hardware technology, the demand for *data stream processing* is further increased. Meanwhile, in the field of database systems, a new demand for *HTAP* (*hybrid transactional and analytical processing*) that integrates the functions of on-line transaction processing (OLTP) and on-line analytical processing (OLAP) is emerging. Based on this background, our group started a new project to develop data stream processing technologies in the HTAP environment in cooperation with other research groups in Japan. Our main focus is to develop new data stream processing methodologies such as fault tolerance in cooperation with the OLAP engine. In this paper, we describe the background, the objectives and the issues of the research.

Keywords: Data stream processing · Fault tolerance
Query processing · OLTP · HTAP

1 Introduction

Over the past 20 years, data stream processing has been one of the important research topics in the research field of databases [1,5,12,16,20,22]. A lot of new algorithms have been developed and various data stream applications and system implementation have also been done.

Today, there are requirements for new technology development in data stream processing as follows:

– *Big data and IoT*: Processing of big data is an important problem not only in the database field but also widely in the research and development of computer systems. *3V* (volume, variety, and velocity) are often mentioned as the features of big data, of which volume and velocity are particularly related to data stream processing. For processing data in an environment where a large amount of data arrives at a high rate, data stream-like processing is inevitably required. In addition, the explosive increase in IoT (Internet of Things) and sensor information processing has brought the situation where large amounts of data arrive continuously even though individual data is small, leading to an increase in the demand for data stream processing.

© Springer Nature Switzerland AG 2018
A. Mondal et al. (Eds.): BDA 2018, LNCS 11297, pp. 3–14, 2018.
https://doi.org/10.1007/978-3-030-04780-1_1

- *Coping with new hardware technologies*: In recent years, we are facing with rapid change in memory capacity and the spread of nonvolatile memories in addition to the remarkable increase of network speed. This trend is believed to last for the time being. Such environmental changes in hardware have a great influence not only on database system technology [3] but also on data stream processing. By using large-scale memories, it is possible to cache more data being processed on the memory, and it is possible to process richer and more complex queries. Furthermore, by utilizing a high-speed nonvolatile memory rather than a magnetic disk with very large delay, even if some data is saved temporarily in a nonvolatile memory, it is possible to continue processing without significant delay. This difference is quite large compared to the past data stream processing which had only to discard data when the load is too heavy. In different words, today's data stream processing system may be able to cope with reliable and resilient processing even in the existence of failures.
- *Fine-grained distributed and parallel processing*: Due to the advance of distributed and parallel systems and multi-core processors, it has become general to process data streams using many nodes and cores. Performance improvement is expected by distributed and parallel processing, but communication overhead among nodes and cores becomes more critical in high-level distributed and parallel environments. For this reason, fine-grained processing is required to fully exploit the advantages of such environments.
- *Fault tolerance and resilience*: Since various applications including critical tasks are processed in a stream manner in today's computing environments, reliable data management is required unlike the assumption of conventional data stream processing. It means that *fault tolerance* (or *resilience*) in data stream processing is required. For example, when a failure occurs in a node in distributed stream processing, the system needs to continue the execution even with the occurrence of a failure, and it is necessary for the node to return to the process promptly after the restoration. In addition, as described below, *state management* in data stream processing is also an important topic in recent years. A state is a set of data in an application that utilizes data stream processing and must be managed at all times in a consistent manner. When trying to execute a full-fledged application on stream-oriented data, a mild error as allowed by conventional data stream processing may not be tolerated in critical applications.

On the other hand, other new developments have also occurred in database systems in recent years. In particular, *HTAP* (*hybrid transactional and analytical processing*) [6,8,9,32] is closely related to data stream processing. Large and high-rate database updates need to be processed at high speed as *on-line transaction processing* (*OLTP*). Meanwhile, the importance of *on-line analitical processing* (*OLAP*), which analyzes data and extracts useful information, is increasing more and more in order to effectively utilize the large amount of data accumulated in the database by OLTP. However, since OLTP and OLAP have quite different properties, it is not easy to execute them at the same time. As a result, HTAP has attracted attention as a new technology, and research has been actively carried out in recent years.

In the assumption of HTAP systems, since data stream-like processing also occurs, coordination of data stream processing and an HTAP system is required. Since data stream processing systems (DSMSs) have been conventionally thought of as independent from database systems, it is a challenge to construct an architecture to cooperate with an HTAP system and to develop data stream processing techniques suitable for such an environment.

Based on this background, our group has just started a project on the data stream processing technology in an HTAP environment jointly with other research groups in Japan. In particular, the followings are important issues in the project:

- Support for fault tolerance and state management that can withstand high reliability requirements in HTAP environments
- Flexible and efficient query processing in the cooperation environment
- Tight coordination between data stream processing and OLAP processing in an HTAP system

The following parts of this paper is organized as follows. In Sect. 2, we describe the trend of fault tolerance and state management support in data stream processing and present open research issues. In Sect. 3, we cover the issues of flexible and efficient query processing. In particular, we focus on approximate query processing and processing of uncertain/probabilistic data streams. In Sect. 4, we mention the notion of an HTAP system and its influence to data stream processing. Section 5 describes our projects, and objectives and issues in the project. Finally, Sect. 6 concludes the paper.

2   Fault-Tolerance and State Management in Data Stream Processing

2.1   Motivation

Today, since big data is processed at a high rate like a data stream, big data processing is treated in a stream-like fashion even in a database. For this reason, applications requiring both a database system function and a data stream processing system function are emerging in today's applications. To cope with the problem, there are basically two approaches:

- To incorporate the data stream processing function in a database system.
- To enhance the facility of a data stream processing system for supporting reliable data management as in a database system. Reliability is one of the requirements for data stream management systems [35].

In this section, we focus on the second approach and show the existing ideas.

2.2 Fault Tolerance in Data Stream Processing

In data stream processing, reliability in processing means that the system supports *fault tolerance* (or *resiliency*). For supporting fault tolerance, there exist several approaches:

- *Replication of data streams*: In this approach, an input data stream is duplicated and processed at multiple nodes. Even when a failure occurs in a node, processing can be continued at other nodes, thus it is effective in terms of fault tolerance support. In the approach called *active standby* [7,34], a backup node performs same computation as the original node and assures reliable processing every time. In another approach called *passive standby*, periodical backups are performed on the backup node [13,24–26]. Although these approaches are effective in terms of resiliency, the cost of duplicated calculation is not always acceptable.
- *Upstream backup and replay*: In *upstream backup*, we temporarily hold incoming data stream items at a upstream node preceding the failure node [4,17,28,38]. In a related approach called the *replay* method, we keep the input data stream item in some reliable storage or some node. In each method, recomputation is performed when a node recovers from the failure. The approach as two problems: (1) the overhead of the storage may not be tolerable depending on the situations, and (2) if the downtime is long, data stream processing may not be able to meet the required service level.
- *Backup-based approach*: In this approach, backup operations are performed for the possible restoration while the process of data stream processing. Basically, there are two approaches, *periodical checkpointing* and *on-demand checkpointing*. For example, the MillWheel system [4] performs periodical checkpointing in the default setting. In the on-demand checkpointing approach, the checkpointing operations are more flexible. In addition, we have several options how to backup the data stream items: non-volatile memories would be a better option compared to HDDs, but we may able to use other nodes for storing backup data.

2.3 State Management in Data Stream Processing

In data stream processing, data is often monitored and queried by incrementally managing *states*. A state corresponds to the data used in an application and the aggregated information which is calculated incrementally. Although it is considered that the state in the conventional data stream processing is not necessarily reliable, proper management of states is becoming more important because today's data stream applications have to deal with more critical situations. Stable management of states shares a common motive with the support of fault tolerance mentioned above, but it comes from the viewpoint of the application side.

Technically, there is tight relationship in state management and fault tolerance, so that the approach as described in the previous section can be applied to the state management. As applications become more complex, new system functions to support higher-level state management would be required.

2.4   Literature Survey

One approach to support fault tolerance is *best-effort fault tolerance*, which is used such as in Storm [36]. This approach tries to support fault tolerance as much as possible, but ignores missing items or lost states if resources are not enough [27]. Although it is efficient, it is not sufficient for building a reliable application because it is not easy to know how much error exists.

Chandramouli et al. [11] compare three methods for fault tolerance support (replay, periodic checkpoint, and on-demand checkpoint). Their contribution is a proposal of a cost model based on NIC bandwidth, and the performance evaluation based on the model is performed. The final conclusion is that the optimal method depends on the underlying condition.

Huang et al. propose the notion of *approximate fault tolerance* for supporting fault tolerance efficiently in the distributed stream processing setting [23]. Since backing up all states and unprocessed data in data stream processing requires a great deal of cost, backups are omitted as long as the required level of reliability is satisfied. In this work, they assume that stream processing is represented as operator sequences. Each node (worker) corresponds to one operator, and each node has its internal state and unprocessed data; backups are performed in per-node basis. Their idea is that they perform backup only when the difference between the ideal value (it is the accurate value when failures do not occur) and the current value exceeds the given threshold. In this sense, their approach belongs to the on-demand checkpoint approach mentioned above.

Fernandez et al. [17] show an approach for scaling out state management for all operators. They perform state decomposition and utilize checkpointing. Apache Samza is a data stream management system that supports scalable state management [31]. It employs a modern system architecture based on the existing research and development efforts. The basic idea is to take the approach of keeping the status of each task in a local DB. In addition, change log (differences between states) for the recovery purpose is registered into a pub-sub system so that efficient recovery on the failure is possible using the change log. IBM Streams manages states based on the pipelined checkpointing method [15]. They propose the two-phase snapshotting protocol to construct a global snapshot in a distributed computing environment. In addition, they provide language abstractions for selective fault tolerance. The state management method in Apache Flink is described in [10]. In the pipelined data stream processing, by managing the state of the application progressively with the lightweight and consistent distributed snapshot mechanism, they support continual execution without significant degradation on failure.

2.5   Open Problems

There are following open problems in supporting fault tolerance:

- *Coping with new hardware technologies*: The progress of high-speed and large-capacity nonvolatile memories would have a positive influence for maintaining fault tolerance. However, there are pros and cons. Since it is possible to load a lot of data on the memory, complex queries can be performed at higher speed. However, it means that the information that needs to be restored when a failure occurs is more complicated and more larger, and it makes system management more difficult. In addition, since the number of nodes in distributed processing and the number of cores in CPUs are further increased in the future, communication overhead during processing would become more critical.

- *Cooperation with DBMS*: When backing up data stream processing, we have an option to write the data into a database system instead of a file system. Such an approach is utilized, for example, in Apache Samza [31]. Since there exists high-speed DBMSs utilizing large-scale memories such as in-memory DBMSs, this is a practical option for the system construction. In this case, the persistence of the backup data is supported by the function of the DBMS.

3 Efficient and Flexible Data Stream Query Processing

To support more complicated data stream processing requirements, further enhancements to existing data stream processing are necessary. In this section, we briefly describe two possible directions.

3.1 Approximate Query Processing for Data Stream Processing

Approximate query processing (AQP) [29] utilizes some kind of approximation in query processing, but allows efficient query processing. AQP is gained interests in terms of the following two perspectives:

- *Supporting interactive queries* [19]: Considering the interaction with the user in database query processing, data analysis, and interactive visualization, it is desirable to present rough results in a short time. Since there is a tradeoff between the accuracy of the results and the response time, it is required to utilize data structures that are as accurate as possible and that can summarizes the data efficiently.
- *Query processing in data stream processing*: This is related with our interest in this paper. In data stream processing, it is necessary to respond to continuous queries for high-speed streams under limited computing resources. Many researches have been conducted on this problem, mainly in the field of algorithms [21].

One commonly used in stream processing is *data synopses* [2,14,30]. It approximates the information of large amounts of data with a composite data structure and there are various types of synopses such as sampling, histograms, wavelets, sketches.

Approximate query processing has been extensively studied in data stream processing in two decades, but there are still remaining problems. In particular, it is considered important to effectively utilize AQP in fault tolerance support mentioned above. There seems to be great possibilities for combining AQP with the idea of approximate fault-tolerance as described in [23].

3.2   Query Processing on Uncertain/Probabilistic Data Streams

Regarding data streams, many research efforts have been done on query process-ing in *uncertain data streams* and *probabilistic data streams* [33,37,39]. Data obtained from sensors, IoT devices, the results of RFID tag recognition, etc. contain noise and errors and have ambiguity. Also, in secondary data obtained by applying a process such as machine learning to sensor data, probability infor-mation may be attached to the data items. Since uncertain and probabilistic data streams also tend to increase, their treatment in the context of the HTAP environment can become an important research issue.

In our project, we consider to combine uncertain/probabilistic data stream processing with the implementation technology concerning fault tolerance. For that purpose, since many techniques are in common with the above-mentioned approach to AQP, we believe that it is possible to construct an integrated frame-work including AQP.

4   HTAP Technology and Data Stream Management

Conventionally, OLTP systems, which are mainly constructed for transaction processing, and OLAP systems, which are targetted for high-level analysis, have been constructed as separate systems. One reason is that the system resource requirements of both systems are significantly different. However, demands for analysis processing in the real world are changing so as to seek more real-time properties. As a result, an integrated system concept called HTAP [6,8,9,32] is born, and prototype systems like IBM Wildfire [8,9] has already been con-structed.

Integration of an HTAP system and data stream processing is an impor-tant research topic. In an HTAP system that processes high-speed data, data processing is performed at the near rate of data stream processing. Although the analysis in an HTAP system performs discovery of new knowledge from data flowing in real time, the process is highly related with event detection in data streams. Considering this point, cooperation of data stream processing and an HTAP system is not only a big technical challenge but also significant in real-world applications. In particular, the OLTP subsystem in an HTAP system handles on-line data items, and there is a possibility of conflict with data stream processing.

Cooperation between transaction processing and data stream processing is one of the challenging issues in this scenario. In particular, there are many prob-lems such as how to share the data being processed in an OLTP subsystem with the data stream processing and how to handle a data stream in the situation where a transaction processing which is related to the data stream is ongoing. For this line of the research, there are existing research/development efforts such as S-Store [28], but there are still remaining issues to be developed in the systems especially in the HTAP context.

5   Objectives and Research Issues of Our Project

In this section, we describe the objectives and the research issues of our project.

5.1   Objectives

Cooperation of Data Stream Processing and OLTP. In the conventional data stream processing, processing of the target data stream(s) is performed alone in many cases, and cooperation with OLTP has not been considered. While OLTP processes data at high speed while guaranteeing persistence and consistency, in data stream processing, it aims to process data arriving in a stream in real time, the reliability and consistency of processing have not been important issues.

However, due to further increases in big data and rapid expansion of IoT and sensor information processing, enormous amounts of data are obtained from time to time, real-time nature is also required for processing and analysis. As one example of cooperation between OLTP and data stream processing, we can consider the situation in which the data that is updated every moment by OLTP is immediately reflected in the processing/analysis like a data stream, and the result is presented in real time. As another example, the data obtained and processed by a data stream processing part can be transferred to the OLTP system in real time for updating the underlying database. That is, in cooperation of OLTP and data stream processing, flexibility to cope with various cooperation requests is required.

Fault Tolerance in Data Stream Processing. Cooperation between OLTP and data stream processing has another viewpoint on how much fault tolerance is guaranteed in data stream processing. Fault tolerance is to withstand system failures and to ensure certain reliability, availability, and consistency. In the conventional data stream processing, the best effort processing is the principle, and correspondence to the system failure has not been generally assumed. However, in the situation of close cooperation between OLTP and data stream processing, requests such as consistency on the OLTP side can spread to data stream processing. For example, when changes in the status of transactions in OLTP change, the correcting data stream processing should reflect the changes. In addition, as a middle of them, cooperation that allows approximate fault tolerance in data stream processing may be required. For example, when integrating OLAP with data stream processing, it would be acceptable to allow a range of error to some extent. By considering approximate fault tolerance, it would be possible to improve the efficiency of data stream processing.

State Management in Data Stream Processing. One of the further related issues is state management in data stream processing. A state is a set of data that we want to reliably manage for a certain period of time and it is necessary for processing and analysis of data streams. Even when a system failure occurs, the state should be managed for recovering the application context on the restart time. State management is closely related to cooperation between OLTP and data stream processing. By providing the state management function in the data stream processing system, it would become more easier to construct an application that combines OLTP and data stream features.

Under the above background, we will study the cooperation method between OLTP and data stream processing in our project, and develop a prototype system based on it. Specific objectives to the above-mentioned subjects are as follows:

1. *Development of cooperation architecture between OLTP and data stream processing*: We need to clarify the functional requirements, the application interfaces, etc. for cooperation between OLTP and data stream processing, and construct a cooperation architecture. It is necessary to consider a flexible cooperation method to process to various cooperation-based requests.
2. *Fault tolerance in data stream processing*: On the premise of cooperation with OLTP, it is quite important to effectively utilize the features of the OLTP system. Corresponding technology development is required according to the cooperation scenario between OLTP and data stream processing. In particular, the development of approximate fault tolerance models and their implementation techniques are important for enhancing cooperation flexibility and efficiency.
3. *Development of state model and state management method*: In order to realize the state management function, we need to enhance the existing solutions for state management in data stream processing. Cooperation with OLTP is also the key point here.
4. *Flexible and efficient query processing on the cooperation architecture*: On the constructed architecture, we have to develop new techniques on AQP and uncertain/probabilistic stream processing.

An abstract image of the data stream managemnt and processing system is illustrated in Fig. 1. The research of our group focuses on data stream management and processing system. The OLTP engine is constructed by collaborators in Japan and its features are are (1) it is a highly distributed and parallel system, and (2) the system is targeted for the near-future hardware technology, etc.

5.2   Research Issues

Cooperation between data stream processing and OLTP is an emerging topic and the analysis of the issue and the existing technologies are not still enough. We think that a detailed analysis of the problem is indispensable for our project.

Item 1 above (development of architecture) is the basis of the project. The key points are effective use of the underlying OLTP engine and tight integration of transaction processing feature in OLTP in the data stream processing system. Since the data stream management and processing system needs close cooperation with the OLTP engine, it would a different approach from the conventional approach of realizing data stream processing engines, and new technologies must be developed.

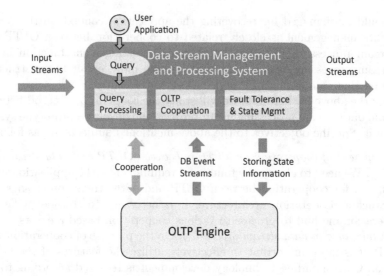

Fig. 1. System architecture

Items 2 (fault tolerance), 3 (state management), and 4 (flexible and efficient query processing) are features constructed over the cooperation architecture and it naturally requires new technologies. Fault tolerance and state management in data stream processing have been addressed in several studies so far, however, tight cooperation with the OLTP system was not performed well.

6   Conclusions

In this paper, we explained the motivation, purpose, and objectives of the new project just started in our research group. Construction of data stream processing features in the HTAP context is the main subject of our research. In particular, cooperation between fault tolerance support in data stream processing and the OLTP function in an HTAP system is a major issue. Over the next few years, our group plans to work on this project. There are important issues other than those mentioned in this paper. For example, automate tuning for dynamic environmental changes by self-reguration system technology [18] would be an important issue to be incorporated in our project.

Acknowledgments. This paper is based on results obtained from a project commissioned by the New Energy and Industrial Technology Development Organization (NEDO) and a project supported by JSPS KAKENHI Grant Number 16H01722.

References

1. Aggarwal, C.C. (ed.): Data Streams: Models and Algorithms, vol. 31. Springer, Heidelberg (2006). https://doi.org/10.1007/978-0-387-47534-9
2. Aggarwal, C.C., Yu, P.S.: A survey of synopsis construction in data streams. In: Aggarwal, C.C. (ed.) Data Streams. Advances in Database Systems, vol. 31, pp. 169–207. Springer, Boston (2007). https://doi.org/10.1007/978-0-387-47534-9_9
3. Ailamaki, A., Liarou, E., Tözün, P., Porobic, D., Psaroudakis, I.: Databases on Modern Hardware. Synthesis Lectures on Data Management. Morgan & Claypool, San Rafael (2017)
4. Akidau, T., et al.: MillWheel: fault-tolerant stream processing at internet scale. PVLDB **6**(11), 1033–1044 (2013)
5. Andrade, H.C.M., Gedik, B., Turaga, D.S.: Fundamentals of Stream Processing. Cambridge University Press, New York (2014)
6. Appuswamy, R., Karpathiotakis, M., Porobic, D., Ailamaki, A.: The case for heterogeneous HTAP. In: CIDR (2017)
7. Balazinska, M., Balakrishnan, H., Madden, S., Stonebraker, M.: Fault-tolerance in the Borealis distributed stream processing system. In: SIGMOD, pp. 13–24 (2005)
8. Barber, R., et al.: Evolving databases for new-gen big data applications. In: CIDR (2017)
9. Barber, R., et al.: Wildfire: concurrent blazing data ingest and analytics. In: SIGMOD, pp. 2077–2080 (2016)
10. Carbone, P., Ewen, S., Fóra, G., Haridi, S., Richter, S., Tzoumas, K.: State management in Apache Flink: consistent stateful distributed stream processing. PVLDB **10**(12), 1718–1729 (2017)
11. Chandramouli, B., Goldstein, J.: Shrink: prescribing resiliency solutions for streaming. PVLDB **10**(5), 505–516 (2017)
12. Chaudhry, N., Shaw, K., Abdelguerfi, M. (ed.) Stream Data Management. Springer, Heidelberg (2005). https://doi.org/10.1007/b106968
13. Cherniack, M., et al.: Scalable distributed stream processing. In: CIDR (2003)
14. Cormode, G., Garofalakis, M., Haas, P.J., Jermaine, C.: Synopses for massive data: samples, histograms, wavelets, sketches. Found. Trends Databases **4**(1–3), 1–294 (2012)
15. da Silva, G.J., et al.: Consistent regions: guaranteed tuple processing in IBM streams. PVLDB **9**(13), 1341–1352 (2016)
16. Ellis, B.: Real-Time Analytics. Wiley, Indianapolis (2014)
17. Fernandez, R.C., Migliavacca, M., Kalyvianaki, E., Pietzuch, P.: Integrating scale out and fault tolerance in stream processing using operator state management. In: SIGMOD, pp. 725–736 (2013)
18. Floratou, A., Agrawal, A., Graham, B., Rao, S., Ramasamy, K.: Dhalion: self-regulating stream processing in heron. PVLDB **10**(12), 1825–1836 (2017)
19. Galakatos, A., Crotty, A., Zgraggen, E., Kraska, T., Binnig, C.: Revisiting reuse for approximate query processing. PVLDB **10**(10), 1142–1153 (2017)
20. Garofalakis, M., Gehrke, J., Rastogi, R. (eds.) Data Stream Management: Processing High-Speed Data Streams. Springer, Heidelberg (2016). https://doi.org/10.1007/978-3-540-28608-0
21. Garofalakis, M., Gibbon, P.B.: Approximate query processing: taming the terabytes! In: VLDB (tutorial) (2001)
22. Golab, L., Özsu, M.T.: Data Stream Management. Synthesis Lectures on Data Management. Morgan & Claypool, San Rafael (2010)

23. Huang, Q., Lee, P.P.C.: Toward high-performance distributed stream processing via approximate fault tolerance. PVLDB **10**(3), 73–84 (2016)
24. Hwang, J.-H., Balazinska, M., Rasin, A., Çetintemel, U., Stonebraker, M., Zdonik, S.: High-availability algorithms for distributed stream processing. In: ICDE, pp. 779–790 (2005)
25. Hwang, J.-H., Xing, Y., Cetintemel, U., Zdonik, S.: A cooperative, self-configuring high-availability solution for stream processing. In: ICDE, pp. 176–185 (2007)
26. Krishnamurthy, S., et al.: Continuous analytics over discontinuous streams. In: SIGMOD, pp. 1081–1092 (2010)
27. Kulkarni, S., et al.: Twitter heron: stream processing at scale. In: SIGMOD, pp. 239–250 (2015)
28. Meehan, J., et al.: S-Store: streaming meets transaction processing. PVLDB **8**(13), 2134–2145 (2015)
29. Mozafari, B., Niu, N.: A handbook for building an approximate query engine. IEEE Data Eng. Bull. **38**(3), 3–29 (2015)
30. Muthukrishnan, S.: Data Streams: Algorithms and Applications. Foundations and Trends in Theoretical Computer Science. Now Publishers, Delft (2005)
31. Noghabi, S.A., et al.: Stateful scalable stream processing at LinkedIn. PVLDB **10**(12), 1634–1645 (2017)
32. Özcan, F., Tian, Y., Tözün, P.: Hybrid transactional/analytical processing: a survey. In: SIGMOD (2017)
33. Ré, C., Letchner, J., Balazinksa, M., Suciu, D.: Event queries on correlated probabilistic streams. In: SIGMOD, pp. 715–728 (2008)
34. Shah, M.A., Hellerstein, J.M., Brewer, E.: Highly available, fault-tolerant, parallel dataflows. In: SIGMOD, pp. 827–838 (2004)
35. Stonebraker, M., Çetintemel, U., Zdonik, S.: The 8 requirements of real-time stream processing. SIGMOD Rec. **34**(4), 42–47 (2005)
36. Storm. http://storm.apache.org/
37. Sugiura, K., Ishikawa, Y., Sasaki, Y.: Grouping methods for pattern matching over probabilistic data streams. IEICE Trans. Inf. Syst. **E-100D**(4), 718–729 (2017)
38. Toshniwal, A.: Storm@twitter. In: SIGMOD, pp. 147–156 (2014)
39. Tran, T.T.L., Peng, L., Diao, Y., McGregor, A., Liu, A.: CLARO: modeling and processing uncertain data streams. VLDBJ **21**(5), 651–676 (2012)

Blockchain-Powered Big Data Analytics Platform

Hoang Tam Vo[1($\boxtimes$)], Mukesh Mohania[1], Dinesh Verma[2], and Lenin Mehedy[1]

[1] IBM Research Australia, Melbourne, Australia
`tam.vo@monash.edu`
[2] IBM Thomas J. Watson Research Center, Yorktown Heights, USA

Abstract. As cryptocurrencies and other business blockchain applications are becoming mainstream, the amount of transactional data as well as business contracts and documents captured within various ledgers are getting bigger and bigger. Blockchains provide enterprises and consumers with greater confidence in the integrity of the captured data. This gives rise to the new level of analytics that marries the advantages of both blockchain and big data technologies to provide trusted analysis on validated and quality big data. Blockchain-based big data is a perfect source for subsequent analytics because the big data maintained on the blockchain is both secure (i.e., tamper-proof and cannot be forged) and valuable (i.e., validated and abundant). Further, data integration and advanced analysis across on-chain and off-chain data present enterprises with even more complete business insights. In this paper, we first discuss a blockchain-based business application for micro-insurance and AI marketplaces, which render blockchain-generated big data scenarios and the opportunity to develop trusted and federated AI insights across the insurers. We then also describe the design of a blockchain-powered big data analytics platform as well as our initial steps being taken along the development of this platform.

1 Introduction

Blockchain (a.k.a. distributed ledger technology [12]) is an emerging technology that is designed to support verification-driven transaction services within a multi-party business network. Blockchain is now being used in several industry applications. A good example is in the Internet of Things (IoT) [7], where blockchain enables IoT devices to send data for inclusion in a shared transaction repository with tamper-resistant records, and enables business parties to access and supply IoT data without the need for central control and management. Recently, there has been a noticeable trend that many industries are forming consortiums and attempting to accelerate efficiency and reduce costs by using blockchain technology. For example, the insurance industry has started the "blockchain insurance industry initiative" (B3i[1]), and the transport/logistics

---

[1] https://b3i.tech/home.html.

© Springer Nature Switzerland AG 2018
A. Mondal et al. (Eds.): BDA 2018, LNCS 11297, pp. 15–32, 2018.
https://doi.org/10.1007/978-3-030-04780-1_2

industry has formed the "blockchain in transport alliance" (BiTA[2]). Specifically, they are experimenting with different technologies and platforms from such as Ethereum [5], Hyperledger Fabric [6] and Corda [3].

Compared to the emerging blockchain technology, enterprise information systems have made great strides over the last several decades toward giving business decision makers the insights they need. The technologies behind these systems, including data warehouses, federated database systems (a.k.a. enterprise information integration) and more recently, data lakes (a.k.a big data technology) are mature and well established in the marketplace. Nevertheless, today's enterprises are facing with an ever-increasing quantity and diversity of data in their organizations from both internal and external sources. With more and more of blockchain networks operating and serving business applications, the amount of quality transactional data as well as business contracts and documents captured within various ledgers are getting bigger and bigger. Blockchain technology brings improvements in all areas related to the quality of captured data including certainty about the origin of data, consensus-driven data version, immutable entries, and audit trails.

Blockchain technology gives enterprises greater confidence in the integrity of the data being collected on business distributed ledgers. Subsequent analytics on these validated data are going to find far more valuable insights than data mining on unvalidated graveyard of big data collected in data lakes of today's enterprises. Furthermore, data integration and advanced analysis across on-chain and off-chain data present enterprises with even more complete business insights. This leads to the new level of analytics that marries the advantages of both blockchain and big data technologies to provide trusted analysis on validated and quality big data, which can drastically impact the way companies around the world do business. In other words, there is a tremendous opportunity for technology companies and enterprises to develop blockchain-powered big data mining platforms. Such analytics platforms are valuable in providing the best trained AI/machine learning model built on top of distributed, transparent and immutable blockchain-generated data.

Further, in many business environments, data required for training the AI model for a given solution may be distributed across multiple organizations or multiple countries. The movement of data across organizations or countries is frequently restricted. The information may contain sensitive details which may not be allowed to be shared across organizational or national boundaries due to prevailing regulations. In other cases, regulations may permit the transfer of data, but the data may be too big to move to a central location, because the network connecting the site where the data is present, and the site where the machine learning is to be done, is relatively slow, has an unacceptable high latency in the amount of time required to transfer the data, or the cost of transferring the data may be excessive. In these cases, the only feasible way to build an AI model is to create partial AI models from each of the sites where the training data is distributed, and to combine these partial models together to create

---

[2] https://bita.studio/.

the eventual AI model that is used. This approach of creating AI models from disparate training data is federated AI. Blockchain technology can play a key part in this approach for improving trust in AI data, models, learning process and outcomes.

In the following, we first discuss a blockchain-based business application for micro-insurance and AI marketplaces, which render blockchain-generated big data scenarios and the opportunity to develop trusted and federated AI insights across insurers. Then, we describe the design of a blockchain-powered big data analytics platform for supporting this application. Finally, we also present our solutions to technical issues when developing this platform including scalable data management and analytics as well as federated AI in blockchain-based business networks.

2 The Case for AI-Driven Blockchain-Enabled Platform for Insurance Industry

In this section, we describe a novel application of blockchain technology to pay-as-you-go automotive insurance marketplace. The motivation for this application stems from the fact that in growth markets like developing countries, up to 35% of 4-wheelers and 70% of 2-wheelers are uninsured [2]. This is because of the hefty premium that one has to pay independent of mileage and driving behaviours and patterns, as well as the lack of transparency in insurance quotes. Additionally, in developed countries, drivers who have an extra car for use on weekends or road trip holidays may find it costly to pay yearly premiums. Several usage-based insurance or micro-insurance start-ups have found their way to the market, e.g., pay as you drive, pay how you drive and mile-based auto insurance. Nevertheless, there is still room for novel solutions in this space.

The above stand-alone solutions from small and medium-sized insurers are developed with various technologies and run independently. Hence, they suffer from a fragmented market and a fragmented view of customers, which potentially leads to an increase in insurance frauds. Using blockchain technology in the micro-insurance domain was demonstrated previously in [15] where each insurer developed their own software stack and driver's mobile application while only sharing data with the blockchain network. This leads to the potential of developing a real-time marketplace platform with a single driver application that collects insurance quotes from multiple insurers competing to insure a trip request from a driver while also leveraging this data economy for sharing, aggregating and analyzing telematics and IoT data to ultimately improve customer experience and business operations.

For example, an insurance company would like to identify the pot holes in the road network of a certain geographical area based on the sensor data from its customers' vehicles so that it can measure the risk of the road segment more accurately. Nevertheless, a single insurance company would face extreme difficulty in developing effective analytics if it relies only on its own customers. This is because it is less likely that it would have a customer base that would uniformly travel through all roads at various weather and environmental conditions for accurate risk analytics for those roads. On the contrary, if multiple such insurance companies agree to collaboratively integrate analytics models (i.e., federated AI) about the road segments that their customers have travelled, they would be able to build a road network analytics service which would benefit every party in the ecosystem towards offering better customer service with a more accurate actuary model.

In addition, this federated AI insight can be further enriched by other niche analytics vendors who are not part of the insurance ecosystem but can provide additional insights related to the insurance customers such as their social behavior, digital identity profile, crime database, health record, etc. These overall federated insights are beyond the reach of nowadays insurance companies (both traditional insurers and emerging insuretechs) and can help strengthen the various insurance models (such as risk, fraud, customer insights, premium computation, etc.) that are presently used. Overall, blockchain technology plays a key part in this platform for improving trust in AI data, models, learning process and outcomes. Specifically, the need for blockchain in this platform includes:

- Blockchain provides a shared distributed database and a system of records that is non-repudiable. Since the proposed platform is an entity that offers a platform for multiple parties to interact seamlessly requiring a no-repudiable system of records for conflict resolutions and settlements, blockchain is a promising solution for such a platform.
- Blockchain enables sharing of data (and AI models) in trusted, accountable, verifiable & auditable fashion where asset owner should be able to author policies with different level of accesses for data sharing and analytics on data.
- Blockchain will also allow track and generate data usage reports as well as track failed data accesses. It can also enable building a data marketplace.
- Blockchain is used as policy enforcer, provenance store and associated analytics. No sharing of data or AI models is possible without verification of policy and access control on blockchain; all access is recorded on blockchain for lineage.
- Meta data on blockchain also certifies data and code for data user as proof of untampered data and model.

3   Architecture of a Blockchain-Powered Big Data Platform

Figure 1 illustrates the overall architecture of a blockchain-powered big data platform. The platform provides data storage and analytics capabilities for supporting three marketplaces:

- Insurance marketplace: This insurance marketplace provides a gateway for customers to request for insurance quotes from multiple insurers and submit an insurance claim to a specific insurer.

- AI marketplace: This AI marketplace gateway allows analytics companies to first build analytics models and then provide these models as analytics services for consumption by insurers. These models can be built using the federated AI approach integrating insights and machine learning models across insurers and non-insurance data providers.
- Value-added service marketplace: This value-added service marketplace provides a gateway for service companies, e.g., car maintenance or roadside assistance, to register and provide their services for insurers and insurance customers.

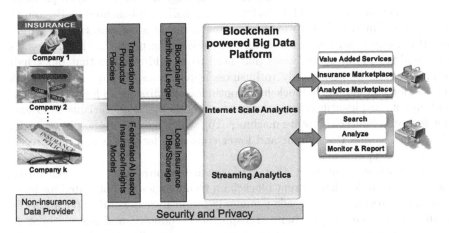

Fig. 1. Blockchain-powered big data platform for micro-insurance and AI marketplaces.

This data platform runs on top of the underlying blockchain and private storages at each participant. Specifically, this data platform enables sharing of data (and AI models) between participants in a trusted, privacy-preserving, accountable, verifiable and auditable fashion.

- Blockchain: This blockchain is the main storage for tamper-proof and auditable data. As there are three marketplaces in this architecture, each of the marketplace could run their own blockchain network where the companies joining these marketplaces can play a role as a peer node in the blockchain network.
- Private storage: Due to confidentiality concern and limited storage capacity of blockchain, the data platform can be designed to use off-chain storage for maintaining actual data whereas the hash of these actual data is stored on the blockchain. This private storage can be implemented either as a centralized component managed by the technology provider or as a decentralized component located at each peer node joining the blockchain network.

The platform also provides an execution environment (e.g., Hadoop, Spark etc.) for two purposes. First, analytics companies use this execution environment to build analytics models based on the data shared on the data platform. Second, insurance companies consume and execute the analytics models offered by these analytics vendors via this analytics platform.

4   Scalable Blockchain Data Management

In the implemented insurance marketplace, a growing number of customers entails an increasing transactional workload for recording insurance policy and trip information, which is challenging to match by today's blockchain platforms. Limited transaction throughput and storage are widely-known problems of blockchain technology. The blockchain originally implemented in Bitcoin [1] uses a "Proof of Work" (PoW) consensus method that is computationally expensive (by design) and requires solving a cryptographic puzzle in the process [8]. This leads to performance issues that need to be addressed so that blockchain technology can be more ready to business applications. The quest for scalable transaction throughput in blockchain continues. Permissioned blockchains where participants are identified use consensus methods based on variants of Byzantine fault-tolerant (BFT) state machines [16], which have been chosen to provide higher transaction throughput and lower consensus latency as shown in recent benchmarks [9].

Nevertheless, even with that performance improvement, the benchmark paper [9] concludes that current blockchain technologies are not suited for large-scale data processing workloads commonly found in real-world applications for finance, insurance, supply chain, transportation industry, and many others. As such, what is needed is a novel mechanism for expanding the throughput of a blockchain, e.g., implicit consensus [11] and sharding consensus [10] that provide high levels of transaction throughput for blockchain. Hence, we develop a method to partition data across multiple blockchains in order to support high transaction throughput and storage capacity while maintaining trust, accountability and traceability.

4.1   Technical Challenges

Data partitioning techniques are commonly employed for conventional databases to address data scalability, data isolation and transaction throughput issues. Nevertheless, none of the current blockchain platforms have proposed any viable solution to scale transaction throughput via partitioning application data across multiple blockchains. This is because data partitioning is not a trivial solution for blockchain based applications and there are several technical challenges that need to be solved [14].

Decentralised and Trusted Routing of Transactions. A mechanism is needed to route transactions based on rules that are trusted by every party in the network. If we adopt a centralized router, the trustworthiness of the whole system depends on this single entity. Therefore, we propose to use another blockchain called "Master chain" to persist and track changes in the partition rules. Once the policy is stored in the "master chain", nodes in the master chain will then be able to route transactions to other different blockchain networks based on the partition rules stored in the master chain.

Handling "Cross-Chain" Transactions. Once the data is partitioned, there may be transactions involving data in several different blockchains, which we call "cross-chain" transactions. To solve this issue, one option is to migrate data from several chains to a particular chain and then execute the transaction on that chain. This approach, however, is not a feasible solution since that would introduce unnecessary delay and instability in the network. Hence, we propose a separate blockchain network called "Mixed chain", which handles "cross-chain" transactions by accessing data from different blockchains and create transactions involving those data separately.

Efficient Retrieval of Transactional Data. Since the data is now partitioned across multiple blockchains, it is critical to devise a mechanism to efficiently retrieve data from these blockchains, especially for queries that are not based on the partitioning attributes. This requires building indexes for efficient query processing.

Modifying Blockchain Node to Perform Routing. The actual routing module, although it uses the partition rules stored in the master chain, needs to be trusted. Therefore, we integrate this routing module with the blockchain node software by extending a special type of smart contract called "system chaincode" as available in Hyperledger Fabric version 1.0. Such "system chaincode" is an integrated part of blockchain software run on a Hyperledger Fabric peer node, and is designed to have system level access, which otherwise is not available to traditional smart contracts that are supposed to run in a sandbox.

4.2   Scaling Blockchain Through Data Partitioning Technique

Figure 2 illustrates the architecture of the scalable blockchain platform that we have developed to address blockchain scalability issues. The platform partitions data across multiple blockchains in order to support data partitioning and cross-chain transactions while maintaining trust, accountability and traceability. In the following, we describe multiple types of sub-blockchains that constitute the entire solution for a scalable blockchain platform.

Master Chain. This blockchain is responsible for three main tasks that are implemented via three corresponding smart contracts.

- "Partitioner" maintains a consensus among parties on a partitioning rule that determines how the entire data domain can be partitioned, e.g., range boundaries of data maintained in each partition of a data domain.
- "Transaction router" routes the incoming transactions to the appropriate partitions that are responsible (according to the partitioning rules) to process these transactions.

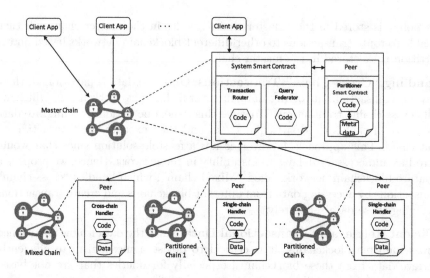

Fig. 2. Scaling blockchain throughput and storage.

– "Query federator" is responsible for handling client queries that retrieve data from one or multiple partitions.

Whereas the "partitioner" is a typical smart contract running inside each peer node in the master chain, the "transaction router" and "query federator" smart contracts are special system smart contracts that have system level capability to access various other software components in the host node, which is however not available to a usual smart contract that runs inside a sandbox.

Multiple Partitioned Chains. These are traditional blockchains, which will maintain different partitions of the entire data domain without knowing the full domain. Each partitioned chain runs a "Single-chain Handler" smart contract that receives transaction and query requests from the master chain and executes these requests.

Mixed Chain. This special blockchain in the system is named after its main responsibility to handle transactions that access data across multiple partitions. This chain runs a "Cross-chain Handler" smart contract that receives cross-partition transaction requests from the master chain and executes these requests.

4.3 Transaction Processing Workflow

Initialization and Setting up Partitioning Rule. In order for the "master chain" to be able to route transactions and queries, it needs to be initialized with information about the endpoints of other chains (i.e., "mixed chain" and "partitioned chains"), as well as partitioning rules which defines how the application data should be split across multiple "partitioned chains". The consensus on partitioning rules between participants can be handled in two ways. In the "off-chain consensus" approach, the participants agree upon partitioning rules based on an off-chain process, then one trusted/delegated participant write this information into the "master chain" via its "partitioner" smart contract. This process is repeated every time there is any change to the partitioning rules.

On the contrary, in the "on-chain consensus" approach, the process for handling every modification to partitioning rules is as follows. First, one participant proposes a change to the partitioning rules by invoking a function in the "partitioner" smart contract. Then, other participants approve or reject the change proposal by invoking corresponding functions in the "partitioner" smart contract. Finally, the "partitioner" smart contract collects sufficient votes for approvals/rejections prior to making the update to the partitioning rules into effect.

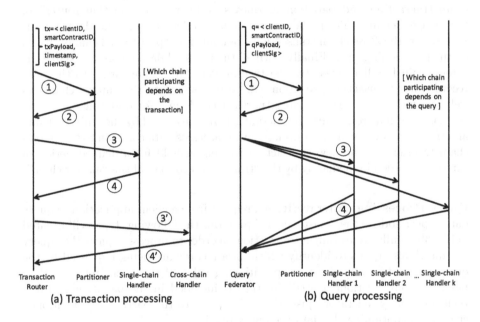

Fig. 3. Transaction and query processing flows.

Transaction Processing in "Master Chain". After receiving a request from a client application to execute a transaction, a peer node in the "master chain" performs the following steps as illustrated in Fig. 3a.

The "transaction router" running in this peer node first identifies the data records to be accessed by this transaction based upon the transaction request. The "transaction router" also looks up the partitioning rules maintained by the "partitioner" smart contract. Note that these partitioning rules can be cached at the "transaction router" for performance optimization purposes so that the "transaction router" does not need to query the "partitioner" smart contract for processing every transaction. This cache can be invalidated and updated when there is any change to the partitioning rules maintained by the "partitioner".

Given the above discussion of the data records to be accessed and the partitioning rules, the "transaction router" then decides which "partitioned chain" the transaction should be forwarded to. In particular, if a transaction only accesses data records within a single partition, then it will be routed to that particular "partitioned chain". Alternatively, transactions that access data records located in multiple partitions will be routed to the "mixed chain" for processing.

Transaction Processing in "Mixed Chain". The workflow to process a transaction in the "mixed chain" is as follows. First, the "cross-chain handler" smart contract of the "mixed chain" receives the information about the cross-chain transaction and partitioning rules sent by the "transaction router" of the "master chain". Then, the "cross-chain handler" fetches the data records accessed by the cross-chain transaction from multiple "partitioned chains" based on the partitioning rules. Finally, with all the required data records available the "cross-chain handler" executes the transaction. Any modification to the data records after transaction execution will be persisted into the "mixed chain" as well as sent back to corresponding "partitioned chains" for updating.

Note that if a cross-chain transaction is being processed by the "mixed chain" and there is a new incoming transaction that accesses data common to that cross-chain transaction, the new transaction will be on hold for a defined period of time or discarded immediately by the "transaction router" of the "master chain".

Query Processing. After receiving a request from a client application to query data, a peer node in the "master chain" performs the following steps as illustrated in Fig. 3b. Similar to routing a transaction, in order to process a query the "query federator" also needs to identify which data records are being requested by the query parameters as well as the partitioning rules. If the query retrieves data based on the partitioning attribute that is included in the partitioning rules, then the "query federator" just routes the query to the particular "partitioned chain(s)" maintaining the data the query requires.

In contrast, if the query retrieves data based on non-partitioning attributes, the "query federator" has no idea which "partitioned chain" is responsible for the queried data. In this case, the "query federator" needs to forward the query to every "partitioned chain" and aggregate the results returned by these chains prior to returning the aggregated query results to the client.

5   Introducing Federated AI for Insurance Marketplace Through Blockchain

We have presented techniques for scaling blockchain systems to support high volume of transactions in pay-as-you-go automotive insurance application. In this section, we describe a federated AI technique that enables insurance vendors to collaboratively enhance various insurance models. Federated AI technique can prove useful in a variety of contexts.

Many insurance companies are a federation of many different agencies, which may not always be able to share information freely across all agencies due to regulatory or network performance reasons. Individual data for some types of insurance may usually not be shareable across national or state boundaries, resulting in a need to train models looking at those data sources within the boundary that is permitted, and then share the models across the boundary using federated learning. In some cases, the insurance company may be working with other suppliers or subcontractors for its information, as an example as re-insurance company working with several primary mortgage insurance service providers. If the reinsurance company wants to build a common AI model for risky loans, it can use federation AI to avoid moving large volumes of data from the primary insurance provides to its own site.

If an insurance company were to track and provide insurance for risky behavior of automobile drivers, or any other insurance which operates on analyzing the personal behavior of an individual, it may have to collect a large amount of personalized information for each individual. Not only does this raise a privacy and regulatory compliance, the amount of data collected can be large, and transporting it may be expensive in some situations (e.g. when automobile data needs to be transferred over cellular networks). In those cases, federation AI can provide a mechanism to build good models which reducing the amount of data transfer that needs to happen.

5.1 Technical Challenges

The environment in which federated AI operates consists of two or more sites (e.g., insurance companies), each of which has a set of data that is to be used for creating a training model. Each of the sites has the ability to train an AI model on their local data. We assume that each of the site has a Federation Agent which can communicate with the machine learning environment that is present at the local site. In general, the sites may use different machine learning environments for training their models. The federation agents communicate with each other via a federation server. The federation server may be co-located at one of the sites, located at another site within the enterprise, or located within a cloud service provider premises. The agents and the servers work together to create a distributed system in which the communication between the agents can be used to create an AI model without moving the data between themselves.

The manner in which data is distributed across different sites, the manner in which the machine learning environments of different sites can be coordinated, and the trust relationships between the different agents and the federation server can vary. The Table 1 lists some of the common situations that can arise in an environment.

In a homogeneous environment, each site is using the same machine learning environment, which makes the practical task of sharing models easier. In a heterogeneous environment, the machine learning environments are different

Table 1. Common situations in a federated AI environment

Environment among sites	Coordination among sites	Data distribution across sites	Trust relationship among sites
Homogeneous	Synchronized	Random	Data Sharing
Heterogeneous	Asynchronous	Skewed	Model Sharing
		Missing Classes	Encrypted Sharing
		Vertically Partitioned	

(e.g. one site may be using Tensorflow, whereas another is using Pytorch), which requires translation of models from one format to another. When sites are coordinated together in a synchronous manner, they can each train a common model at the same time, and can exchange model parameters at period intervals with the server. This allows a concurrent training of the model across all of the training data sites. In some enterprises, such synchronization may not be feasible. Different sites can be training model at different times, and only completely trained models can be exchanged among sites.

Training data and its distribution among sites can have a strong impact on how different models ought to be combined. In some cases, the data across all of the sites can be viewed as a random distribution across the overall collective data that is available. More frequently, the data is skewed with some sites having more data than others, and the size of the available data at each site results in models with different level of accuracy and fidelity. In some cases, some classes of data may only be available at a single site, and be missing from other sites, which require special mechanisms for handling them. In other cases, the data may be vertically partitioned, i.e. some attributes of a type of class of data may be present at one site, while other attributes may be present at other sites.

Each combination of the different attributes described in the table above creates a new flavor of federated AI algorithms. Even without considering the heterogeneity among machine learning environments, we need 24 different flavors of federated AI to be supported among the agents and the servers. The good news is that we have invented algorithms that can handle each of the flavors with a fidelity that comes close to that of centralized learning - where all the datasets are collected at a single site and used to train a machine learning model.

The server acts as the central point for coordination for any session in which a federated machine learning model is built. It can be used by a user to configure the options which are to be used for the federation process, resulting in a unique federation session number. Once this federation session number is provided to the federation agent, the agent can join the federation session and participate in the model building exercise. The actions of the agent and the servers are selected so as to enable the particular flavor of federation that is enabled.

5.2 Initial Solution

An insurance company could enhance its actuarial modelling by contacting the AI marketplace to get quotes for the federated actuary model integrated by an analytics company, referred to as fusion manager who combines multiple local sub-optimal insight models originally built by individual insurance companies (cf. Fig. 4) using the federated AI approach. In this approach, insurance providers collaborate and integrate their sub-optimal analytics models based on their silo customer data into a more complete AI model that benefits everyone in the ecosystem. Here, the insurance companies do not share their customer data, but they instead share with the blockchain network metadata related to the sub-optimal AI model built locally at their location. The role of blockchain platform in this solution is to secure and validate the data, AI models, the learning process and outcomes.

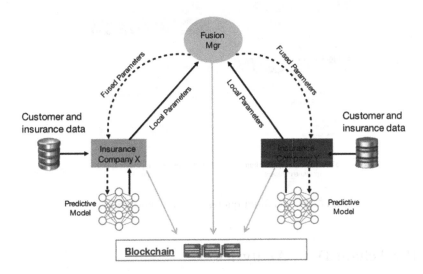

Fig. 4. Federated AI for insurance industry running on top of blockchain platform.

Federation AI requires not just the algorithms at the federation agents and the federation server, it also requires a communication protocol among them for the transfer and storage of model parameters during different stages of the federation process. One can use either a prevailing protocol, e.g. a REST protocol for this communication, or use blockchain for the purpose of communication among different nodes. Blockchain is a good protocol to maintain a reliable distribution ledge of transactions, but could be inefficient due to the need to build consensus among different agents.

There is an important role for blockchain to play, not in the main process of federation, but in determining how the federation process is tracked and registered. During the building of a federated AI model, some sites may introduce erroneous data or models, either maliciously or inadvertently due to poor maintenance or a bug. It will be useful to keep a provenance record of which sites contributed what new model parameters as the federation process progresses, and the impact of that parameter on the overall model. For keeping this provenance information, block chain provides a handy mechanisms. In our initial implementation of the system [13], we have used a private efficient protocol between the federation server and the agents, but used block chain to provide the provenance and trust information among the different agents.

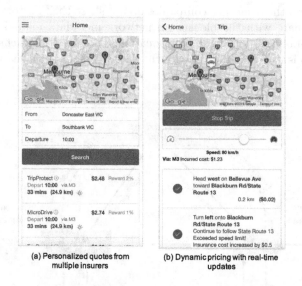

(a) Personalized quotes from
multiple insurers

(b) Dynamic pricing with real-time
updates

Fig. 5. Personalised quotes and dynamic pricing.

6   Blockchain Data Analytics

The data maintained in blockchain-powered systems possess two interesting characteristics. Firstly, it is a pool of curated data that is timestamped. Consequently, analyzing temporal data residing in those systems would be a common task. Secondly, data stored in a business blockchain network is usually process oriented, i.e., log of processes acted by business parties in the network. Hence, rule and process based analytics for compliance checking would be much useful in a blockchain-powered data system. In this section, we describe how analytics and insights learnt from the data and analytics models shared in the blockchain can be used to improve customer satisfaction as well as business operations, which is hard to achieve in traditional insurance industry approaches where each small and medium-sized insurer only has a fragmented view of a customer.

6.1   Improving Customer Satisfaction

Finding Optimal Routes with Personalised Insurance Cost. The customer, i.e., driver, wants to quickly find alternative routes from a source to a destination and their corresponding insurance costs at an expected departure time. Inputs for this request are entered by the customer via the mobile application. After receiving the user request, the routing engine (using Google Direction Service [4]) of the marketplace gateway finds alternative routes between the source and destination at the planned travel time.

These routes are then passed to the pricing engine of multiple insurers for obtaining quotes. Each insurer may develop its own risk models. For example, the risk of travelling along these routes is calculated based on route characteristics—extracted from historical data on the blockchain—such as traffic statistics, accident statistics, type of roads and speed limitation. This route risk is then combined with other risk factors such as time frame risk, personalised driver risk (based on their previous driving behaviour and claim history recorded on the blockchain), vehicle risk, and driving difficulty (based on weather forecasts) in order to estimate the likelihood of accidents and degree of damage, which determine the insurance cost for each route. These alternative routes with their insurance quotes offered by multiple insurers are returned to the mobile application (cf. Fig. 5a) for the user to select.

Transparent Bookkeeping of Trip History on Blockchain. Presented with several possible routes and insurance quotes for their trip request, the customer decides which route is best suited for their personal preference based on multiple criteria such as total travel time and insurance cost. Some customers may choose the fastest route which may come with the most expensive insurance cost due to higher risk travelling on a main road. In contrast, another customer may prefer the cheapest insurance cost and choose a route that takes longer time but utilises safer roads.

The customer selects their preferred route on the mobile application and this pay-as-you-go insurance application changes to travel mode, i.e., the road segments travelled by the insured car are periodically recorded and sent to the back-end system for persisting on the blockchain. This data is transparent to both the customer and the insurance company, and will be used to verify if the customer has consistently travelled along the agreed path in the case there is an insurance claim made. Every time the new trip information has been successfully persisted on the blockchain, the mobile application marks a tick on the road segment that the insured car has just travelled so that the customer is confident that the insurance company has acknowledged their actual trip (cf. Fig. 5b).

Dynamic Pricing Alert. The customer is expected to consistently follow the original selected route as well as the speed limit along the route. In the case the insured car diverts from the original path or exceeds a speed limit, the insurance premium may be recalculated. In particular, at some point in the middle of the journey the customer might speed up exceeding the road speed limit due to being in a hurry. Their speeding is detected by the mobile application and sent to the back-end system for further processing. First, this speeding event is persisted onto the blockchain. Second, the pricing engine of the insurer is notified to recalculate insurance cost given real-time conditions and alert the changes to the user through the mobile application (cf. Fig. 5b). Apart from speeding events, our pay-as-you-go insurance application is also able to provide premium recalculation based on other dynamic variables, e.g., a driver's diversion from original routes, weather events and traffic conditions.

Real-Time Road Assistance. Real-time trip verification, i.e., determining whether the user has so far consistently followed the agreed route, is used to detect any path diversion and notify new insurance cost. This capability is also useful for other scenarios such as providing real-time assistance during the trip, in particular, in an unlucky event the insured car may break down in the middle of a journey. The customer can activate a special function provided in the mobile application to trigger an emergency assistance request.

Receiving this request, the marketplace gateway system invokes a smart contract to handle emergency requests from the customer. The smart contract first verifies whether the customer has followed the agreed path and in the case of positive check it will automatically notify the emergency assistance team of the insurance company to provide the user being in trouble with prompt help. This automated and effective emergency response would increase customer satisfaction while reducing business cost.

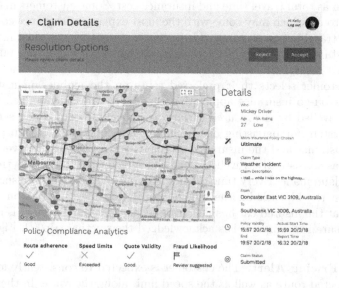

Fig. 6. AI-assisted claim processing.

6.2 Enhancing Business Operations

Having shown the use of analytic models built from the data shared in the blockchain for improving customer satisfaction, we now describe how insights extracted from the blockchain data can enhance the business operations of the insurers as well. Assume that the driver reports a car collision during a trip and make an insurance claim. The data related to this particular trip and the driver's claim history retrieved from the blockchain become an indisputable source of evidence to assist the insurer in claims processing.

In particular, policy compliance analytics (cf. Fig. 6) based on the data recorded on the blockchain can be performed automatically by the system and shown to the insurance staff responsible for resolving the claim. For instance, the system can verify the route adherence and speed compliance during the trip, i.e., whether tamper-proof blockchain data confirms that the driver followed the original selected route or actually diverted from the route or exceeded speed limits prior to the collision. These analytics assist the claim resolving staff to make decision whether to accept or deny this claim.

In addition, patterns of insurance claims extracted from historical claim data on the blockchain can also help flag fraudulent claims when there is a likelihood that the driver reported an event that did not actually occur. Specifically, for opportunistic fraud, i.e., the exaggeration of otherwise legitimate claims, and premeditated fraud, i.e., deliberate fabrication of a claim, a technique for detecting these types of fraud is pattern matching on the blockchain data, i.e., the same patterns that are used in multiple claims across insurers by the same customer or other customers can be a red flag that requires further fraud investigation.

Furthermore, historical claims stored on a shared blockchain help reduce fraudulent non-disclosure, i.e., misrepresentation of facts material to the insurance policy such as a driver's failure to disclose their history of claims and accidents. A driver will find it difficult to hide their bad record with one insurer when moving to another insurer. Even though the fraudulent driver may try to modify their identity when insuring with different insurers, information retrieval techniques such as entity matching can be used to match the customer information with the existing identity managed in the blockchain.

7   Conclusion

In this paper, we have described the design of a blockchain-powered big data analytics platform and its novel application to pay-as-you-go automotive insurance and AI marketplace. In this platform, blockchain is being used for creating the marketplace, and we believe federated AI can play a big role in creating collaboration between the parties. We also present technical solutions for scalable blockchain data management and federated AI/analytics services based on the data and AI models shared in the platform.

Acknowledgement. We would like to thank Dain Liffman, Ziyuan Wang, Josh Andres, Nick Waywood, John Wagner and Ermyas Abebe for their helpful discussion about application of blockchain technology to insurance industry.

References

1. Bitcoin: A Peer-to-Peer Electronic Cash System (2008). https://bitcoin.org/bitcoin.pdf
2. Growth Insurance Market (2016). http://www.timeslive.co.za/thetimes/article1508818.ece
3. Corda (2017). https://github.com/corda/corda
4. Direction Service (2017). https://developers.google.com/maps/documentation/directions
5. Ethereum (2017). https://www.ethereum.org
6. Hyperledger (2017). https://www.hyperledger.org
7. Atzori, M.: Blockchain-based architectures for the internet of things: a survey (2017). https://ssrn.com/abstract=2846810
8. Bonneau, J., Miller, A., Clark, J., Narayanan, A., Kroll, J.A., Felten, E.W.: SoK: research perspectives and challenges for bitcoin and cryptocurrencies. In: Proceedings of IEEE SSP, pp. 104–121 (2015)
9. Dinh, T.T.A., Wang, J., Chen, G., Liu, R., Ooi, B.C., Tan, K.L.: Blockbench: a framework for analyzing private blockchains. In: Proceedings of SIGMOD, pp. 1085–1100 (2017)
10. Kokoris-Kogias, E., Jovanovic, P., Gasser, L., Gailly, N., Ford, B.: Omniledger: a secure, scale-out, decentralized ledger. Cryptology ePrint Archive, Report 2017/406 (2017)
11. Ren, Z., Cong, K., Pouwelse, J., Erkin, Z.: Implicit consensus: blockchain with unbounded throughput. CoRR abs/1705.11046 (2017)
12. Tschorsch, F., Scheuermann, B.: Bitcoin and beyond: a technical survey on decentralized digital currencies. IEEE Commun. Surv. Tutor. **18**(3), 2084–2123 (2016)
13. Verma, D.C., Calo, S.B., Cirincione, G.: Distributed AI and security issues in federated environments. In: Proceedings of ICDCN Workshops, pp. 4:1–4:6 (2018)
14. Vo, H.T., Kundu, A., Mohania, M.: Research directions in blockchain data management and analytics. In: Proceedings of EDBT, pp. 445–448 (2018)
15. Vo, H.T., Mehedy, L., Mohania, M., Abebe, E.: Blockchain-based data management and analytics for micro-insurance applications. In: Proceedings of CIKM, pp. 2539–2542 (2017)
16. Vukolić, M.: The quest for scalable blockchain fabric: proof-of-work vs. BFT replication. In: Camenisch, J., Kesdoğan, D. (eds.) iNetSec 2015. LNCS, vol. 9591, pp. 112–125. Springer, Cham (2016). https://doi.org/10.1007/978-3-319-39028-4_9

Humble Data Management to Big Data Analytics/Science: A Retrospective Stroll

Sharma Chakravarthy[(✉)], Abhishek Santra, and Kanthi Sannappa Komar

Information Technology Laboratory (IT Lab),
Computer Science and Engineering Department, University of Texas at Arlington,
Arlington, TX 76019, USA
sharma@cse.uta.edu

Abstract. We are on the cusp of analyzing a variety of data being collected in every walk of life in diverse ways and holistically as well as developing a science (Big Data Science) to benefit humanity at large in the best possible way. This warrants developing and using new approaches – technological, scientific, and systems – in addition to building upon and integrating with the ones that have been developed so far. With this ambitious goal, there is also the accompanying risk of these advancements being misused or abused as we have seen so many times with respect to new technologies.

In this paper, we plan on providing a retrospective bird's-eye-view on the approaches that have come about for managing and analyzing data over the last 40+ years. Since the advent of Database Management Systems (or DBMSs) and especially the Relational DBMSs (or RDBMSs), data management and analysis have seen several significant strides. Today, data has become an important tool (or even a weapon) in society and its role and importance is unprecedented.

The goal of this paper is to provide the reader an understanding of data management and analysis approaches with respect to where we have come from, motivations for developing them, and what this journey has been about in a short span of 40+ years. We sincerely hope this presentation provides a historical as well as a pedagogical perspective for those who are new to the field and provides a useful perspective that they can relate to and appreciate for those who have been working and contributing to the field.

Keywords: Data management · Relational databases
Data warehouses · Event and stream data processing
Data mining · Video situation analysis · Big data analytics/science

1 Introduction

This is a position paper on big data analytics and science that is intended to capture the steps and journey that has led to this stage. In this paper, we will start

© Springer Nature Switzerland AG 2018
A. Mondal et al. (Eds.): BDA 2018, LNCS 11297, pp. 33–54, 2018.
https://doi.org/10.1007/978-3-030-04780-1_3

with the humble beginnings of how and why data management became important, approaches developed over the last 40+ years driven by new technologies (both hardware and software) and science (algorithms, theories, abstractions, etc.) and primarily motivated by user or application needs. We are also familiar with cases where technologies/ideas that have come way before their time and have not survived. Hence, it is important that many things – requirements, technologies, as well as readiness of the users to adapt – have to come together in a timely and synergistic manner for success.

Specifically, in this paper, we identify various milestones in managing and analyzing data and briefly describe how they came about, their purpose, and their impact. Relational Database Management Systems (or RDBMSs) became the *de facto* standard for managing data (considered large at that time) for real-world applications whose requirements gave rise to many new abstractions. Theory of concurrency control, recovery (the so called ACID properties - Atomicity, Consistency, Isolation, and Durability), and query optimization provided the underpinnings for industrial-strength data management systems. These systems were optimized high throughput in mind for short transactions termed OLTP or On Line Transaction Processing. A non-procedural query language SQL (or Structured Query Language) provided ease of querying from an end user perspective.

Then the need for more complex analysis of data and the need for combining data from various source or operational databases led to the development of data warehouses, data marts and more recently we have the notion of data lakes (mainly as storage of very large amounts of data) in conjunction with Big data analysis! The need to analyze large amounts of data in ways beyond the capabilities of an RDBMSs – especially collections that provided a broader understanding of the business – gave rise to OLTP or On Line Analytical Processing, subject-oriented dimensional analysis, and ETL (Extract, Transform, and Load) tools for maintaining and managing the archival collections. SQL was extended to support dimensional analysis as well. ROLAP (Relational OLAP), MOLAP (Multi-dimensional OLAP), and HOLAP (Hybrid OLAP) followed.

Data mining is the process of automatically discovering useful information from large data repositories. Data collected and validated (termed labeled) has been used for generating models (supervised approach) that can be used for predicting outcomes for new data. Data has also has been processed in a number of ways to glean patterns without using labeled data (e.g., clustering, termed unsupervised). Data mining is different from querying or analyzing different types of reports generated using the approaches outlines above. Starting with text mining, a number of traditional mining techniques have been developed (decision trees, clustering, neural nets, etc.) Both supervised and unsupervised approaches have shown to be needed for different applications. Not too long ago, graph mining and association rule mining (and its flavors) became possible with the advances in data representation, availability of large real-world data, and relevant technical advances. Data Mining, as we have it today, became even more important from a business perspective (similar to Data warehouses, but with different requirements) when we progressed in our ability (storage, networking, processing, and algorithms) to handle vast amounts of *real business data* (in contrast to samples of representative data) for identifying non-intuitive nuggets with certain confidence for driving business goals.

The advent and the availability of inexpensive and small (even intelligent) sensors (as compared to bulky sensors such as moats) generated requirements not satisfied by existing data management systems leading to the development of event and stream processing approaches for continuous data. As specification of latency requirements, throughput, and memory needs along with processing data coming at different rates were not supported by DBMSs, new ways of satisfying them had to be developed.

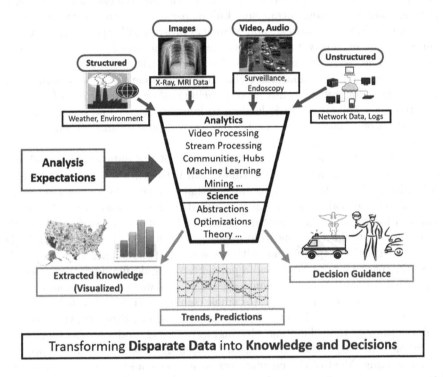

Fig. 1. High level big data analytics/science architecture

This was followed by the need for processing very large amounts of data once with a specific goal (as compared to management and analysis of schema-based data over its lifetime) which gave rise to the Map/Reduce abstraction combined with transparent fault tolerance (which is different from ACID properties, and commits of DBMSs). Minimal user input was considered important leading to the support of sorting and shuffling in a transparent manner. The requirement was to achieve both data and computation scalability in an arbitrary manner to deal with ever increasing data. Around the same time, NoSQL systems were developed for applications that do not require all the bells and whistles of a traditional DBMS, and to model different types of data closer to their native representation (instead of shoe horning them into a single model.) Instead of ACID, CAP (Consistency, Availability, and Partition tolerance) was introduced.

It is important to note that these strides are not sequential although discussed in that way. Some have been sequential (for example, RDBMSs to Data Warehouses, RDBMSs to Stream Data Processing) building upon the strengths and weaknesses of previous capabilities whereas others have happened in an overlapping manner. For example, stream data processing, Association rule mining, and data warehouses have overlapped. For these, our choice of order of discussion is arbitrary.

Finally, we are staring at even more ambitions requirements for holistically analyzing (and developing formalisms for) disparate data that corresponds to 4V's (Volume, Velocity, Variety, and Veracity) or even 5V's (plus Value) which is the challenge currently faced by the community addressing big data analysis/science. The basic premise here is that dealing with each of the V's individually or in small combinations (what has been done up to this point to a large extent) is not sufficient, but need to include *all or combinations* of them as warranted for a holistic analysis leading to inferring better and concise knowledge for decision making. Towards this end, we will present some of *our* ongoing contributions towards big data analysis.

Figure 1 shows, at a high level, the problem of big data analysis and science. As shown in the figure, the ultimate goal is to synthesize meaningful and beneficial knowledge with good confidence that can be used for decision making (what humans call wisdom which is culled from data and events based on a combination of nature and nurture as biologists put it) by using all available relevant disparate data coming from a variety of sources. The inverted triangle shows the reduction aspect along with some of the technologies that are available today (used for analytics), and a partial list of underpinnings (i.e., science) using which we develop these technologies. The way we see big data analysis is that instead of addressing each V (or small combinations of V), a holistic approach is the desired goal driven by analysis expectations. However, the problem is still the same as that of culling, filtering, aggregating, and inferring nuggets of (actionable) knowledge that can be used for decision making including real-time decision making. Most of the current approaches have addressed a small subset of this problem.

Although not explicitly shown in the figure, techniques for the visualization of data as well as the derived knowledge is quite important. Pictorial representation and multi-dimensional subject-oriented analysis of the results are also very useful for understanding the results of analysis.

Personalized health care is a good example of how it is critical to avail and process all types of data related to a person over a period of time (lab results, X-rays, EKG, endoscopy video etc.) to make a meaningful decision for that individual rather than using average cases which is how it is done today. Similar applications include climate change studies, fusion applications, and others.

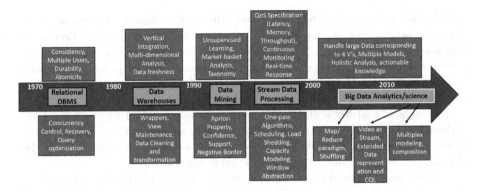

Fig. 2. Data management time line leading to big data analytics/science

Figure 2 shows an approximate time line of the major advances for dealing with data management as technologies and requirements changed. Boxes above the time line capture the high level requirements for each advancement and the boxes below show the abstractions that were developed/used for those advancements. This is an approximate time line mainly to show the maturation of various components leading to these advancements.

As we can glean from the rest of the paper, many of the advances discussed deal with a single or a few types of data. That is a critical step before we can embark on combining all available types of data and be able to holistically synthesize knowledge from them. As we have done some work in each of the fields discussed in this paper, we will briefly identify our contributions in each section and its relevance towards the overall goal of big data analysis.

The remainder of the paper is organized as follow. Within each section, we will identify our contributions to the topic under discussion. Section 2 discusses early developments of data base management systems focusing on the relational model. Section 3 elaborates on the need for combining data from multiple source DBMSs and its flexible analysis. In Sect. 5, we briefly trace the approaches to mining leading to the current status. Section 4 elaborates on the differences between event & stream processing systems and RDBMss as well as the extensions developed. In Sect. 6, we discuss several topics and their relevance to big data analysis. We mainly discuss our contributions towards big data analysis in this section. Finally, Sect. 7 has conclusions.

2   Traditional Data Management

The need for an integrated way to manage data (for payroll, banking, etc.) was recognized very early on (in the 1960s) as the usage of file-based systems was becoming unmanageable. Customized applications had to be executed using master and slave sets of files as there was no recovery. Applications were heavily dependent of the formats of files used. One of the authors was part of the team in India that processed payroll in the early 1970s using file-based systems – that too using tape drives as there was not enough online disk storage to keep data that was used once a month. The fear always was the failure, while processing the payroll completely (having no atomicity guarantees), at any point in time and corrupting the files. Hence a backup copy (termed slave) was always maintained for that eventuality.

Hierarchical and Network DBMSs were developed in the 1960s as a solution. Although concurrency control and recovery were available, there was no theoretical underpinning. Queries had to be written as programs still aware of the abstraction, and to some extent physical representation.

Traditional relational database management systems (RDBMSs), consisting of a set of persistent relations, a set of well-defined operations, B+ tree and hash indexing, and highly optimized query processing and transaction management components (both concurrency control and recovery) with theoretical underpinnings were developed in the 1970s. Relational DBMSs have been researched for over forty years and are used today for a wide range of applications. It is one of the most-widely used systems either as a stand-alone system or behind a web site. Typically, data processed by a DBMS is less frequently updated, and a snapshot of the database is used for processing queries. Abstractions derived from the applications for which a DBMS [47,48] is intended, such as consistency, concurrency, recovery, and optimization have been used in many other systems.

Queries that are processed by a traditional DBMS are termed *ad hoc* (also OLTP) queries. They are specified, optimized, and evaluated *once* over a snapshot of a database. Industrial strength relational DBMSs were developed and has become the de-facto standard for data management applications for structured data. Even other types of data (e.g., graph data) have been modeled and processed using relational DBMSs as other choices were not available until recently. We have used RDBMSs for mining subgraphs and have encountered the problem of self-joins not efficiently optimized by RDBMSs. The burden of designing a schema and populating the DBMS meant that its use was expected over a life span that amortized the cost and effort needed for building and using it. For applications that required formalized approaches to concurrency, recovery, query optimization, and in addition provided several utility tools, RDBMSs remained a preferred choice.

Both the data model and the widely popular SQL[1] have undergone several extensions over the last few decades. SQL92 with minimal aggregate operators and group by capabilities have been revised several times, each time adding features that have enhanced its expressiveness significantly. Now we have table expressions, abstract data types, triggers to mention a few. Similarly, XML has been added natively in almost all DBMSS with the accompanying path queries and their optimization. In addition, object-oriented features have been added to compete with native object-oriented DBMSs. These days RDBMSs are competing with not object-oriented DBMSs, but NoSQL systems and cloud computing.

---

[1] Prof. Michael Stonebraker fondly refers to SQL as the inter-galactic data speak. Others may see it differently.

Of course, the application space has also expanded significantly (not every application needing ACID properties) and perhaps there is room for more than one model. Only time will tell.

2.1 Our Contributions

We have made a number of contributions for the traditional DBMS with respect to multiple query [9,15,49,50] and semantic query optimization [14]. In addition, our prior research has contributed event-condition-action rules for supporting active database capability [13,24,25] in relational and object-oriented data models [2,7,12,14]. This is further discussed in Sect. 4. Several abstractions were developed (e.g., extended Tx model for rule processing, event algebra) as part of this work.

3 Data Warehouses

With the wide usage of RDBMSs, information increasingly became an important corporate asset, with consequential requirements for not only its efficient collection, organization, and maintenance, but also for leveraging it for business decision making. Hence the term BI (or Business Intelligence) and decision support systems came about to exploit data as an asset. For analysis leading to enterprise decision making, it was necessary to integrate/combine needed data from a number of sources. These sources were, typically, the so called operational databases belonging to the same enterprise. They could also include external sources, in addition to operational databases. Some of these external sources may be web sites or read-only databases. Note that several databases (even from different vendors) were being used within an organization by different departments. Also, mergers and acquisitions only exacerbated this problem.

In addition, it was also important to analyze data using different subject-oriented dimensions (e.g., year, product, and zip code as dimensions) which was either not possible or very difficult using the extant RDBMSs. Dimensional analysis had to be done at different granularities in order to understand the business better. This drilling down of the dimensional analysis further for a detailed understanding and slicing of data in different ways (using different combinations of dimensions) to get understanding from different perspectives became an important business requirement. Data warehouse was one of the approaches to satisfy these new business requirements.

A data warehouse (DW) can be seen as a selective integration and analysis of distributed, heterogeneous information. These are typically operational databases (essentially RDBMSs.) They had been optimized for OLTP rather than complex data analysis. Also, executing business analysis queries could interfere with or slow down the operational databases that had to provide quick response time. This was undesirable. Also, the business analysis was not done in real-time and hence even if the data was not very fresh (i.e., archived), it was not an issue. Hence the approach was to collect **only data needed for analysis** from multiple operational databases and create a "data warehouse" that has

required data using a simpler schema and use it for analysis. Before we introduce
the data warehouse architectures, it is useful to understand the functionality differences between operational databases and data warehouses or decision support
systems as shown in Table 1. Other requirements were the schema must be simple, data should be clean, consistent, and accurate, allow complex analysis, and
refresh the data warehouse within a given time window.

Table 1. Functionality of operational DBMS vs. data warehouse

Function	Operational DBMS	Data warehouse
Data content	Current values	Archival data
Data organization	Application-based	Subject area across enterprise
Nature of data	Dynamic	Static until refreshed
Data structure, Format	Complex: suitable for operational Computation (normalization)	Simple: Suitable for business Analysis (star, snowflake)
Random access	High	Moderate to low
Data update	Field-by-field basis	Accessed and read: no update
Usage	Structured, repetitive	Highly unstructured analytical processing
Response time	Sub-second to 2 to 3 s	Seconds to minutes, hours

Note that, as a separate strand of research, schema integration, global schema
development, and related research had been ongoing for quite a number of years
by this time. So, the approach to data warehouses was very different – and more
operational and pragmatic – than the ones researched for a providing a *single
logical view* of multiple databases). It is important to mention that federated and
multi-database approaches for integrating and/or using different data models
have been researched.

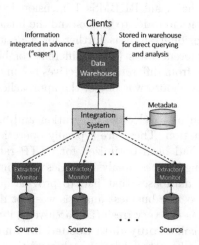

Fig. 3. Data warehouse: stored data architecture

Based on the differences in
purpose and the decoupling of
operational databases from Data
warehouses, at least two architectures were considered. In the
first one, an integration system would be used with its
own meta data, and analysis
queries would be directly processed by the integration system using data accessed from
operational databases (through
a wrapper.) This was termed
'lazy' or 'data-on-demand' approach that did not satisfy the

decoupling, and other requirements such as interference. The second approach (shown in Fig. 3) was to build an explicit data warehouse where data was collected by the integration system from operational databases (periodically) with its own schema and analytical processing capabilities.

This approach was closer to the requirements and decoupled the operational databases. Several issues had to be solved, such as how and how often to refresh the data in the data warehouse, impact of refreshing on the consistency of both the operational as well as data warehouse. In addition, SQL had to be extended to include dimensional analysis operators and new ways of indexing (e.g., inverted index, bloom filters) as well as optimization. Data warehouse schema was purposely kept simple (e.g., star and snowflake) and some views were materialized as part of optimization. Cube and rollup operators were introduced for dimensional analysis and a number of techniques were developed for multi-dimensional analysis. In addition a number of ETL (Extract, Transform, and Load) tools based on new techniques were developed for fulfilling the requirements of Data Warehouse approach to data management.

3.1 Our Contributions

For this data management approach, there were several theoretical issues that were addressed by researchers. The Data Warehouse maintenance problem was characterized and shown to be complex and relevant. In characterizing the problem, three main components were identified: policies, QoS (or Quality of Service) criteria, and source capabilities. Our work in this space has made contributions [27–30] to an understanding of each of these components. In addition, dependencies between them have been identified and analyzed. This has resulted in straightforward heuristics for policy selection, and a tool based on a cost model for detailed comparison of policies.

4 Event and Stream Data Processing

Complex event processing (CEP) systems developed in the mid-eighties [8,25], Postgres [55], and ETM [26] and stream processing (SP) developed in the late nineties converged as inexpensive sensors became smaller in size, cheaper, and pervasive. The initial drivers of so called 'active databases' were to provide capability to identify a class of situations based on operations on a DBMS to automatically generate notifications avoiding the need for manually running scripts periodically for doing that. For this to happen, the changes required to the architecture and functionality of a DBMS were explored. Later, this notion of event detection and processing was generalized to any application rather than limiting it to a DBMS.

However, the drivers for stream processing or DSMS (Data Stream Management Systems) were somewhat different. Table 2 summarizes the differences between a traditional DMBS and a Stream Data processing system.

Table 2. Back-of-the envelope comparison of a DBMS with a DSMS

DBMS	DSMS
Persistent relations	Transient streams
One-time (*ad hoc*) queries	Continuous queries
Random (disk) access	Sequential (in-memory) access
Unbounded disk storage	*Bounded* main memory
Snapshot or current state used	Arrival order, `window` important
Relatively low update frequency	Varying and bursty input rates
No QoS support	QoS support critical
Requires precise results	May tolerate approximate results
Computes all results on a snapshot	Computes `window`-based results incrementally
Transaction management critical	Transaction management not critical

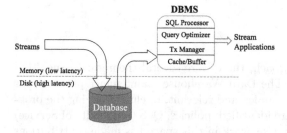

Fig. 4. Data stream processing using a DBMS

It is clear from Table 2 that applications that process stream data do not readily fit the traditional DBMS model and its processing paradigm, since DBMSs were not designed to manage high-frequency updates (in the form of data streams) and to provide continuous computation and output for queries. Hence, the techniques developed for DBMSs were re-examined to meet the requirements of applications that use stream data. This re-examination has given rise to a paradigm shift along with new approaches and extensions to new techniques for query language and optimization. Capacity modeling, scheduling, and load shedding that were not part of the traditional DBMSs architecture were developed for the new paradigm. The differences in architecture and processing are captured eloquently in Figs. 4 and 5.

We briefly describe the seminal works in stream data processing. *Aurora* [56] is a system for managing data streams for monitoring applications. Continuous queries in Aurora are specified in terms of a dataflow diagram consisting of boxes and arrows (using a GUI). Optimization and limited

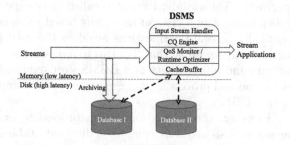

Fig. 5. Data stream processing using a DSMS

load shedding were incorporated into Aurora. Borealis [32] is a second generation system that addressed distributed stream processing.

Stanford Stream Data Manager (or *STREAM*) is a general-purpose data stream management system that introduced Continuous Query Language (CQL) [3] which included the notion of windows, chain scheduling algorithm that optimized memory, and its variants for handling different types of constraints.

4.1   Our Contributions

MavStream [33–35, 46], as our contribution to the area of stream data processing, addressed stream processing holistically including capacity modeling which modeled each relational operator and combinations of operators using a queueing model to determine expected latency for a given processing capacity. Path capacity scheduling was developed for optimizing latency as opposed to memory, and a suite of scheduling approaches to trade off latency with memory. Load shedding was addressed with respect to placement of load shedders and determining an optimal selection of them based on error tolerance. Stream processing system was also integrated with an earlier developed event processing system synergistically into a single architecture termed MavEStream. A number of other systems addressed stream processing which are not discussed here due to paucity of space.

5   Data Mining or Knowledge Discovery in Databases

Data mining aims at discovering important and previously unknown patterns from the datasets. Although not explicitly termed data mining and might not have used real-world business data, the concept of understanding data in ways that are different from querying and analysis that was available for RDBMSs and data warehouses pre-dates them. Classification, clustering, prediction, deviation analysis, and neural networks were used by many businesses for selective marketing, credit card transaction approval, and mortgage and other types of lending. Supervised and unsupervised approaches were developed and multi-fold cross validation was widely used for establishing the accuracy of models. A number of algorithms were developed, some couched in expert systems used by businesses. Due to the limitations of storage and processing, the sizes of the data sets used were small and often statistically representative samples were used for processing (instead of all available data) and results extrapolated (or generalized) for larger data sets.

The rapid improvement in the size of the storage devices along with the associated drop in the cost in the 1990s, and increase in the computing power as well as the wide use of statistical approaches for processing data gave rise to the field of data mining as we know it today. Suddenly, it became feasible for organizations to store unprecedented amounts of organizational data and process it. These organizations, though having a gold mine of data, were not able to fully capitalize on its value mainly because the algorithms and approaches had to be scaled to very large data sizes. Typically, the data captures the business trends over a period of time and hence using real-world business data (rather than samples) became the goal. However, the nuggets of useful knowledge hidden were not so easy to discern. To compete effectively, decision makers felt they needed to identify and utilize "nuggets of knowledge" buried in the collected data and take advantage of the high return opportunities in a timely manner.

While these developments were ongoing, although graph theory has been around for a very long time, graphs for data representation were not that popular. A few researchers were using graphs for representing data in specific domains and were trying to identify patterns in graphs as businesses were trying to do the same using their transactional data. One of the early work on graph mining was Subdue [19] which developed main memory algorithms for identifying substructures in graphs (or forests) that were "interesting" based on some metric. They use a information theoretic metric termed minimum description length (or MDL) for this purpose. The data sets were drawn from chemical representations, CAD circuits.

Data mining became a hot research area with the advent of association rule mining and graph mining. Association rule mining [1] started with market basket analysis for identifying items bought together with given support and confidence from actual point-of-sales data. These data sets were huge (for example, Walmart's point-of-sales data around that time was estimated to be around 1Gb per day) and multiple years of data could not be held in main memory for analysis. For association rule ming of this data, novel data structures (e.g., hash tree) and approaches to reduce the number of passes on the data (to minimize I/O's) were developed. As the search space was prohibitively large, *a priori* and other properties were identified to reduce the number of item sets carried over from iteration to the next one. A large amount of work followed resulting in a number of mining systems marketed by almost every major vendor.

Along the same lines, the expressiveness of graphs became important with the advent of Internet and graph mining and its importance for identifying important and useful graph patterns became apparent. Frequent subgraphs, identification and counting of triangles and other substructures in very large graphs became important in addition to interesting substructures. Today, there is renewed interest in using graphs for modeling and analysis of complex data as discussed in Sect. 6.3.

5.1 Our Contributions

We have contributed to both association rule mining [37, 38, 41–43] and substructure discovery. For association rules, we have extensively evaluated performance of database algorithms on different RDBMSs and compared them. This helped us to identify some of the quirks in the optimization of SQL queries by different vendors and understand the difficulties of optimizing queries with 10 to 20 joins, a large number of them being self joins. RDBMSs query optimization were not designed with those number of joins in mind.

On the graph mining side, we have tried to scale the main memory substructure mining algorithm of Subdue using a number of alternative approaches. The first one [4,10,45] mapped graphs into relations and the substructure discovery algorithm to SQL in order to leverage the built-in capabilities of a DBMS (buffer manager, query optimizer) instead of re-inventing them for mining. This allowed us to scale the size of the graphs to millions of nodes and edges. This approach has certain limitations due to large number of joins as well DBMS's inability to order a relation using columns. Recently, we have been able to successfully scale this algorithm even further to arbitrary sizes using the map/reduce paradigm as discussed in Sect. 6.1.

6   Approaches to Big Data Analytics/Science

To achieve the goals of big data analytics/science, a number of perspectives on how to analyze a data set as well as a number of approaches and their combinations need to be taken into consideration. Research is ongoing by a large number of scientists with a broad brush covering data sets from a variety of domains. In this section, we present some of the approaches that we are working on to address the big data analysis problem in a small way. Although they do not right now solve the big data analysis problem as is posited in Fig. 1, they address components whose solutions are likely to contribute to the overall solution.

6.1   Data and Processing Scalability Using Map/Reduce

Scalability of algorithms has been a research focus for a long time, whether it is in terms of increased CPU cycles or large amounts of data. This is not a new problem but the need for the scalability of *both* data and computation poses new challenges. From a data management perspective, relational databases and data warehouses have tried to address this issue to some extent, but the ever increasing data sizes and their concomitant analysis needs beg for a more general solution. Sometimes this is referred to as horizontal (increasing data sizes by adding more disks) and vertical (increasing processing by adding more processors) scalability. In the traditional data management scenario where SQL is predominantly used, its optimization (e.g., joins and other operators) and generation of good plans need to be cognizant of data distribution characteristics as well as processor characteristics. Further, concurrency control (usually based on locking), buffer size, among others play a role in the parallelism possible or achieved. This makes it difficult to add resources arbitrarily to a DBMS and hence scalability is a big issue. Several novel indexes as well as column-based systems have been developed for dealing with scalability of data and processing.

There are many instances where new approaches and techniques were developed as extant data management systems were not able to handle the application's requirements. The development of BigTable [16] is a clear example of the difficulties faced for indexing web pages and retrieving them with an acceptable response time. More recent NoSQL systems are other examples which are used by many newer applications that needed to support complex data models and computations/traversals (e.g., HBase and Hadoop/hive used by Facebook) over those models. Finally, Map/reduce paradigm itself resulted from the need to process unstructured log data without having to design a schema and loading it into a DBMS as the processing was done very infrequently (or only once.)

Although the Map/Reduce paradigm itself is simple – that of divide and conquer using hashing, what has made it practically useful is the couching of that abstraction as part of a system that supports sorting, shuffling, and fault tolerance in a *transparent* manner. Use of distributed file systems, minimal burden on the user (only two functions), and the ability to use arbitrary number of processors (by specifying as a parameter) has made this approach a viable one for many big data applications.

Our Contributions: We have achieved data scalability using divide and conquer with and without using map/reduce. We have developed generic Map/Reduce based algorithms for horizontal scalability of substructure discovery that can work with any partitioning strategy. The basic components of graph mining - subgraph expansion, duplicate removal and counting of isomorphic substructures were incorporated into the algorithms for the Map/Reduce paradigm by carefully orchestrating new representations. Vertical scalability was achieved by showing that these algorithms produce the same results (loss-less property) irrespective of the number of partitions. Experiments validated the advantage of using Map/Reduce based substructure discovery to scale to arbitrarily large graphs [20–22]. In an effort to analyze the partitioning strategies and associated algorithms from a performance standpoint, we have done a component cost analysis of substructure discovery in a distributed framework. The cost analysis identified places for improvements in using the range-based partitioning strategy over its counterpart. Theoretical justification along with experimental evaluation of the improvements were verified by varying a number of user parameters. The cost analysis also pointed out the portability of our algorithms to a different paradigm such as Spark to reap similar benefits.

Our approach to query processing on graphs developed a cost-based plan generator by defining and using a catalog that is relevant to graphs [23,31]. To process a plan on large graphs, partitioning of graphs was used for processing the plan on each partition separately and combining the results in a loss-less manner. Several heuristics were developed for optimizing the number of partitions loaded for this purpose [5,6].

6.2   Stream-Based Video Situation Analysis

This project, underway at Information Technology Laboratory (or IT Lab) of UT Arlington, is one of the examples of combining different strands of research to obtain a result that is greater than the sum of its parts. The basic idea is to use the current state of the art approaches from image and video analysis and event and stream data processing and to extend them in appropriate ways to combine them synergistically.

Image and Video Analysis (IVA) has been ongoing for several decades and has developed an impressive array of techniques [17] for image and video processing: object identification and re-identification, event & activity modeling, and summarizing a video to name a few. A large number of techniques (both pattern recognition-based and machine learning-based) for processing video frames to characterize objects have been developed to deal with camera angles, lighting effects, color differences, as well as object identification and re-identification.

Our Contributions: Based on our understanding of the two domains (refer to Sect. 4 for Stream data processing [11]), we believe that stream data processing (both CEP and SP) can be effectively adapted for analyzing videos in a different way than how it is being done currently.

The focus of this work is to bring the advantages of general-purpose window-based querying and aggregation for analyzing video contents. This is a significant departure from the traditional way of analyzing images and videos. This approach, we believe, will augment the current image and video analysis capabilities. For certain types of videos, queries expand the scope of analysis that can be performed. In order to accomplish this, first, we extract the contents of each frame of a video to the extent possible using state of the art techniques. These are stored in a representation that is much richer than the relational model (e.g., store bounding boxes, feature vectors, etc.) and process this data as a stream using extended CQL. We have built upon and extended the query language AQuery [39] and the arrable data representation. We have introduced new operators for comparing two objects for similarity (instead of exact match), compress multiple occurrences of objects in meaningful ways, and vector operations.

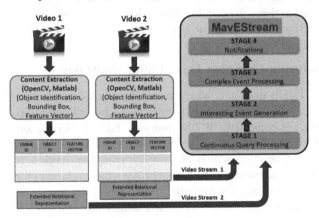

Fig. 6. MavVStream architecture

Figure 6 shows the architecture of MavVStream which extends the MavStream architecture and combines with feeding streams coming out of video pre-processing. With this approach, we have been able to demonstrate the processing of some of the queries listed below. Assuming two streams of videos from two cameras – one for entry and one for exit – the following classes of continuous situation queries can be formulated on the video footage[2].

---

[2] This could be for a building, mall, check post, or a parking lot etc.

[Situations – Set 1]

1. How many people/cars enter the building/parking lot every hour on the hour or between a specified interval or on a particular day?
2. Identify the slowest/busiest hour of the day (in terms of number of people entering).
3. List people/cars who stayed for less than an hour in the building/parking lot.
4. List the average duration a person/car stayed at the property on Mondays/weekends.
5. Did two individuals (images given) enter **or** exit the building within 5 min of each other?

The above queries are of aggregation types where information from the video is aggregated along temporal or other dimensions. It is also possible to ask queries that include spatial dimension as well or both spatial and temporal dimensions. For example, it is possible to extend the above class of queries to:

[Situations – Set 2]

1. Did a person go up to the check post/entrance but turn around (i.e. did not cross/enter)?
2. Did two people bump into each other?
3. Was a person picked up by a vehicle?

Note that the above queries are some what different from the earlier ones and require identification and extraction of various types of actions, such as walking, turning around, carrying an object, getting rid of an object etc. We have formulated some of the actions needed for **Situation 2** in terms events that are detected over primitive events (e.g., object's location in a frame).

6.3   Modeling and Analyzing Complex Data Using Multiplexes

Modeling Alternatives: IMDb data set, which most of our readers may be familiar with, captures movie/tv episode, actor, directors and other related information. This is a large data set consisting of 70+ years of data. This data set can be modeled and analyzed in multiple ways for different purposes. It can be modeled as a relational database on which we can ask complex queries. It can also be modeled as a data warehouse (using star or snowflake schema) for performing dimensional analysis using, for example, year, actor, director, and genre as dimensions. The analysis that can be performed using these two models will produce exact, aggregate information, as well as dimensional analysis from various perspectives. It is also possible to generate association rules for this data set by using each movie as a basket and features, such as director, actor, genre, year as items.

However, the above analysis is different from an analysis where you are interested in identifying trends or groupings. For example, identifying a group of
directors who have worked with the strongest group of actors who have acted
in comedy and drama genres is not possible with a query-based or mining approach. Also, identifying the way in which genres have changed over time is also
difficult (if not impossible) using the previous models. Several queries may have
to be posed and results used for obtaining those results.

The IMDb data set can also be modeled using graphs. With this approach,
instead (or in addition to) of querying, you can perform different types of analysis, such as interesting substructure detection, frequently occurring subgraphs,
community detection to identify groups of vertices that are more connected to
each other than to other vertices in the graph [18, 40], or hub detection for identifying nodes that have more connectivity (or influence) than others using different
measures [44]. In order to do this, the data set need to be modeled as a graph.

In order to model a data set as a graph, nodes and edges need to be identified
that correspond to the analysis being performed. This is an important and also
difficult part of modeling. In a straightforward graph model, actors can be used
as nodes and each attribute (e.g., type of genre) can be used to connect two
nodes (i.e., an edge) if two actors (nodes) have acted in at least one (or k in
general) movies classified with that genre (note that movies may belong to multiple genre) the nodes are connected by an edge with the label of the genre. This
straightforward modeling creates a single graph (termed a monoplex) that will
have multiple edges between nodes (based on different relationships) for a large
data set. Analysis of such a graph needs to separate graphs based on labeling
if one wants to only analyze actors with respect to a particular genre. As the
number of features increase, the complexity of the graph as well as its understanding & processing will increase. For the same data set, there are several ways
of doing this modeling. Although community detection and hub detection algorithms exist, they typically do not handle multiple edges forcing one to analyze
the monoplex as mentioned above.

6.3.1 Our Contributions

Multiplex-Based Modeling: Instead of modeling complex data as a monoplex, multiplexes have been proposed [36, 51]. In a multiplex, instead of creating
a single graph with colored nodes and/or edges, a number of graphs are created,
each representing one aspect or feature or perspective. For example, Fig. 7(a)
shows such a multiplex. In each layer, actors are nodes and one genre is used
to create edges. We believe this model is better than a monoplex as the graph
in each layer is smaller, will not have multiple edges, and easier to understand
semantically.

Note that in Fig. 7(a), each layer has the same set of nodes (actors in this case). Hence, there is no need to connect nodes across layers. This type of multiplex is termed a *homogeneous multiplex*.

It is also possible to model this data set in alternative ways using graphs. For example, actors could be linked using movies in

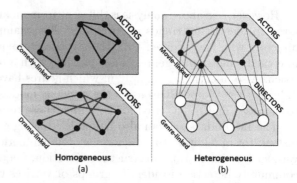

Fig. 7. Multiplex modeling of IMDB dataset

which the two actors have worked together. In a different layer, directors can be linked if two directors have directed a movie of a specific genre type. This multiplex is shown in Fig. 7(b) where the two layers do not have the same set and type of nodes. In this model, it is possible to capture a third relationship between nodes of the two layers. For example, directors who have directed an actor can be linked as shown in Fig. 7(b). This type of multiplex is termed a *heterogeneous multiplex* based on the node types of the layers. Given these two types, it is possible to have *hybrid multiplexes* as well.

Efficient Analysis of Multiplexes: Our ongoing research towards big data analytics includes modeling and efficient analysis of multiplexes. Applying community and hub detection algorithms for a multiplex is being explored by the research community [54,57]. This, in our view, is not the best way as the complexity of community and hub detection increases and the decomposition of the problem into its components (layers) is not leveraged.

For analyzing multiplexes in a holistic and flexible way, we propose **composition** as the approach of choice. The basic idea of composition is to analyze **individual layers** using extant algorithms and *compose* the results of individual layers to obtain results for any combinations of layers in a loss-less manner. As part of [52], we have proposed an intersection-based approach to compute communities of any AND-based feature combination in a loss-less manner, if the layer-wise communities are self-preserving. Based on the analysis expectation, our approach allows computing communities for the required subsets of layers without having to generate the combined layer. Thus, for an exhaustive community analysis of a multiplex with N layers, instead of generating $O(2^N)$ combined layers we just need to re-use the N sets of layer-wise communities. We have extended this flexible analysis approach to **hub detection** of homogeneous multiplexes as well [53].

In summary, as shown in Fig. 1 and in this section, analysis expectations drive the choice of models and the types of algorithms used for analysis. The goal should be to develop a large suite of modeling and analysis tools so appropriate ones can be chosen for analyzing a given problem.

7   Conclusions

In this retrospective stroll, we have outlined the relevance and need for data management from its humble beginnings to the current expectations and status. This is purely our perspective and we believe that a number of technological and other advances as well as application/user expectations coming together symbiotically in a timely manner has resulted in the current status. Needless to stay that this evolution is still in its infancy and will continue with additional contributions as the data sizes become even larger, the variety of data types even broader, data characteristics even wider (than 4Vs), and expectations on the kinds of knowledge inferred for decision support even greater.

Acknowledgment. We would like to thank Dr. Sanjukta Bhowmick on her collaboration with us on the multilayer network analysis.

References

1. Agrawal, R., Imielinski, T., Swami, A.: Database mining: a performance perspective. IEEE Trans. Knowl. Data Eng. **5**(6), 914–925 (1993)
2. Anwar, E., Maugis, L., Chakravarthy, S.: A new perspective on rule support for object-oriented databases. In: SIGMOD Conference, pp. 99–108 (1993)
3. Arasu, A., Widom, J.: A denotational semantics for continuous queries over streams and relations. SIGMOD Rec. **33**(3), 6–12 (2004)
4. Balachandran, R., Padmanabhan, S., Chakravarthy, S.: Enhanced DB-subdue: supporting subtle aspects of graph mining using a relational approach. In: Ng, W.-K., Kitsuregawa, M., Li, J., Chang, K. (eds.) PAKDD 2006. LNCS (LNAI), vol. 3918, pp. 673–678. Springer, Heidelberg (2006). https://doi.org/10.1007/11731139_77
5. Bodra, J., Das, S., Santra, A., Chakravarthy, S.: Query processing on large graphs: scalability through partitioning. In: Ordonez, C., Bellatreche, L. (eds.) DaWaK 2018. LNCS, vol. 11031, pp. 271–288. Springer, Cham (2018). https://doi.org/10.1007/978-3-319-98539-8_21
6. Bodra, J.D.: Processing Queries Over Partitioned Graph Databases: An Approach And It's Evaluation. Master's thesis, The University of Texas at Arlington, May 2016
7. Chakravarthy, S., Anwar, E., Maugis, L., Mishra, D.: Design of sentinel: an object-oriented DBMS with event-based rules. Inf. Softw. Technol. **36**(9), 559–568 (1994)
8. Chakravarthy, S., et al.: HiPAC: A Research Project in Active. Time-Constrained Database Management. Technical report, Xerox Advanced Information Technology, Cambridge (1989)
9. Chakravarthy, S.: Divide and conquer: a basis for augmenting a conventional query optimizer with multiple query proceesing capabilities. In: ICDE, pp. 482–490 (1991)
10. Chakravarthy, S., Beera, R., Balachandran, R.: DB-subdue: database approach to graph mining. In: Dai, H., Srikant, R., Zhang, C. (eds.) PAKDD 2004. LNCS (LNAI), vol. 3056, pp. 341–350. Springer, Heidelberg (2004). https://doi.org/10.1007/978-3-540-24775-3_42
11. Chakravarthy, S., Jiang, Q.: Stream Data Management: A Quality of Service Perspective. Springer, Boston (2009)

12. Chakravarthy, S., Krishnaprasad, V., Anwar, E., Kim, S.: Composite events for active databases: semantics, contexts and detection. In: VLDB, pp. 606–617 (1994)
13. Chakravarthy, S., Nesson, S.: Making an object-oriented DBMS active: design, implementation, and evaluation of a prototype. In: Bancilhon, F., Thanos, C., Tsichritzis, D. (eds.) EDBT 1990. LNCS, vol. 416, pp. 393–406. Springer, Heidelberg (1990). https://doi.org/10.1007/BFb0022185
14. Chakravarthy, U.S., Grant, J., Minker, J.: Logic-based approach to semantic query optimization. ACM Trans. Database Syst. **15**(2), 162–207 (1990)
15. Chakravarthy, U.S., Minker, J.: Multiple query processing in deductive databases using query graphs. In: VLDB, pp. 384–391 (1986)
16. Chang, F., et al.: Bigtable: a distributed storage system for structured data (awarded best paper!). In: 7th Symposium on Operating Systems Design and Implementation (OSDI 2006), 6–8 November 2006, Seattle, WA, USA, pp. 205–218 (2006). http://www.usenix.org/events/osdi06/tech/chang.html
17. Chellappa, R.: Frontiers in image and video analysis NSF/FBI/DARPA workshop report. In: Workshop, p. 120 (2014). www.umiacs.umd.edu/~rama/NSF_report.pdf
18. Clauset, A., Newman, M.E., Moore, C.: Finding community structure in very large networks. Phys. Rev. E **70**(6), 066111 (2004)
19. Cook, D.J., Holder, L.B.: Substructure discovery using minimum description length and background knowledge. J. Artif. Intell. Res. **1**, 231–255 (1994)
20. Das, S.: Divide and Conquer Approach to Scalable Substructure Discovery: Partitioning Schemes, Algorithms, Optimization And Performance Analysis Using Map/reduce Paradigm. Ph.D. thesis, The University of Texas at Arlington, May 2017
21. Das, S., Chakravarthy, S.: Partition and conquer: map/reduce way of substructure discovery. In: Madria, S., Hara, T. (eds.) DaWaK 2015. LNCS, vol. 9263, pp. 365–378. Springer, Cham (2015). https://doi.org/10.1007/978-3-319-22729-0_28
22. Das, S., Chakravarthy, S.: Duplicate reduction in graph mining: approaches, analysis, and evaluation. IEEE Trans. Knowl. Data Eng. **30**(8), 1454–1466 (2018). https://doi.org/10.1109/TKDE.2018.2795003
23. Das, S., Goyal, A., Chakravarthy, S.: Plan before you execute: a cost-based query optimizer for attributed graph databases. In: DaWaK 2016, Porto, Portugal, 6–8 September 2016, pp. 314–328 (2016)
24. Dayal, U., et al.: The HiPAC project: combining active databases and timing constraints. SIGMOD Rec. **17**(1), 51–70 (1988)
25. Dayal, U., Buchmann, A.P., Chakravarthy, S.: The HiPAC project. In: Active Database Systems: Triggers and Rules for Advanced Database Processing, pp. 177–206. Morgan Kaufmann (1996)
26. Dittrich, K.R., Kotz, A.M., Mulle, J.A.: An event/trigger mechanism to enforce complex consistency constraints in design databases. SIGMOD Rec. **15**(3), 22–36 (1986)
27. Engström, H., Chakravarthy, S., Lings, B.: A systematic approach to selecting maintenance policies in a data warehouse environment. In: Jensen, C.S., et al. (eds.) EDBT 2002. LNCS, vol. 2287, pp. 317–335. Springer, Heidelberg (2002). https://doi.org/10.1007/3-540-45876-X_22
28. Engström, H., Chakravarthy, S., Lings, B.: Implementation and comparative evaluation of maintenance policies in a data warehouse environment. In: Eaglestone, B., North, S., Poulovassilis, A. (eds.) BNCOD 2002. LNCS, vol. 2405, pp. 90–102. Springer, Heidelberg (2002). https://doi.org/10.1007/3-540-45495-0_14

29. Engström, H., Chakravarthy, S., Lings, B.: A heuristic for refresh policy selection in heterogeneous environments. In: ICDE, pp. 674–676 (2003)
30. Engström, H., Chakravarthy, S., Lings, B.: Maintenance policy selection in heterogeneous data warehouse environments: a heuristics-based approach. In: DOLAP, pp. 71–78 (2003)
31. Goyal, A.: QP-SUBDUE: Processing Queries Over Graph Databases. Master's thesis, The University of Texas at Arlington, December 2015
32. Hwang, J.H., Cha, S., Çetintemel, U., Zdonik, S.B.: Borealis-R: a replication-transparent stream processing system for wide-area monitoring applications. In: SIGMOD Conference, pp. 1303–1306 (2008)
33. Jiang, Q., Adaikkalavan, R., Chakravarthy, S.: NFM^i: an inter-domain network fault management system. In: ICDE, pp. 1036–1047 (2005)
34. Jiang, Q., Adaikkalavan, R., Chakravarthy, S.: MavEStream: synergistic integration of stream and event processing. In: International Conference on Digital Communications, p. 29 (2007)
35. Jiang, Q., Chakravarthy, S.: Queueing analysis of relational operators for continuous data streams. In: CIKM, pp. 271–278 (2003)
36. Kivelä, M., Arenas, A., Barthelemy, M., Gleeson, J.P., Moreno, Y., Porter, M.A.: Multilayer networks. CoRR abs/1309.7233 (2013). http://arxiv.org/abs/1309.7233
37. Kona, H., Chakravarthy, S.: An SQL-based approach to incremental association rule mining. Found. Comput. Decis. Sci. J. (2006). Special issue
38. Kona, H., Chakravarthy, S.: Partitioned approach to association rule mining over multiple databases. In: Kambayashi, Y., Mohania, M., Wöß, W. (eds.) DaWaK 2004. LNCS, vol. 3181, pp. 320–330. Springer, Heidelberg (2004). https://doi.org/10.1007/978-3-540-30076-2_32
39. Lerner, A., Shasha, D.: Aquery: query language for ordered data, optimization techniques, and experiments. In: Proceedings of the 29th International Conference on Very Large Data Bases, vol. 29, pp. 345–356. VLDB Endowment (2003)
40. Leskovec, J., Lang, K.J., Dasgupta, A., Mahoney, M.W.: Community structure in large networks: natural cluster sizes and the absence of large well-defined clusters (2008)
41. Mishra, P., Chakravarthy, S.: Performance evaluation and analysis of k-way join variants for association rule mining. In: James, A., Younas, M., Lings, B. (eds.) BNCOD 2003. LNCS, vol. 2712, pp. 95–114. Springer, Heidelberg (2003). https://doi.org/10.1007/3-540-45073-4_9
42. Mishra, P., Chakravarthy, S.: Performance evaluation of SQL-OR variants for association rule mining. In: Kambayashi, Y., Mohania, M., Wöß, W. (eds.) DaWaK 2003. LNCS, vol. 2737, pp. 288–298. Springer, Heidelberg (2003). https://doi.org/10.1007/978-3-540-45228-7_29
43. Mishra, P.: Performance Evaluation and Analysis of SQL-based Approaches for Association Rule Mining. Master's thesis, The University of Texas at Arlington, December 2002
44. Newman, M.: Networks: An Introduction. Oxford University Press Inc., New York (2010)
45. Padmanabhan, S.: HDB-Subdue: A Relational Database Approach to Graph Mining and Hierarchical Reduction. Master's thesis, The University of Texas at Arlington, December 2005
46. Qingchun, J.: A Framework for Supporting Quality of Service Requirements in a Data Stream Management System. Ph.D. thesis, The University of Texas at Arlington, August 2005

47. Ramakrishnan, R.: Database Management Systems. WCB/McGraw-Hill (1998)
48. Elmasri, R., Navathe, S.B.: Fundamentals of Database Systems, 2nd edn. Benjamin/Cummings, Redwood City (1994)
49. Rosenthal, A., Chakravarthy, S., Blaustein, B.T., Blakeley, J.A.: Situation monitoring for active databases. In: VLDB, pp. 455–464 (1989)
50. Rosenthal, A., Chakravarthy, U.S.: Anatomy of a mudular multiple query optimizer. In: VLDB, pp. 230–239 (1988)
51. Santra, A., Bhowmick, S.: Holistic analysis of multi-source, multi-feature data: modeling and computation challenges. In: Reddy, P.K., Sureka, A., Chakravarthy, S., Bhalla, S. (eds.) BDA 2017. LNCS, vol. 10721, pp. 59–68. Springer, Cham (2017). https://doi.org/10.1007/978-3-319-72413-3_4
52. Santra, A., Bhowmick, S., Chakravarthy, S.: Efficient community re-creation in multilayer networks using boolean operations. In: International Conference on Computational Science, ICCS 2017, 12–14 June 2017, Zurich, Switzerland, pp. 58–67 (2017). https://doi.org/10.1016/j.procs.2017.05.246
53. Santra, A., Bhowmick, S., Chakravarthy, S.: HUBify: efficient estimation of central entities across multiplex layer compositions. In: 2017 IEEE International Conference on Data Mining Workshops, ICDM Workshops (2017)
54. Solé-Ribalta, A., De Domenico, M., Gómez, S., Arenas, A.: Centrality rankings in multiplex networks. In: Proceedings of the 2014 ACM Conference on Web Science, pp. 149–155. ACM (2014)
55. Stonebraker, M., Hanson, E., Potamianos, S.: The POSTGRES rule manager. IEEE Trans. Softw. Eng. **14**(7), 897–907 (1988)
56. Zdonik, S.B., Stonebraker, M., Cherniack, M., Çetintemel, U., Balazinska, M., Balakrishnan, H.: The aurora and medusa projects. IEEE Data Eng. Bull. **26**(1), 3–10 (2003)
57. Zhang, H., Wang, C.D., Lai, J.H., Philip, S.Y.: Modularity in complex multilayer networks with multiple aspects: a static perspective. Appl. Inform. **4**, 7 (2017)

Fusion of Game Theory and Big Data
for AI Applications

Praveen Paruchuri$^{(\boxtimes)}$ and Sujit Gujar

Machine Learning Lab, IIIT Hyderabad, Hyderabad, India
{praveen.p,sujit.gujar}@iiit.ac.in

Abstract. With the increasing reach of the Internet, more and more people and their devices are coming online which has resulted in the fact that, a significant amount of our time and a significant number of tasks are getting performed online. As the world moves faster towards more automation and as concepts such as IoT catch up, a lot more (data generation) devices are getting added online without needing the involvement of human agents. The result of all this is that there will be lots (and lots) of information generated in a variety of contexts, in a variety of formats at a variety of rates. Big data analytics therefore becomes (and is already) a vital topic to gain insights or understand the trends encoded in the large datasets. For example, the worldwide Big Data market revenues for software and services are projected to increase from 42 *Billion* USD in 2018 to 103 *Billion* in 2027. However, in the real-world it may not be enough to just perform analysis, but many times there may be a need to operationalize the insights to obtain strategic advantages. Game theory being a mathematical tool to analyze strategic interactions between rational decision-makers, in this paper, we study the usage of Game Theory to obtain strategic advantages in different settings involving usage of large amounts of data. The goal is to provide an overview of the use of game theory in different applications that rely extensively on big data. In particular, we present case studies of four different Artificial Intelligence (AI) applications namely Information Markets, Security systems, Trading agents and Internet Advertising and present details for how game theory helps to tackle them. Each of these applications has been studied in detail in the game theory literature, and different algorithms and techniques have been developed to address the different challenges posed by them.

1 Introduction

The online world has been growing at a tremendously fast pace due to the increasing reach of the Internet, enabled by the penetration of devices like mobile phones and other smart devices along with the significantly increasing bandwidth. This aided with the explosion of social networking sites, image and video sharing websites along with an ever-expanding online storage space has resulted in orders of magnitude increase of publicly available data. The availability of

© Springer Nature Switzerland AG 2018
A. Mondal et al. (Eds.): BDA 2018, LNCS 11297, pp. 55–69, 2018.
https://doi.org/10.1007/978-3-030-04780-1_4

these large datasets has completely altered the way Internet-based companies operate. The companies no longer charge money from customers for the services they provide but instead are very interested to know more details of the customers so they can perform data analytics to obtain useful information that is of commercial value. Given the large datasets involved here, we refer to this process as *Big Data Analytics* [62].

Similarly, increasing computerization of almost all processes we can think of has led to tremendously big databases or data in other storage formats. While the dataset available for these companies are private, the critical challenge remains that these companies need to analyze large data sets to deduce facts and information that can give them an edge over their competitors. In both the scenarios, the need has been the ability to analyze big data sets and perform deductions for different purposes. While the processing speeds have roughly kept in line with Moore's law [32] the data available now is orders of magnitude larger and hence led to the introduction of distributed processing frameworks such as Hadoop [51] and its companion tools such as Spark, Hive, etc.

Big data is characterized by the three V notation: Volume, Velocity, and Variety [62]. What this implies is that not only a lot of data is getting generated but the generation is being done at a fast pace in multiple formats from different sources. As presented in [10], the worldwide Big Data market revenues for software and services are projected to increase from $42B(Billion)$ USD in 2018 to $103B$ in 2027, attaining a Compound Annual Growth Rate (CAGR) of 10.48%. As part of this forecast, Wikibon estimates the worldwide Big Data market is growing at an 11.4% CAGR between 2017 and 2027, growing from $35B$ to $103B$. Please find the summary in Fig. 1 (Source: Wikibon and reported by Statista).

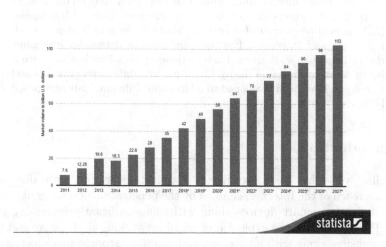

Fig. 1. Revenue forecast (Source: Wikibon and reported by Statista).

In this paper, we illustrate the idea that game theory is an additional useful tool in the hands of big data researchers and practitioners and big data is relevant in designing game theoretic systems. *Game theory* is a mathematical tool that studies and analyzes strategic interactions between rational decision makers [33]. When we look at the big picture for why big data analytics is needed, apart from gaining insights or understanding trends, the analysis many times needs to be operationalized in the real world which leads to strategic advantages. Game theory can help to do this process in a combined manner as opposed to viewing the analysis part as one step and then identifying which idea or piece of data or information can lead to strategic advantages. The goal of this paper is to illustrate how cross-disciplinary research between game theory and big data is enabling newer Artificial Intelligence (AI) applications.

The rest of the paper is structured as follows: We first provide a brief introduction to game theory in Sect. 2. We then analyze in detail how game theory can provide strategic advantages in specific applications namely Information Markets (Sect. 3), Security systems (Sect. 4), Trading agents (Sect. 5) and Internet Advertising (Sect. 6) and how these applications generate and/or use big data.

2   Game Theory Overview

According to Myerson, a Nobel laureate 2007, *"Game theory is the study of mathematical models of conflict and cooperation between intelligent, rational decision-makers"*. Game theory analyzes and predicts the behavior of strategic agents (players) with conflicting interests. Besides this, it also suggests the strategies to play. In this section, we briefly introduce critical concepts in game theory.

To analyze a game, we need to define what the elements of a game are. Elements of a game include:

- **Players** identify the agents playing the game.
- **States** of the game.
- **Actions** that change the state of the game.
- **Knowledge** (beliefs) of the state and actions.
- **Outcome** of the players' actions e.g., payoffs for each player.
- **Payoff or Utility** that each player derives from the outcome (based on actions of all the players).

- **Assumptions about a game**
    • Every player acts rationally so as to maximize its own payoff.
    • Information about game is *common knowledge*

The key question here is, how will the players strategize in a game? To answer this, we first define what a strategy is. *Strategy* is an algorithm or rule by which each player chooses an action, i.e., a complete contingent plan explaining what a player will do in every situation (state). In *Pure strategy*, for each state (or believed state), the action is chosen in a deterministic way whereas, in *Mixed strategy*, a player is allowed to randomize across its pure strategies.

In game theory, the main outcome we look for is equilibrium of a game. An equilibrium is a strategy profile in which all the players prefer to follow the strategies recommended to them as deviation does not help them. There are different types of equilibrium notions, predominantly *dominant strategy equilibrium* and *Nash equilibrium* which we describe in the next subsection.

2.1   Equilibrium

Any finite game can be described using tuple, $\Gamma =< N, (S_i), (u_i) >$ where N is the set of the players, the S_is are possible strategies on how to act across all the states of the game and u_i are the utilities obtained by the players. First, we define what are dominated strategies.

Definition 1. *Given a game $\Gamma =< N, (S_i), (u_i) >$, a strategy $s_i \in S_i$ is said to be weakly dominated if there exists another strategy $s_i' \in S_i$ such that*

$$u_i(s_i, s_{-i}) \leq u_i(s_i', s_{-i}) \forall s_{-i} \in S_{-i}$$

with strict inequality for at least one s_{-i}. In such a case, we say strategy s_i' weakly dominates strategy s_i.

Intuitively, in the above case, a rational player will always prefer to play s_i' over s_i. If there is a unique such strategy for each player, we say the game possesses a dominant strategy equilibrium.

Definition 2. *A strategy $s_i^* \in S_i$ is said to be a* weakly dominant strategy, *for player i if it weakly dominates every other strategy $s_i \in S_i$.*
That is, $\forall s_i \neq s_i$,

$$u_i(s_i, s_{-i}) \leq u_i(s_i^*, s_{-i}) \forall s_{-i} \in S_{-i}$$

with strict inequality for at least one s_{-i}.

However, note that, given a game, most of the games need not possess a dominant strategy equilibrium. In the absence of dominant strategy equilibrium, the most celebrated concept in game theory comes into the picture, *Nash equilibrium*.

Definition 3. *Pure Strategy Nash Equilibrium: A strategy profile $(s_1^*, s_2^*, \ldots, s_n^*)$ is called as Pure Strategy Nash Equilibrium (PSNE), if for each player i, s_i^* is a best response strategy to s_{-i}^*.*

That is, $\forall i$

$$u_i(s_i^*, s_{-i}^*) \geq u_i(s_i, s_{-i}^*) \ \forall \ s_i \in S_i$$

Again, we can easily construct the games where no pure strategy Nash equilibrium exists. It should be noted that the players can also randomize across their strategies, that is, they can play mixed strategies. We therefore define a mixed strategy Nash equilibrium as follows.

Definition 4. *Mixed Strategy Nash Equilibrium: A strategy profile* $(\sigma_1^*, \sigma_2^*, \ldots, \sigma_n^*)$ *is called as Mixed Strategy Nash Equilibrium, if for each player* i, σ_i^* *is a best response strategy to* σ_{-i}^*.

That is, $\forall i$

$$U_i(\sigma_i^*, \sigma_{-i}^*) \geq U_i(\sigma_i, \sigma_{-i}^*) \,\forall\, \sigma_i \in \Delta(S_i)$$

Notice the difference between u_i and U_i. John Nash showed in his celebrated work that given any finite game, it consists of at least one mixed strategy Nash equilibrium [34]. For more details about game theory, an interested reader can refer to [33].

Many times, in the real world, we need to induce a game among the players such that at equilibrium they strategize in a manner we would like them to. For example, data is essential in many AI applications, and this data may be available with strategic players who need not reveal it truthfully to the (AI) system designer unless they are offered proper incentives. This leads to *Mechanism Design Theory*.

2.2 Mechanism Design – A Reverse Game Engineering Aka Incentive Engineering

As pointed out in the previous subsection, agents may have certain private information. To study such situations, *Bayesian* games or *games with incomplete information* are proposed. A strategic form game with incomplete information is defined as a tuple $\Gamma = < N, (\Theta_i), (S_i), (p_i), (u_i) >$ where

- $N = \{1, 2 \ldots, n\}$ Set of players
- Θ_i Set of types of Player i, $i = 1, 2, \ldots, n$
- S_i Set of strategies of Player i, $i = 1, 2, \ldots, n$
- $p_i : \Theta_i \rightarrow \Delta(\Theta_{-i})$. $p_i(.|\theta_i)$ specifies probability distribution over Θ_{-i}. $p_i(.|\theta_i)$ represents Player i's belief about types of other players if its type is θ_i.
- $u_i : \Theta \times S \rightarrow \mathbf{R}$ payoff/utility function for Player i

In mechanism design, the designer, aka social planner in game theory literature, offers certain outcomes (X) based on the actions of the players. The mapping from Θ_is to X is called a *social choice function*. The goal is to design a mechanism, that is define rules of the game such that, at equilibrium, the players report their private information truthfully to the social planer. This is referred to as incentive compatibility.

Definition 5. *Dominant Strategy Incentive Compatibility (DSIC): A social choice function* $f : \Theta_1 \times \ldots \times \Theta_n \rightarrow X$ *is said to be dominant strategy incentive compatible DSIC (or truthfully implementable in dominant strategies) if the direct revelation mechanism* $\mathcal{D} = ((\Theta_i)_{i \in N}, f(\cdot))$ *has a (very weakly) dominant strategy equilibrium* $s^*(\cdot) = (s_1^*(\cdot), \ldots, s_n^*(\cdot))$ *in which* $s_i^*(\theta_i) = \theta_i, \forall \theta_i \in \Theta_i, \forall i \in N$.

That is, SCF f is DSIC if

$$u_i\left(f(\theta_i, \theta_{-i}), \theta_i\right) \geq u_i\left(f(\theta'_i, \theta_{-i}), \theta_i\right)$$

$$\forall\, \theta'_i \in \Theta_i,\ \forall\, \theta_i \in \Theta_i;\, \forall\, \theta_{-i} \in \Theta_{-i},\ \forall\, i \in N$$

Mechanism design is very rich in concepts and results. For a more detailed tutorial on mechanism design, the interested readers can refer to [19,20].

3   Information Markets

Suppose a social planner is interested in building a pollution map for the city of Zurich. Jutzeler *et al.* proposed models for data analytics on Zurich ultra-fine particle database [27] for such activities. To build this, we need to gather information at various locations. In addition to data collection, an underlying assumption is that the data provided is valid data. Similarly, if one wants to build some health-care analytics based on data of citizens of a metro, we need trusted data.

Reliable data can be produced by aggregating and processing crowd-sensed information obtained by agents. There are three main challenges to aggregate information in this way. The first challenge is that the agents should be willing to gather and provide data which may not happen, unless sufficient incentives are offered. The second challenge is to ensure that the agents are truthful in providing the data. The third challenge is that agents should be prevented from colluding to provide the same data and accumulate incentives. We call such marketplaces where information can be traded as *Information Markets*. There are two types of game-theoretic mechanisms are deployed in information markets: (i) *prediction Market based*, and (ii) *Peer Prediction based*

A *Prediction Market* seeks to predict the outcome of a future event. The market prices can indicate what the crowd thinks the probability of the event is. The main purpose of prediction market is to elicit aggregate beliefs over an unknown future outcome. Traders with different beliefs trade on contracts whose payoffs are related to the unknown future outcome and the market prices of the contracts are considered as the aggregated belief. These mechanisms use strictly proper scoring rules to incentivize agents for truthful revelation e.g., logarithmic scoring rule [24]. Prediction markets are used in data analytics, e.g., [12,44].

Peer Prediction markets use an interesting approach. Let's say we want to gather information about pollution level at a particular location in Zurich at 6:30pm. We can ask multiple crowd agents about this. A strategic agent may report fake information. In peer prediction mechanisms, the same task is assigned to multiple agents and each agent is matched with another randomly selected agent and paid only if both the answers match. The incentive offered is inversely proportional to the likelihood of the answer. There are multiple challenges to resolve here and hence is a very active area of research.

[40] proposed Bayesian Truth Serum (BTS) mechanism which does not require knowledge of any common prior information. However, it is applicable

only for a large number of agents. This seminal work is one of the first to enable information elicitation without common knowledge assumption and inspired several new mechanisms. [60] and [61] propose mechanisms that neither assume any common prior information nor large number of agents but are robust to private beliefs of agents. However, these mechanisms suffer from *temporal separation.* It requires one report before and one after executing the crowd-sensed task.

Riley *et al.* [43] proposed minimalistic mechanism under the assumption that all the agents with the same outcome have the same posterior expectations. Faltings *et al.* [15] introduced Peer Truth Serum (PTS) which is a minimalistic mechanism but assumes prior belief model. The mechanism also admits uninformed equilibria where agents do not perform measurements. Radanovic *et al.* [41] propose a improved version of [15], that introduces Logarithmic Peer Truth Serum (LPTS) by eliminating the dependency on prior belief model. The mechanism produces worse payoff than truthful reporting for uninformed equilibria and against misbehaving agents acting on collusion strategies. For more details about game theory and data collection, we refer an interested reader to the book [16].

4   Stackelberg Security Games (SSGs)

Protecting national infrastructure such as airports, historical landmarks, or locations of political or economic importance is a challenging task for police, military and other security agencies, a challenge that is exacerbated by the threat of terrorism [39]. Security agencies across the world therefore have the unenviable task of needing to provide security 24*7 across the length and breadth of countries. Police typically need to follow a multi-pronged approach involving usage of cameras, placing checkpoints, deploying physical patrol units including canine units, intelligence collection, patrolling with robots or drones and others. However, limited security resources prevent full security coverage at all times, hence the need to make optimal usage of available resources. The key idea behind the line of work on SSGs is that one agent (the leader) must commit to a strategy that can be observed by the other agent (the follower or adversary) before the adversary chooses its own strategy. Such a formulation is called Stackelberg game and has been used or studied in different contexts [5, 11, 38, 54].

SSGs have received wide attention in literature [55] and led to a number of deployed systems for a variety of settings (a number of them initiated at the Teamcore group of University of Southern California). The initial deployed system in this line of work, the ARMOR system [39] modeled the security domain as a general purpose Bayesian Stackelberg game and used the DOBSS algorithm to compute the optimal leader strategy [37]. Successful deployment of this system resulted in tremendous interest in SSGs that led to development of significantly enhanced systems for a wide variety of domains. The next system developed in this line of work was IRIS [56], a game-theoretic scheduler for randomized deployment of the U.S. Federal Air Marshal Service (FAMS) that has been in use since 2009. It was followed by the PROTECT system [50] which got deployed for generating randomized patrol schedules for the U.S. Coast Guard in Boston, New York, Los Angeles, and other ports around the United States. As mentioned in [52], SSGs have also found applications in cybersecurity [48,53], robot patrolling [4], drug design against viruses [36], traffic enforcement [45], adversarial learning [59], and many others. Green security games [17,18] focus on defending against environmental crimes. In recent times, SSGs are being used to solve a lot more complex problems such as protecting biodiversity in conservation areas that span over 2500 square kilometers [17] and screening 800 million airport passengers annually throughout USA [6]. A detailed overview of SSGs is presented in [52].

Figure 2 shows the global presence of security games efforts. As shown in the figure the efforts in this direction span numerous countries and continents. Source for the figure: Tutorial titled "Advances in Game Theory for Security and Privacy" presented in ACM-EC 2017.

Global Presence of Security Games Efforts

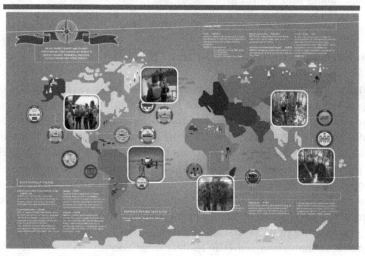

Fig. 2. SSGs Worldwide [Source: From tutorial titled "Advances in Game Theory for Security and Privacy" presented in ACM-EC 2017]

The question that naturally arises is how is all of this connected to big data? And the high level answer for this is, apart from the fact that the nature of domains are vastly different which throws up a different set of constraints and challenges, the size of the solution space has in general have increased significantly. As stated in [52], the defender strategy space in recent applications like TSG (Threat Screening Game) is larger than 10^{33}. In the road networks security problem, the adversary space is of the order 10^{18}. Note that this is not the raw data that needs to be stored but is generated due to combinations of raw data that result in strategy space. However this still doesn't mean the problems are any easy and will need smart ways in which the strategy space gets built and pruned in real time that allows these algorithms to be used in practice. Many of these are deployed systems which means that inspite of having huge strategy spaces game theoretic models have been able to solve real world problem sizes of this order. The exact details of each domain and the specific algorithms used to solve them have been presented in detail in literature. We therefore skip the details here apart from noting that the as SSGs get even more used in practice the strategy spaces are going to get bigger which means smarter game theoretic techniques would need to be developed to handle them.

5 Trading Agents

As the world gets increasing automated, electricity markets are not immune to change. The concept of smart grid is gaining attention and there is a lot of interest in making different aspects of it a reality. Smart grid includes a variety of operational and energy measures including smart meters, smart appliances, renewable energy resources, and energy efficient resources [47]. To take advantage of these properties of smart grid, governments around the world are trying to convert their electricity grid into a smart grid with supporting retail market infrastructure and customer participation in power markets through demand side management and distributed generation [57]. Power TAC [28,29] provides a low overhead simulation of a future smart grid environment with renewable energy production, smart metering, autonomous agents acting on behalf of customers and retailers, state-of-the-art customer models, and realistic market designs. The Power Trading Agent Competition (Power TAC) [29] held annually using the Power TAC simulation environment encourages development of autonomous broker agents that aim to maximize profit by buying and selling energy in the wholesale and retail markets, subject to some fixed costs and constraints. The broker would therefore need to design competitive tariffs to offer in retail market while carefully balacing the demand and supply by trading in a wholesale market. Since it is a real-time supply demand balancing problem, they can also make up the difference with a significant penalty from balancing market which makes advanced and accurate planning a lot more profitable.

Past tournaments have shown that brokers can be built using a variety of techniques including Game Theoretic, Machine Learning and Search/Planning based strategies which have been found to be pretty useful to dynamically price tariffs and predict customer usage while simultaneously placing bids in wholesale auctions. Brokers that used game theoretic modeling in combination with other techniques include [9,30,57,58]. Brokers that used MDPs to model strategies include [13,42,57,58,63]. Note that some of these works use MDPs along with game theoretic modeling since developing a broker agent for power TAC includes developing strategies for wholesale, retail and balancing markets that work in tandem. Brokers that employed genetic algorithm, fuzzy-logic and tailored heuristics include [30,35,46]. The Power TAC 2018 Finals had 7 brokers from research groups across the globe. The tournament had a total of 324 games, with all possible combinations of 7-broker games (100 games), 4-broker games

(140 games; 80 games for each broker), and 2-broker games (84 games; 24 games for each broker). The tournament ran 24*7 for close to two weeks and a large amount of data was generated in the process. As happened with data from prior years, the newly generated data can help with data driven modeling wherein the generated data can be used for designing new or improving the design of existing bots whether using a game theoretic approach or otherwise.

Figure 3 shows the architecture diagram for the power TAC simulator. As shown in the figure there are a number of players involved across the wholesale, retail and balancing markets and a lot of information gets exchanged in the process. Source for the figure: power TAC website http://www.powertac.org/ (October 2018).

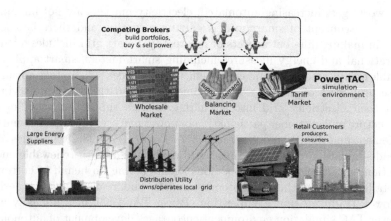

Fig. 3. Power TAC architecture [Source: power TAC website http://www.powertac. org/ (Retrieved October 2018)]

6   Internet Ad Auction

Internet advertising is becoming popular due to advances Artificial Intelligence (AI) and the increasing reach of internet. AI, especially data analytics, plays a crucial role in identifying appropriate target consumers for ads, leading to a paradigm shift in the advertising industry. In 2016, advertisers spent more than 100 Billion USD on Internet advertising, which accounts for more than 25% of the budget across all the modes of advertising industry.

Advertisers being strategic agents, may manipulate the underlying AI (in particular predictive data-analytics) algorithms which learn about the consumer behavior. For example, consider advertising on search engines such as Google, Bing, etc. Whenever a search engine's user searches for a specific keyword(s), the search engine along with the search results, also displays relevant advertisements. Typically, the selection of ads to be displayed is based on the valuation per click that advertisers perceive they obtain. These valuations are collected in the form of bids through an auction called sponsored search auction (SSA).

In SSA, not only the bid but the probability of an ad receiving a click, also known as click-through rate (CTR) of the ad plays an important role. In the beginning, neither the search engine nor the advertisers know CTRs. The search engine can use Multi-Armed-Bandit (MAB) algorithms to learn CTRs by displaying each ad multiple times and observing clicks. However, a naive implementation of MAB algorithms can fail when advertisers misreport their valuation per click. Hence, one needs to design a manipulation resistant MAB algorithm which is called a MAB Mechanism. In short, there is a need to fuse game theoretic techniques with the machine learning while designing such AI algorithms inspired by modern web applications.

In Multi-Armed Bandit (MAB) problem, there is a decision maker who has to pull an arm from a given set of arms having unknown stochastic rewards. The dilemma for the decision maker is to choose an arm that has the highest expected reward based on historical observations (exploit) or choose another arm to learn about its expected reward (explore). MAB problem was first studied by Robbins in 1952. Based on this seminal work, MAB problems have been extensively studied for regret analysis and convergence rates. Readers are referred to [1] for regret analysis in finite time MAB problems. The regret of a MAB algorithm is measured as expected loss in the rewards as compared to the case when the expected rewards are known.

In Sponsored search auctions over search engines, CTRs of the advertisers are not known initially. The value per click is private information of the advertisers. The decisions are based on both the stochastic rewards as well as such private information, and strategic advertisers may manipulate the MAB algorithm. To prevent manipulations in SSA, computer scientists proposed *MAB mechanisms*. [3] derived a characterization for truthful MAB mechanisms in the context of sponsored search auctions if there is only a single slot for advertisement. They have shown that any truthful MAB mechanism must have at least $\omega(T^{2/3})$ regret in T rounds. In known valuations settings (or non-strategic agents who report valuations truthfully) the regret is $O(T^{1/2})$ for the same set of agents. [14] have also addressed the problem of designing truthful MAB mechanisms for sponsored search auctions with a single sponsored slot. Though they have not explicitly attempted the characterization of truthful MAB mechanisms, they have derived similar results on regret as in [3]. [64] considered the learning of CTRs with agents who can reincarnate themselves. [64] uses a weaker notion of incentive compatibility, namely Bayesian Incentive Compatibility to optimally learn the CTRs and avoid re-incarnation by the advertisers with a new identity. However, the paper considers only a single slot case and does not perform any regret analysis.

A general transformation which outputs a randomized truthful mechanism with any monotone allocation rule is designed in [2]. As an application of this transformation, a MAB mechanism for single slot sponsored search auction that is ex-post DSIC with regret of $O(T^{1/2})$ is obtained. The problem of incentivizing agents to explore the arms in order to maximize the expected reward is considered in [31]. [49] characterizes, what are called all MAB mechanisms for multi-slot sponsored search auctions that are dominant strategy incentive compatible while [21] designs one such auction. [22] proposes to use Thompson sampling-based approach to design auction in MAB settings.

Apart from MAB mechanism design approach, there is vast amount of literature on use data analytics techniques in Internet Advertising, e.g., [7,8,23,25,26] and the references cited therein.

7 Conclusions

In this paper, we present an overview of the use of game theory in different applications that rely extensively on big data. In particular, we present case studies for four different AI applications namely Information Markets, Security systems, Trading agents and Internet Advertising and present details for how game theory helps to tackle them. Our survey of these applications shows that fusion of game theory and big data indeed provides significant advantages that may not be available otherwise. For example, in Information markets, game theory is used to incentivize agents to (truthfully) report the data they possess. The availability of large data sets in Security games allows for a more accurate modeling of the game matrices and opponent modeling which helps to identify better security strategies. In the Trading agents application, there is a large scale generation of data that happens independent of whether the broker is game theoretic or not. This large scale data helps or has helped with a data driven refinement of the trading agents for better performance. Note that different agents have different strategies and hence different ways in which the data gets used for refinement. Finally in the Internet Advertising application, big data plays a crucial role for targeted advertisement through a better design of auctions.

References

1. Auer, P., Cesa-Bianchi, N., Fischer, P.: Finite-time analysis of the multiarmed bandit problem. Mach. Learn. **47**(2–3), 235–256 (2002)
2. Babaioff, M., Kleinberg, R.D., Slivkins, A.: Truthful mechanisms with implicit payment computation. In: Eleventh ACM Conference on Electronic Commerce, pp. 43–52. ACM (2010)
3. Babaioff, M., Sharma, Y., Slivkins, A.: Characterizing truthful multi-armed bandit mechanisms: extended abstract. In: Tenth ACM Conference on Electronic Commerce, pp. 79–88. ACM (2009)
4. Basilico, N., Gatti, N., Amigoni., F.: Leader-follower strategies for robotic patrolling in environments with arbitrary topologies. In: AAMAS, pp. 57–64 (2009)

5. Brown, G., Carlyle, M., Salmeron, J., Wood., K.: Defending critical infrastructure. Interfaces **36**(6), 530–544 (2006)
6. Brown, M., Sinha, A., Schlenker, A., Tambe., M.: One size does not fit all: a game-theoretic approach for dynamically and effectively screening for threats. In: AAAI (2016)
7. Budhiraja, A., Reddy, P.K.: An improved approach for long tail advertising in sponsored search. In: Candan, S., Chen, L., Pedersen, T.B., Chang, L., Hua, W. (eds.) DASFAA 2017. LNCS, vol. 10178, pp. 169–184. Springer, Cham (2017). https://doi.org/10.1007/978-3-319-55699-4_11
8. Cheng, H., Cantú-Paz, E.: Personalized click prediction in sponsored search. In: Proceedings of the Third ACM International Conference on Web Search and Data Mining, pp. 351–360. ACM (2010)
9. Chowdhury, M.M.P., Kiekintveld, C., Son, T.C., Yeoh, W.: Bidding strategy for periodic double auctions using Monte Carlo tree search. In: AAMAS, pp. 1897–1899 (2018)
10. Columbus., L.: 10 charts that will change your perspective of big data's growth (2018). https://www.forbes.com/sites/louiscolumbus/2018/05/23/10-charts-that-will-change-your-perspective-of-big-datas-growth/#3c7b5c0d2926. Accessed Oct 2018
11. Conitzer, V., Sandholm., T.: Computing the optimal strategy to commit to. In: Proceedings of ACM EC, pp. 82–90 (2006)
12. Cowgill, B., Wolfers, J., Zitzewitz, E.: Using prediction markets to track information flows: evidence from google. In: Das, S., Ostrovsky, M., Pennock, D., Szymanksi, B. (eds.) AMMA 2009. LNICST, vol. 14, p. 3. Springer, Heidelberg (2009). https://doi.org/10.1007/978-3-642-03821-1_2
13. Cuevas, J.S., Rodriguez-Gonzalez, A.Y., Cote, E.M.D.: Fixed-price tariff generation using reinforcement learning. In: Modern Approaches to Agent-based Complex Automated Negotiation, pp. 121–136 (2017)
14. Devanur, N.R., Kakade, S.M.: The price of truthfulness for pay-per-click auctions. In: Tenth ACM Conference on Electronic Commerce, pp. 99–106 (2009)
15. Faltings, B., Li, J.J., Jurca, R.: Incentive mechanisms for community sensing. IEEE Trans. Comput. **63**(1), 115–128 (2014)
16. Faltings, B., Radanovic, G.: Game theory for data science: eliciting truthful information. Synth. Lect. Artif. Intell. Mach. Learn. **11**(2), 1–151 (2017)
17. Fang, F., Nguyen., T.: Green security games: apply game theory to addressing green security challenges. In: ACM SIGecom Exchanges (2016)
18. Fang, F., Stone, P., Tambe., M.: When security games go green: designing defender strategies to prevent poaching and illegal fishing. In: IJCAI (2015)
19. Garg, D., Narahari, Y., Gujar, S.: Foundations of mechanism design: a tutorial part 1-key concepts and classical results. Sadhana **33**(2), 83–130 (2008)
20. Garg, D., Narahari, Y., Gujar, S.: Foundations of mechanism design: a tutorial part 2-advanced concepts and results. Sadhana **33**(2), 131–174 (2008)
21. Gatti, N., Lazaric, A., Trovò, F.: A truthful learning mechanism for contextual multi-slot sponsored search auctions with externalities. In: Thirteenth ACM Conference on Electronic Commerce, pp. 605–622 (2012)
22. Ghalme, G., Jain, S., Gujar, S., Narahari, Y.: Thompson sampling based mechanisms for stochastic multi-armed bandit problems. In: Proceedings of the 16th Conference on Autonomous Agents and MultiAgent Systems, pp. 87–95. International Foundation for Autonomous Agents and Multiagent Systems (2017)
23. Ghose, A., Yang, S.: An empirical analysis of sponsored search performance in search engine advertising. In: Proceedings of the 2008 International Conference on Web Search and Data Mining, pp. 241–250. ACM (2008)

24. Hanson, R.: Logarithmic markets coring rules for modular combinatorial information aggregation. J. Prediction Markets **1**(1), 3–15 (2012)
25. Hillard, D., Schroedl, S., Manavoglu, E., Raghavan, H., Leggetter, C.: Improving ad relevance in sponsored search. In: Proceedings of the Third ACM International Conference on Web Search and Data Mining, pp. 361–370 (2010)
26. Joshi, A., Motwani, R.: Keyword generation for search engine advertising. In: Sixth IEEE International Conference on Data Mining Workshops ICDM Workshops 2006, pp. 490–496. IEEE (2006)
27. Jutzeler, A., Li, J.J., Faltings, B.: A region-based model for estimating urban air pollution. In: Proceedings of the 28th Conference on Artificial Intelligence (AAAI), pp. 424–430 (2014)
28. Ketter, W., Collins, J., Reddy., P.: Power tac: a competitive economic simulation of the smart grid. Energy Econ. **39**, 262–270 (2013)
29. Ketter, W., Collins, J., Weerdt, M.: The 2018 power trading agent competition (2017)
30. Liefers, B., Hoogland, J., Poutré, H.L.: A successful broker agent for Power TAC. In: Agent-Mediated Electronic Commerce. Designing Trading Strategies and Mechanisms for Electronic Markets, pp. 99–113 (2014)
31. Mansour, Y., Slivkins, A., Syrgkanis, V.: Bayesian incentive-compatible bandit exploration. In: Proceedings of the Sixteenth ACM Conference on Economics and Computation, pp. 565–582. ACM (2015)
32. Moore, G.E.: Cramming more components onto integrated circuits. Electron. Magaz. (1965). Accessed Oct 2018
33. Narahari, Y.: Game theory and mechanism design. World Scientific **4** (2014)
34. Nash, J.: Non-cooperative games. Ann. Math. **54**, 286–295 (1951)
35. Özdemir, S., Unland, R.: AgentUDE17: a genetic algorithm to optimize the parameters of an electricity tariff in a smart grid environment. In: Advances in Practical Applications of Agents, Multi-Agent Systems, and Complexity: The PAAMS Collection, pp. 224–236 (2018)
36. Panda, S., Vorobeychik., Y.: Stackelberg games for vaccine design. In: AAMAS, pp. 1391–1399 (2015)
37. Paruchuri, P., Pearce, J.P., Marecki, J., Tambe, M., Ordonez, F., Kraus., S.: Playing games for security: an efficient exact algorithm for solving bayesian stackelberg games. AAMAS **2**, 895–902 (2008)
38. Paruchuri, P., Pearce, J.P., Tambe, M., Ordonez, F., Kraus., S.: An efficient heuristic approach for security against multiple adversaries. In: AAMAS (2007)
39. Pita, J., et al.: Deployed armor protection: the application of a game theoretic model for security at the los angeles international airport. In: AAMAS Industry Track, pp. 125–132 (2008)
40. Prelec, D.: A bayesian truth serum for subjective data. Science **306**(5695), 462–466 (2004)
41. Radanovic, G., Faltings, B.: Incentive schemes for participatory sensing. In: Proceedings of the 2015 International Conference on Autonomous Agents and Multiagent Systems, pp. 1081–1089. International Foundation for Autonomous Agents and Multiagent Systems (2015)
42. Reddy, P.P., Veloso, M.M.: Strategy learning for autonomous agents in smart grid markets. In: IJCAI, pp. 1446–1451 (2011)
43. Riley, B.: Minimum truth serums with optional predictions. In: Proceedings of the 4th Workshop on Social Computing and User Generated Content (SC14) (2014)

44. Ritterman, J., Osborne, M., Klein, E.: Using prediction markets and twitter to predict a swine flu pandemic. In: 1st International Workshop on Mining Social Media, vol. 9, pp. 9–17 (2009)
45. Rosenfeld, A., Maksimov, O., Kraus, S.: Optimizing traffic enforcement: from the lab to the roads. In: Rass, S., An, B., Kiekintveld, C., Fang, F., Schauer, S. (eds.) Decision and Game Theory for Security. GameSec 2017. LNCS, vol. 10575. pp. 3–20. Springer, Cham (2017). https://doi.org/10.1007/978-3-319-68711-7_1
46. Rúbio, T.R.P.M., Queiroz, J., Cardoso, H.L., Rocha, A.P., Oliveira, E.: TugaTAC broker: a fuzzy logic adaptive reasoning agent for energy trading. In: Rovatsos, M., Vouros, G., Julian, V. (eds.) EUMAS/AT -2015. LNCS (LNAI), vol. 9571, pp. 188–202. Springer, Cham (2016). https://doi.org/10.1007/978-3-319-33509-4_16
47. Saleh, M.S., Althaibani, A., Esa, Y., Mhandi, Y., Mohamed., A.A.: Impact of clustering microgrids on their stability and resilience during blackouts. In: ICSGCE, p. 195–200 (2015)
48. Sengupta, S., et al.: A game theoretic approach to strategy generation for moving target defense in web applications. In: AAMAS, pp. 178–186 (2017)
49. Sharma, A.D., Gujar, S., Narahari, Y.: Truthful multi-armed bandit mechanisms for multi-slot sponsored search auctions. Curr. Sci. **103**, 1064–1077 (2012)
50. Shieh, E.A., et al.: Protect: an application of computational game theory for the security of the ports of the united states. In: AAAI (2012)
51. da Silva Morais, T.: Survey on frameworks for distributed computing: hadoop, spark and storm. In: 10th Doctoral Symposium in Informatics Engineering-DSIE (2015)
52. Sinha, A., Fang, F., An, B., Kiekintveld, C., Tambe., M.: Stackelberg security games: looking beyond a decade of success. In: IJCAI, pp. 5494–5501 (2018)
53. Sinha, A., Nguyen, T.H., Kar, D., Brown, M., Tambe, M., Jiang., A.X.: From physical security to cybersecurity. J. Cybersecur. **1**, 19–35 (2015)
54. Stengel, B.V., Zamir., S.: Leadership with commitment to mixed strategies. Technical Report LSE-CDAM (2004)
55. Tambe., M.: Security and Game Theory: Algorithms, Deployed Systems, Lessons Learned. Cambridge University Press, Cambridge (2011)
56. Tsai, J., Kiekintveld, C., Ordonez, F., Tambe, M., Rathi., S.: Iris-a tool for strategic security allocation in transportation networks. In: AAMAS Industry Track (2009)
57. Urieli, D., Stone, P.: TacTex'13: a champion adaptive power trading agent. In: AAAI, pp. 465–471 (2014)
58. Urieli, D., Stone, P.: Autonomous electricity trading using time-of-use tariffs in a competitive market. In: AAAI (2016)
59. Vorobeychik., Y.: Adversarial ai. IJCAI, pp. 4094–4099 (2016)
60. Witkowski, J., Parkes, D.C.: Peer prediction with private beliefs. In: Proceedings of the 1st Workshop on Social Computing and User Generated Content (SC 2011) (2011)
61. Witkowski, J., Parkes, D.C.: Peer prediction without a common prior. In: Proceedings of the 13th ACM Conference on Electronic Commerce, pp. 964–981. ACM (2012)
62. Xiao.: Impact of clustering microgrids on their stability and resilience during blackouts. In: CSGCE, pp. 195–200 (2015)
63. Yang, Y., Hao, J., Sun, M., Wang, Z., Fan, C., Strbac, G.: Recurrent deep multiagent Q-learning for autonomous brokers in smart grid. In: IJCAI, pp. 569–575 (2018)
64. Zoeter, O.: On a form of advertiser cheating in sponsored search and a dynamic-VCG solution. In: Proceedings of TROA (2008)

Financial Data Analytics and Data Streams

Distributed Financial Calculation Framework on Cloud Computing Environment

Rao Casturi[1(✉)] and Rajshekhar Sunderraman[2]

[1] Risk Reporting and Technology, Voya Investment Management,
Atlanta, GA 30327, USA
Rao.Casturi@voya.com
[2] Department of Computer Science, Georgia State University,
Atlanta, GA 30303, USA
raj@cs.gsu.edu

Abstract. Even though the recent technological innovations in cloud computing, distributed data base architecture grew to a point where they can now address Big Data process, still most of the companies are struggling to implement their solutions on cloud computing environment. This is mainly due to lack of proper case studies and application frameworks available on a public research domain. Most of the implementations are vendor provided, high cost, consulting type and long drawn projects which consume valuable Business and IT resources for an organization. In this paper we propose a simple to implement (parallel data load and any aggregation calculation framework), easy to maintain and flexible architecture framework which can be adapted as a tool for small to mid-size investment organization in implementing a Distributed Cloud Computing Architecture. Our framework is an extension of traditional Distributed Database Design with the horizontal partition of the relations to parallelize the computation on Azure SQL instance and materialize the aggregated results with SQL Views for users, Business Intelligence (BI) Reporting, Data Mining and Knowledge Discovery applications. The solution is implemented on Azure SQL Cloud Computing platform to build the financial calculation framework.

Keywords: Big data · Mortgage factors · Azure Cloud

1 Introduction

The raw data (factor data) for Mortgage pools [1, 2] is released on a monthly basis by Mortgage Agencies like Ginnie Mae (GN), Freddie Mac (FH) and Fannie Mae (FN). The individual financial institutions either use third party vendor provided solutions or build their own data gathering applications to collate this information. The data set is around 2 TB every month and grows from month over month as the loans and pools tend to increase as time progresses. Currently we are only targeting a small portion of the mortgage pool universe as the current architecture can't even handle bigger data sets in our existing calculation framework. With our proposed new architecture, we can target full universe of mortgage pool data or factor data. For smaller institutions the cost in implementing third party vendor for Big Data solutions are expensive and can't

© Springer Nature Switzerland AG 2018
A. Mondal et al. (Eds.): BDA 2018, LNCS 11297, pp. 73–88, 2018.
https://doi.org/10.1007/978-3-030-04780-1_5

justify the costs. After going through various cost benefit scenarios, we in our organization decided, to build a flexible framework which not only capture the raw mortgage factor data in timely manner but also build our user de-fined calculation analytical engine to aggregate factor data to support future Portfolios Analytical Decision Systems (PADS) [3] with the help of latest Microsoft Cloud Computing Technology (Azure). Our approach and successful implementation of our proposed solution gave our portfolio managers the ability to analysis huge amount of data in a very short period compared to the legacy EXCEL based application. The EXCEL model was severely limited to the amount of information we can analyze in a given day. This is due to the technical limitations of processing Big Data in EXCEL. Excel can be effective as an analysis tool for small amount of data but when it comes to Big Data analysis, it is severely limited. Subsequent paragraphs, however, are indented.

The contribution of our paper is a flexible calculation framework implemented successfully on AZURE SQL Cloud Computing [4] Environment by "distributing the raw Big Data among multiple SQL relations (tables) and build a parallel execution meth-od based on Node based architecture harnessing the computational power of Azure SQL Cloud based implementation which can save huge costs to the organization in terms of hardware and software. The architecture on Azure is expandable to meet the organization needs when it comes to computational power. The "pay as you use" operation model for a long time in service sectors and that concept is now introduced and implemented successfully with Microsoft Azure Cloud Computing. We can scale up or down on our computational power depending on our needs.

Our contribution and research in creating a node-based architecture on Azure SQL Database [5] can be used by any organization as a case study to implement their data driven financial calculation applications which deal with large or Big Data and utilize industry strength cloud platforms (Microsoft, ORACLE, AMAZON etc.).

The current paper is organized into seven sections. Section 2 introduces the preliminary definitions and background information. Section 3 highlights the problem, presents related work done on migration of traditional database models to cloud based models in academia, and introduces our proposed solution. In Sect. 4, we present the implementation of our solution over the current mortgage database. Section 5 compares the benefits from existing architecture to the newly proposed framework. Section 6 is our conclusion and Sect. 7 gives our future works including Data Mining initiatives and Hadoop and MapReduce project initiatives.

2 Preliminaries and Background Information

An Investment Portfolio can consist of Fixed Income Bonds [1] and or Equity securities. In the day and age of big data, the key analytical indicators [2] of these investment assets drive lot of portfolio decisions. Calculating and picking up trends in these analytical measures is a challenge when we need to go through millions of mortgage loans. For our discussion, we will focus on the mortgage backed securities. The Mortgage Backed Securities (MBS) can be further classified into two broad categories by their defining attributes (Agency and Non-Agency mortgage-backed). The pool of residential mortgages can be thousands of individual loans put together.

Individual or institution investors can buy this MBS. The Fig. 1 is a very high level of how an investor or an institution can invest into a MBS depending on their risk appetite.

Fig. 1. High-level Mortgage Backed Securities life cycle

The mortgage pools are made up of individual loans. The data size of these pools can be viewed as Big Data. The main characteristics of Big Data is defined by the main three pillars by which we categorize the underlying data set. They are called 3Vs. [9].

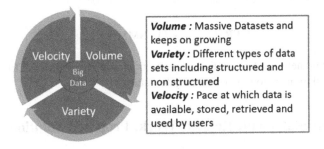

Fig. 2. Big Data categorization diagram

The Fig. 2 shows 3Vs of Big Data in our study are Volume (Terabytes of data) in raw format with the Variety (mixed data values) and Velocity (Changing daily) which makes it a best candidate for our study and implement our flexible calculation framework which can be leveraged for other investment data related decision systems.

The details of MBS [8] security or the structure is out of scope for this paper discussion. To make it simple if we take the loan 1 from Fig. 3 we can see what is outstanding on the mortgage as a balance and what is the rate at which the loan is being serviced and the how may months are remaining on the loan. Along with it also give the % weight of the loan in a given pool where it belongs. This is key for an investor when they are looking to invest in this pool of securities. From these basic measures we compute more complicated measures to use as indicators for investing.

Loan	Outstanding mortgage balance	Weight in pool	Mortgage rate	Months remaining
1	$125,000	22.12%	7.50%	275
2	$85,000	15.04%	7.20%	260
3	$175,000	30.97%	7.00%	290
4	$110,000	19.47%	7.80%	285
5	$70,000	12.39%	6.90%	270
Total	$565,000	100.00%	7.28%	279

Fig. 3. Example of a pools of loans in an MBS pass through security [1]

There are several calculations done on various dimensions (range bound). E.g. Showing the WAC over 0–2, 2–5, 5–7, 7–10 and above 10. Another example of a measure is Conditional Prepayment Rate (CPR). CPR is a sub calculation on Single Month Mortality Rate (SMMR) which is calculated for each month of the mortgage and use that SMMR to calculate the CPR for a specific month. The Eq. (2) shows the calculation for one-month CPR. SMM is calculated with as shown in Eq. (1). Some of the mortgage calculations are recursive in nature.

$$\text{SMM}(t) = \text{Prepayment in month }(t)/(\text{beginning mortgage balance for month }(t)$$
$$- \text{ scheduled principal payment in month }(t)) \tag{1}$$

$$\mathbf{CPR}(1) = 1 - (1 - \text{SMM}(1))12 \tag{2}$$

Prior to our new implementation, the solution was based on EXCEL with VBA code for calculation various fixed aggregate values.

3 Problem Statement, Related Work, Proposed Solution

3.1 Problem Statement

Excel is a tool for analysis for a limited data set but not for a Big Data analytical analysis. There are three major issues we face in our current EXCEL based tool. Extract Transform and Load (ETL) [7] the raw agency mortgage files are not sustainable and is not possible as each file can contain millions of tuples. The second issue of EXCEL as a calculation tool limits the ability of flexibility for any new metric calculations which are frequently needed on either ad-hoc basis or on a permanent basis. Due to the nature of links and look ups and no control on editing in the existing EXCEL solution, it opens a huge operational risk to the organization. The third constraint is the Storage of historical data [19] for trend analysis and data mining. Excel can't handle a 2 TB data set or aggregated data [14] of 5 to 10 million rows in EXCEL for one month depending on number of calculations needed and performed. For Data Mining we do need historical data set to see patterns and further extend it to AI and Machine Learning algorithms, our current solution can't be used.

Summarizing the issues, we have three main problem segments. (1) ETL is only possible for one deal at a time and is time consuming. (2) User defined calculations are not easily possible. (3) Trend and Data Mining [20] capability is not available. Solving these three issues will increase the productivity of our investment teams giving more time for analysis rather than working on data collection and code manipulation. Our contribution to the research community is a perfect use case for a financial calculation framework using the Cloud Computing solution for our Big Data problem. Our Mortgage Factor Raw data can be viewed as Big Data and a good candidate for our "Capture, Transform, Calculate and Visualize" (CTCV) framework which can be leveraged for other investment data related decision systems.

3.2 Related Work

There is lot of academic research and vendor products available in the domain space ETL in terms of processing multiple files efficiently. There are several architectural frameworks which can be implemented on our problem. In academia as well as in industry there is major research work carried out in terms of supporting large files coming from various source systems. During 2002 and 2003 Google came up with their proprietary file system Google File System (GFS) [12] which led the way to several other research groups to come up with their architecture to support large file systems. Google also published their MapReduce [13] programming model which can process terabytes of data on thousands of machines. MapReduce program was distributed over large clusters of commodity machines. During this time, Apache open-source developed Hadoop architecture and called it as Hadoop Distributed File Systems (HDFS) and introduced their MapReduce programming model. The MapReduce architecture uses a map function specified by users that processes key-value pair to generate a set of intermediate key-value pairs, and a reduce function that merges all the intermediate values with the intermediate key [11]. Even though our implementation is not on Hadoop platform we are researching the possibility of our implementation of raw factor file load via Hadoop-MapReduce architecture.

As part of the research for our project, we came across several Distributed Processes (DP) [19] or Distributed Computing (DC) methods and papers. DP and DC are inter-linked and to some extent distributed processing is already implemented in our modern computer architecture. For example, in single-processor based computing systems, the central processing unit (CPU) and the input and output devices and operations (I/O) are separated and overlapped where it is possible. Parallel computing is further defined by Flynn's Taxonomy of Parallel Architectures by data and instruction set execution. Single-Instruction, Single-Data (SISD), Multiple-Instruction, Single-Data (MISD), Single-Instruction, Multiple-Data (SIMD) and Multiple-Instruction, Multiple-Data (MIMD). Each one of the four (SISD, MISD, SIMD, MIMD) are studied extensively and implementations are very widely used in research as well as in industry. There are various ELT tools adapted and incorporated all four types of architectural designs. Example Microsoft SQL Server Integration Tool (SSIS) can handle multiple data files in parallel by using the For-Loop Container [7]. The design is to create Parallel Task and Child Task and copy the Parallel Task to achieve the parallelism by copying the child task inside the Parallel Task Container.

Bootstrapped to SSIS package the SQL Procedure we came up to calculate our user defined calculations is a perfect solution for the parallel execution of the code.

We evaluated several industry products in the Big Data and Cloud Computing space including Oracle Cloud Platform, Microsoft Cloud Technology (AZURE) [10] and Amazon Web Services (AWS). The best suited for our needs is Microsoft platform minimizing the business disruption. Figure 4 gives us a high-level view of various products the vendors offer in the Cloud Computing Space.

Cloud Product	Short Description and services
SaaS: Software as a Service	Centrally hosted and managed for the end users. Scalable to meet customer demand. E.g.. Like any utility company
PaaS: Platform as a Service	Vendor provides platform and Customer provides applications. E.g. Virtual Machines, SQL Serves , Web Servers etc.
IaaS: Infrastructure as a Service	Vendor provides all the support and maintenance of server farms, running virtualization software.

Fig. 4. Brief description of major services by Cloud Computing providers

The gaps we found in our research literature is how to implement a *user defined Cloud Computing Architecture for a Financial Calculations*. This is not done or not available to research community as to design and build such kind of applications needs a lot of financial acumen along with a very strong Database Design and Development knowledge. The vendor products available are very expensive and are focused to a very narrow needs of business. Some example of such vendors is in MBS space are KDSGlobal, MDS etc. We could not find any relevant research material on this topic so we decided to research and find a solution which can be ported easily to other financial calculation not just restricting to our Mortgage Factor Aggregation (MFA). The complex calculations should be handled by a flexible user defined rules to accomplish the MFA.

3.3   Proposed Solution

Our approach is to use the Azure SQL Database Server to host our raw factor data for mortgage pools (Big Data) and run distributed user defined calculations on the raw data files sourced to SQL stage tables in parallel and materialize the results for BI Tools. The actual data load can be accomplished in parallel using Microsoft SQL Server Integration Services (SSIS) Looping techniques. This is not the focus for our paper as parallel loading is done many a times using many vendor tools like Microsoft SSIS and Hadoop - Map Reduce Techniques. The challenge for us is to design a database schema to accommodate our goal of providing easy, flexible and cost-effective user defined calculation engine on a Cloud Platform using distributed database design patterns. The focus of this paper is on the second part of the overall problem statement of parallel execution of user defined calculations on Azure SQL Cloud Computing environment.

At high level the architecture of the proposed solution is shown in Fig. 5. The idea was tried several times and implemented.

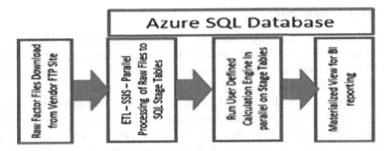

Fig. 5. Proposed Distributed Calculation on Data Copies (SQL Stage Tables)

The proposed architecture is a flexible data base design using the standard Relational Schema Architecture on Microsoft SQL Server Database. The execution of the calculation engine runs in parallel as multiple instances with different data set. The actual creation of the Azure SQL Database is out of scope for this paper.

The architecture in layman terms is like a single cash counter at a store checkout verses multiple cash counters. The customers with the products they bought can be treated as our raw factor files and each cash counter is a node designed to calculate the total amount to be charged per each customer and updates one main register with the customer's payment. This is concept of parallel processing is not a new concept but adapting to financial calculations on cloud computing environment gives re-search community to use it in a public domain. Technical details are discussed in the implementation section.

4 Implementation

Our proposed implementation is on Microsoft AZURE SQL Server with a S3 standard (100 DTUs). For this implementation we created a non-elastic database pool and with 250 GB database. The actual creation of the AZURE SQL Server Database is out of scope for this paper but can be referenced to prior paper by the same authors [21]. Along with the Azure implementation, we also implemented the same solution on on-premises SQL database on Windows server with 8 CPUs and 20 GB RAM with SQL 2012 database. The on-premises solution serves two purposes. The on-premises solution serves as a benchmark to our Azure and serial processes of Big Data and secondly the on-premises solution gives us an opportunity to propose an institutional solution for Azure applications for future investment frameworks and paves path for "Global Access" of our applications.

Our IT department was able to procure a user license for our research on Azure Cloud Platform providing the tools needed (Visual Studio for SSIS and Microsoft SQL Management Studio Client) in order to connect and test our proposed concept of user defined calculation database schema. The key it to utilize the database design and ability of Azure Cloud database architecture to drive.

4.1   Current Architecture

The current architecture can only process one factor file a given time. There is a Data Load Form which will ask the user to input the deal number and then the EXCEL Macro (VBA Code) will be executed to get the data from raw factor files from the Vendor Data Source (VDS) depending on the user data license.

Even though we can modify the EXCEL Solution code to take in multiple deals at a time, still the raw factor file sourcing, parsing and loading will be an issue in EXCEL sheet. Some of these deals will have many loans (in millions) which EXCEL sheet can't handle. The EXCEL application is built to extract the data, transform the data and also calculate the user defined calculations one deal at a time. Modification of the VBA Code is very manually. The EXCEL version served as a business use case for us to build more robust and enterprise solution using the standard tools and the utilize new cloud computing technology platform.

4.2   Proposed Architecture

The proposed architecture is based on AZURE SQL Server with RDBM [9] Architecture. The core architecture is based on the main assumption that user defined calculation rules can be dynamic. Users can come up with another set of new calculations which are critical to the business and the system should able to handle without any code rewrites. If there is a specific calculation which our calculation engine can't handle, we can incorporate it easily in our main calculation engine by extending the existing calculation classification category. The calculation engine is based on calculation classification category as a primary driver for extendable code. To have a dynamic calculation and reporting framework, we designed a flexible database schema, giving the users the ability to edit the core calculation rules. A calculation rule can be defined and saved in a SQL Relation as following in terms of relational algebra shown in the next couple of lines shown as C(i)

$$C(i) = \pi R(\text{reportdate}, \text{collateralgroupname}, \text{agencytype})$$
$$(\text{reportdate}, \text{collateralgroupname}, \text{agencytype}) FSUMobal, SUM \text{ olnsz } (\text{StageTabe R}) \qquad (3)$$

C(i) is the user defined calculation rule
π is the selection of the attributes from the Relation (Stage Table R)
F is the aggregate function to be used on the set group by attributes

The calculation "C" can be set up by the user and that we capture as an individual calculation. The database schema [17, 18] consists of 5 major relations (SQL tables). They are (1) User Calculation Rule Master also called Calculation Field Master (2) Range Definition Master (3) Distinct Values Data Set (Key-Value Pair) (4) Result Materialization and (5) Raw Factor Data. With few supporting relations we were able to build a dynamic user defined and easy to implement architecture which is portable to any other data sets. We enhanced the calculation framework to parallelize by adding a "Node" identifier when a calculation is initiated. As shown in Fig. 6. The Raw Factor

Files are sourced through the ETL process to the SQL Stage Tables. Once the raw factor file is loaded the SSIS Package will initiate "Master SQL Procedure" (MSP) with a Node ID and "SoureTableName" (STN) as parameters. MSP Will work on the STN going through all the user defined calculations and save the results to "Results" Relation and marking the file process complete in the Process Log Table. The MSP is written in anticipation of platform independent by using standard SQL code.

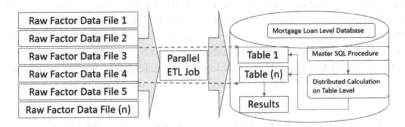

Fig. 6. Proposed Architecture (Parallel computational execution on Table 1..n)

SQL Table Description: Calculation Field Master (CFM) will hold the calculations to be done on different attributes. Example of a user set up is shown in the Fig. 7. This table has 16 attributes and to save space we are showing only few attributes. The Field Name denotes the actual Calculation to be done and rest of the fields will give more details about the calculation. Where there are weighted average or any specific calculations using a denominator and numerator, we can specify that in the CFM table. There are fields available for custom numerator and denominator.

Calc ID	Field Name	Calculation type	Where Condition	Custom Calculation	Range ID
1	CBal	SUM	AND Olnsz <> 0	(100*(1-POWER(SUM(CBal)/ NULLIF(SUM(Sbal1),0),12)))	NULL
2	Olnsz Repo	MAX	NULL		NULL
3	AGE	DYNAMIC RANGE	NULL		NULL
4	FICO	RANGE	NULL		1

Fig. 7. Calculation field master table layout

Range Master (RM): The RANGE calculation is used to set up an aggregation over a range. Example is FICO where the data should be aggregated by a specific bucket using the Range Values. There are two types of Range Values. One is a static range value like FICO and other type is dynamic. For a dynamic range the calculation engine will determine the range (min, max) with in each deal during the execution and the calculation engine will generate the ranges for the given attribute into four buckets.

Range ID	Field ID	Range Description	Min Value	Max Value
1	4	FICO > 0 AND FICO < 500	0	500
1	4	FICO > = 500 AND FICO < 600	500	600
1	4	FICO > = 600 FICO < 650	600	650

Fig. 8. Range Master Table Layout

A sample user defined calculation is to find the FICO Score by the shown in the sample FICO Range Set up in the RANGE MASTER Table in Fig. 8. The CalcId = 4 from the Field Master has a relationship with the FieldId in Range Master table giving the calculation engine the needed information to group by the field ranges defined. The Type "RANGE" in the Field Master table give the calculation further information to fetch the ranges for the given field. This is all done dynamically during the actual execution of the calculation.

Distinct Values (DV) or Key-Value: Holds all the distinct domain values for the attributes which we use for the calculation. Example for Servicer attribute, the do-main values could be any qualified servicer to service the loan. The key value for the servicer "1 ST STATE BANK" is currently 12 and for AUBURNBANK is 34. The KeyID will be used to save the final results. This Key-Value table is called Distinct Value Table (DVT) for easy reference. The Fig. 9. shows the Distinct Values Table Layout.

Key ID	Attribute	Value
9	Collateral Group Name	2014-28-G5
10	Loan Prefix	GN
11	Loan Prefix	FN
12	Servicer	1 ST State Bank
24007	Field Display	AtIssuance_SATI_Units_CBRs

Fig. 9. Distinct Values Table (DVT) Layout (Key-Value Pair)

Attribute Master: This determines what domain values are needed to build the Key-Value table which is used by the calculation. This table has 3 fields Attribute Id, Attribute Name, Attribute active flag which can be used to suppress if we don't need the specific attribute in future. All the tables are few other tables and all are designed with Relational Database Design architecture.

With our new proposed architecture, the Vendor File Extract is built on Microsoft SQL Integration Services (SSIS) which is an industry standard ETL tool. Individual file is parsed to an individual table (Stage table) by providing the source data copy for each calculation node. Node ID give the details in a supporting table which file is passed as variable to the SQL Stored Procedure. Process Flag "N" Indicates that the file is not assigned any Node or not ready for calculation yet.

Date Key	Asof Date	Process Flag	Node ID
1	201801	Y	1
1	201801	Y	2
1	201801	N	NULL
1	201801	N	NULL

Fig. 10. Master Node Table (Adhoc Date Table)

Once the source raw file is available the entry in the Master Node Table, the entry will be updated with the Source Table Name and the Node ID. Example if a source file (remic_FN_aa.txt) is available for the node 1 then the entry will look like the first row in the Fig. 10. With the source file name added to the table.

Summary Data: The summary table holds the final calculations done on the raw files and materialized the final aggregated results. The sample table layout is shown in the Fig. 11. The Result table is designed as Key-Value table following standard STAR [9] Schema. The Summary Data table was designed for the maximum flexibility and adaptability. The L1, L2, L3 and L4 will define the grouping for the Field ID (translated to the actual field using the Key-Value Pair). Date Key Id will bring back the data for which the calculation is been done. Like ASOF Data of the raw file. The N1 defines the values for that grouping level. We designed four numeric value fields in Summary Data table to accommodate all the calculation needs. We don't need any more than four numeric values to capture any MBS calculation. There is no need to modify this table design for any future expansions in storing any kind of summarized data with our design. The view created for the users and for BI applications bring the actual data values for reporting needs.

FACT ID	DATE KEY VALUE	FIELD ID	L1	L2	L3	L4	N1
1	1	1	5	10	2054	0	16343223.2
2	1	1	6	10	2054	0	48287178.7
3	1	1	7	10	2054	0	35042135.3
4	1	1	8	10	2054	0	19986496.1
5	1	1	9	10	2054	0	44080898.3
857	1	1	83259	10	2054	0	34710340.4
871	1	1	83249	10	2054	0	3214011.47
879	1	1	83245	10	2054	0	0
1036	1	1	58108	10	2054	0	0
1040	1	1	58112	10	2054	0	689229.569

Fig. 11. Fact Table Result with Key Value Pair for various aggregation sets

The interpretation of the above Fact Table if we convert them to actual values will be as for Field Cbal (Field ID) the SUM value will be 1634223.2 (N1) for the collateral group name (L1) 2013-67-FN, Agency as FN (L2). L3 and L4 are used for Top values, CBRs, CPRs and CRRs. The L3 and L4 are multipurpose fields but values are functionally dependent on the calculation filed master calculation field set up.

If a calculation has a range bound value then the range id will be saved in L4 of the Fact Table. In the traditional calculation frameworks these ranges or any range bound values are created as additional attributes in the base table and used in the calculation (aggregation functions). By our methods of dynamically generating these ranges at the calculation time, reduces the storage needed to store these as a new attributes and also give the ability to build new ranges without any DDLs (ALTER Table) to extend the existing schema. The range can be defined with a CASE statement as shown in following code snippet for each tuple. This statement is generated dynamically by the calculation engine and saves an extra attribute to be created in the source raw factor table.

CASE   WHEN (FICO > 0 AND FICO < 500) THEN 1
WHEN (FICO >= 500 AND FICO < 650) THEN 2
WHEN (FICO >= 650 AND FICO < 700) THEN 3
WHEN (FICO >= 700 THEN 4 **END** AS **FICO_BUCKET**

In the sample SQL {FICO_BUCKET} is replace by the CASE Statement at the execution time of the calculation engine as following SQL Statement to calculate the Sum of Original Balance (OBAL).

SELECT ASOF_DATE, LOAN_ID, {FICO_BUCKET}, SUM(OBAL) AS OBAL
FROM dbo.LOAN_LEVEL_DATA
GROUP BY ASOF_DATE, LOAN_ID, FICO_BUCKET

New Proposed Algorithm and Process: The new proposed algorithm is encompassed into two SQL Stored Procedures (SP). The two steps (Distinct Values, Aggregation) are done in memory to minimize the IO operation. The high level process flow in Fig. 13 and algorithm shown in Fig. 12.

The main algorithm shown in the Fig. 13, between (Start) and (End) the steps are performed on each Factor Raw Files (FR). The calculation engine is activated by providing an identifiable Node ID and a Source File Name on which the calculations are to be performed.

The main calculation is encompassed into a single SQL Stored procedure which has 2 parameters. The Fig. 12 shows the name of the procedure and the 2 parameters. The first parameter is the source file table name (Stage Table) and the next parameter is the node id. One raw factor file contains more than one deal. A deal is not split across multiple raw factor files. The Raw Data has 73 attributes for each file and there are 97 different calculations involved for each loan. The SQL Stored Procedure is built for flexibility to expand or add any new Calculation Types which are not currently available in the exiting code. The main procedure which goes through each row in the Calculation Master to pick up the Calculation Type and generates a complex dynamic SQL based on all the different attributes from Calculation Master Table and executes

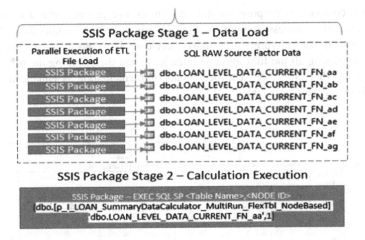

Fig. 12. Parallel execution of SSIS Package with the Node Attached Calculation

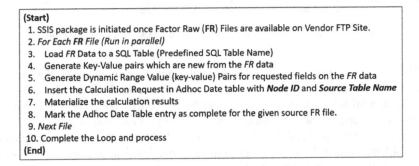

Fig. 13. High-level Algorithm for the New Distributed Calculation Engine

the calculation on the stage data source. In-memory SQL tables are used to speed up the calculations and dropped once the calculation is performed and results are materialized.

The reasons why we used the SQL Stored Procedure for calculation implementation with distributed stage tables having the local data copy, over SSIS package implementation was discussed in our conclusion section.

5 Results

Our implementation is a big step up from existing application based on EXCEL version. As we see the EXCEL version can't handle some of the larger files as these files are contain millions of rows which EXCEL can't handle nor build for such kind of application or data processing. Our way of with simple and open architecture design and using the best in class (SSIS, SQL SPs) enabled the complete process to be done for approximately under an hour for 2,440 loans having 14,895,727 rows with total of 7 files. For a comparison purpose we did benchmark our process with a bench-mark process (sequential process) implemented on a standalone on-premises SQL database before migrating that to Azure SQL Database in cloud. The timing on the same data set but with processing all the files into one huge table and then running the user defined calculation on that one table. The benchmark implementation is enhanced to have INDEX keys on the single table.

			INDEX KEYS			Calculation				
ID	Scenario Type	Achitecture	Avaliable	# of Keys	Time (minutes)	# of Calculations	Time (minutes)	Total Rows	Type	Total time (minutes)
1	Factor Raw Files load to a single SQL Table	On-premisis	No	NA	NA	97	171	14,895,727	Single Instance	171
2	Factor Raw Files load to a single SQL Table	On-premisis	Yes	17	21	97	28	14,895,727	Single Instance	45
3	Factor Raw Files Loaded to a separate SQL Stage Table	On-premisis	No	NA	NA	97	4.5*	14,895,727	Parallel - Multiple Instances	4.5*
4	Factor Raw Files Loaded to a separate SQL Stage Table	On-premisis	Yes	17	12	97		14,895,727	Parallel - Multiple Instances	16.5*
5	Factor Raw Files Loaded to a separate SQL Stage Table	Azure Cloud SQL DB	No	NA	NA	97	51*	14,895,727	Parallel - Multiple Instances	51*

Note : *Average time (Min : 1 minute Max 6 minutes depending on the Factor Raw file size

Fig. 14. Various Scenarios to test to propose a best-fit for organization

The benchmark implementation (Sequential Standalone on-premises SQL implementation) took approximately 3 h to run calculations on the 14MM tuples (ID #1 in Fig. 14). In total we marked around 171 min for the benchmark process to complete. We then compared to our new process which the longest calculation path took 6 min which has 3.2 mm tuples and the table has NO INDEX Keys. The average completion time of 4.5 min which shows in the ID #3 (parallel multiple instances of calculation) in the Fig. 14. Without any INDEX KEYS still the overall process took less time than the one we created for benchmark process. The ID #5 in the results table Fig. 14. shows our Azure implementation with No Index Keys still outperform the single file single calculation instance with a speed up of approximately 58%. The Azure implementation did show a longer execution time than the on-premises parallel (ID #3) process. This is due to the contract and service we have on Azure which is not totally comparable to the on-premises environment. Even though Cloud computing process is slower than the on-premises process, makes our case stronger to migrate to Azure due to the ability to grow the memory and the CPUs in the cloud. Our point is over a period of time, we do need more computational power and with cloud we can achieve that without any overhead on our existing infrastructure. The time saving is due to the distributed nature of the calculation engine which can be run in parallel over multiple source files.

For the results it is very clear that ID #3 which is our preferred implementation outperforms any other implementation. The Azure implementation ID #5 can outperform our preferred implementation of on-premises distributed implementation if we had gone to a higher service contract. Having higher DTU on Azure Cloud we can conclude that our implementation of distributed framework can outperform on-premises data architecture and also save money by not having hardware on-site.

6 Conclusion

As expected, the new method has a superior performance compared to ad-hoc approach on EXCEL based VBA. The EXCEL can't even be a solution in an enterprise environment. This can be a serious problem from an operational risk. Even though the cloud computing did not outperform the parallel and distributed solution we had as benchmark over 8 CPU, 20 GB RAM, it gave us a very good understanding of the future needs for Azure Cloud DB requirements. Example to match our current needs we would go to the next level of Azure Subscription will cost around $4.056 per one-hour computational need which at our current parallel timing, we will spend less than $100 per monthly calculations. Our standard license of S3 on Azure which gives a limited number of CPUs and RAM memory. With more resources (DTU and CPUs) we can achieve a higher speedup using Azure. With more computational power we can bring down the final calculation time to near 4 min to match the on-premises. The implementation is to spread our calculations on distributed files (tables) and running the calculations in parallel on the Azure SQL Cloud Environment.

There are very strong reasons why we implemented our user defined calculation through a SQL Stored procedure rather than a SSIS ETL [6] package. The first reason (a) SSIS is efficient when the calculation is at a tuple level and complicates at a group level. Second reason (b) In SSIS it is not possible to build a dynamic range (MIN, MAX). Third (c) the portability of the SSIS is limited as it needs Microsoft Developer tools. SQL Stored procedure is portable to any standard SQL instance running with ANSI Standard SQL. This is major win as we can port our code to any environment either on premises SQL instance or Cloud Computing SQL instance. The fourth (d) Maintaining a SQL Stored procedure is simpler than an SSIS and the skill set required to enhance a SQL Stored Procedure is much widely available than SSIS coding. In future we are going to implement the ETL load with Hadoop and MapReduce to have the portability for ETL also.

Even though the on-premises implementation showed better results, the disadvantage of this is an overhead of hardware and software costs on our organization to maintain this solution. We recommend the Azure solution with a higher subscription to gain the performance speedup and reduce overall technology costs in maintenance hardware and software on-premises.

7 Future Work

We are extending our architecture for Data Mining [9] to find trend and identify market events and alert our Risk and Portfolio Managers. This we are calling it as "Early Warning Analytical Trigger System (EWATS)". Data aggregation is a key building block in our Data Mining [16] and Machine Learning financial applications [19]. The biggest work we are undertaking is to implement the data load process using the Big Data ETL tools using Hadoop and MapReduce. Even though we wanted to try and use the Massively Parallel Processing (MPP), due to time constraints we were not taking advantage of MPP. This is slated as future development requirement to give us more computational power by implementing over Hadoop [15] and MapReduce. Our implementation did give the foundation for future Distributed Cloud Computing Framework for Financial Applications.

References

1. Fabozzi, F.: Fixed Income Analysis, CFA Institute Investment Series, 2nd edn. Wiley
2. Fabozzi, F.: The Hand Book of Fixed Income Securities, 7th edn. McGraw-Hill
3. Han, J., Kamber, M., Pei, J.: Data Mining Concepts and Techniques, 3rd edn. Morgan Kaufmann
4. Microsoft Press: Cloud Application Architecture Guide
5. Microsoft Azure Documentation: Load data from flat files into Azure SQL database. Microsoft
6. Microsoft SQL Server 2012: Analysis Services – Multidimensional Modeling, E-book publication, June 2012
7. Microsoft SQL Server 2012 Integration Services: Tok, Wee-Hyong, Parida, Rakesh; Masson, Matt; Ding, Xiaoning; Srivashnmugam, Kaarthik; Microsoft Press E-Book publication
8. Jorion, P.: Financial Risk Manager Handbook, 5th edn. GARP, Wiley
9. Elmasri, R., Navathe, S.: Fundamentals of Data Base Systems, 7th edn. Pearson
10. Microsoft Azure Documentation: Overview of Azure Data Lake Store, Microsoft
11. Apache Hadoop. http://hadoop.apache.org/
12. Ghemawat, S., Gobioff, H., Leug, S.-T.: The Google file system. In: SOSP 2003, 19–22 October 2003, Bolton Landing, New York, USA (2003)
13. Gemayel, N.: Analyzing Google file system and Hadoop distributed file system. Res. J. Inf. Technol. **8**, 66–74 (2016)
14. Agarwal, S., et al.: On the computation of multidimentional aggregates. In: Proceedings of the International Conference on Very Large Databases, pp. 506–521 (1996)
15. White, T.: Hadoop: The Definitive Guide, 3rd edn. O'Reilly Media, Inc., Sebastopol (2012)
16. Yen, S., Chen, A.: An efficient data mining technique for discovering interesting association rules. IEEE (1997)
17. Codd, E.F.: A relational model of data for large shared data banks. Commun. ACM **13**(6), 377–388 (1970)
18. Ozsu, M., Valduriez, P.: Principles of Distributed Database Systems, 3rd edn. Springer, New York (2011). https://doi.org/10.1007/978-1-4419-8834-8
19. Witten, I., Frank, E., Hall, M.: Data Mining Practical Machine Learning Tools and Techniques, 3rd edn. Morgan Kaufmann
20. Eaton, C., Deroos, D., Deutsch, T., Lapis, G., Zikopoulos, P.: Under-standing Big Data Analytics for Enterprise Class Hadoop and Streaming Data. McGraw-Hill. ISBN 978-0-07-179053-6
21. Casturi, R., Sunderraman, R.: Script based migration toolkit for cloud computing architecture in building scalable investment platforms. In: Elloumi, M., et al. (eds.) DEXA 2018. CCIS, vol. 903, pp. 46–64. Springer, Cham (2018). https://doi.org/10.1007/978-3-319-99133-7_4

Testing Concept Drift Detection Technique on Data Stream

Narinder Singh Punn[(⊠)] and Sonali Agarwal

Indian Institute of Information Technology Allahabad, Allahabad, India
{pse2017002, sonali}@iiita.ac.in

Abstract. Data mutates dynamically, and these transmutations are so diverse that it affects the quality and reliability of the model. Concept Drift is the quandary of such dynamic cognitions and modifications in the data stream which leads to change in the behaviour of the model. The problem of concept drift affects the prognostication quality of the software and thus reduces its precision. In most of the drift detection methods, it is followed that there are given labels for the incipient data sample which however is not practically possible. In this paper, the performance and accuracy of the proposed concept drift detection technique for the classification of streaming data with undefined labels will be tested. Testing is followed with the creation of the centroid classification model by utilizing some training examples with defined labels and test its precision with the test set and then compare the accuracy of the prediction model with and without the proposed concept drift detection technique.

Keywords: Concept drift · Data stream testing · Drift detection techniques Supervised/unsupervised learning

1 Introduction

As today an enormous amount of information is coming from various sources also known as data streams, arriving sequentially and at high speed. Some of this information can serve as the input data for the classification models which is a two-phase process involving Training phase: build a classification model with the given labelled objects and Testing phase: given a model, predict the labels of unknown objects. Since data streams are not always stationary, the change in information can occur in such a way that it changes the prediction behaviour of the classification model which is also known as the concept drift. In non-stationary contexts [1], the classifier will suffer from the concept drift due to data evolution and new data families, i.e. class values. This concept drift is a serious issue, and it must be handled in such a way that the performance of the model does not degrade. Thus making it necessary to test the concept drift detection technique on the data stream.

To build a sustainable model, it is necessary to detect the concept drift as a means to track the performance of the model. To address this problem, first, develop the classification model using some of the labelled data which will serve as training phase of the model and remaining data can be used in the testing phase of the model. In this paper, the centroid-based classification model is used, and to this, concept drift

© Springer Nature Switzerland AG 2018
A. Mondal et al. (Eds.): BDA 2018, LNCS 11297, pp. 89–99, 2018.
https://doi.org/10.1007/978-3-030-04780-1_6

detection technique is added to check for concept drift before making the prediction and retraining the model. And finally, with the help of experimental results, it is shown that how the accuracy of the model is affected by the use of proposed concept drift detection technique.

2 Related Work

Data stream is the continuous flow of data in real time which is coming from various sources at high speed. Due to this, there are many challenges like infinite length, concept-drift, concept-evolution and feature-evolution, which are faced in Data stream [2]. A lot of research is afoot to address these challenges [3].

In the field of the data stream, concept drift is one of the critical challenges to be dealt with. As for now, there is no specific literature available for testing of the concept drift on the data stream. There have been many ongoing types of research on the concept drift detection techniques [4]. There are many well-known algorithms for the detection of concept drift, and the majority of these algorithms [5, 6] are classified under the category of tree-based or rule-based [4].

In this paper, the focus is given on the statistical approach [7] to detect the concept drift and test the same approach for how it affects the accuracy of the model. The significance of this approach is that it helps to understand multiple models which are strategically analyzed and combined to solve the problem.

3 Problem Formulation

Given the classification model, some threshold: 'λ' is defined where $\lambda \in [0, 1]$ for proposed detection process, which must be satisfied by the test data points to belong to some class. If no such class exists then, the conclusion can be made that data point may belong to some new family of class, and it is more likely to cause concept drift in the proposed model.

Now more parameters: '$\mu_1, \mu_2, \mu_3, \ldots, \mu_n$' are computed where $\mu \in [0, 1]$ and 'n' is number of classes; for any new object whose class label is unknown. These parameters are calibrated by proposed concept drift detection procedure whose value defines how closely new data object is related to different classes.

If $\mu_n > \lambda \ \forall \ n =$ number of classes then the conclusion can be made that it is concept drift.

From the above relation, it can be stated that threshold value affects the sensitivity of the proposed detection procedure; smaller the value of threshold more sensitive is the technique to detect the concept drift.

4 Concept Drift Detection Procedure

This procedure describes how to calculate the parameter 'μ', compare its value with the threshold and make the decision for concept drift.

Approach: Initially start by building the classification model in the training phase. Before predicting the unlabeled data, it is passed through concept drift detection procedure and check how it relates to the current behaviour of the model.

At first, find the centroids of all the available classes in the classification model. These centroids, training data with which the model was developed, Test data whose class needs to be predicted and Threshold act as input to the proposed concept drift detection technique. The parameter 'μ' measures the closeness of the test data with the classes present in the model. It can be calculated as: first compute the distance of test object from the centroid of the class and then make count of objects which are in vicinity of test object i.e. all those data points (by which the model was trained) whose distance from the centroid is less than the distance between the test object and centroid. Then the ratio of this count and total data points of the class gives the value of 'μ'.

After the 'μ' values are computed for test object concerning each class, compare it with the threshold whose value defines the sensitivity of the detection. Smaller the threshold value the more it is sensitive to detect the concept drift. If all the values of 'μ' corresponding to the test object concerning all the classes are greater than the threshold, then it can be concluded that it would lead to concept drift.

Algorithm: Concept Drift Detection

Input: Threshold, Trained data, Centroid and Test data whose label is unknown.
Output: 1/0: Concept Drift/No Concept Drift.
1. Get the centroid of each class:
    a. centroid[number_of_classes] ← []
    b. for group in data:
        i.   temp ← [size_of_features]
        ii. temp[] ← $\sum$(features)/Data_Length
        iii. centroid[group] ← temp
    c. return centroid
2. Finding the 'μ':
    a. for group in centroid:
        i.   for features in centroid[group]:
             1. euclidean_distance1[group] ← norm((features),(test_data))
        ii. for features in trainded_data[group]:
             1. euclidean_distance2 ← norm((features),(test_data))
             2. if (eculidean_distance2 <= euclidean_distance1[group][0])
                a. countIn ← countIn+1
             3. total ← total+1
        iii. μ[group] ← (countIn/total)
3. Comparing 'μ' with 'λ':
    a. if (μ[group] > λ):
        i.   return 1
    b. else:
        i.   return 0
4. end.

For instance consider three-class classification model consisting of green, red and yellow coloured data points belonging to the class of different shapes as shown (Fig. 1). Black coloured shapes indicate the centroids of the classes.

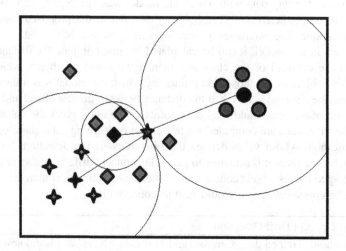

Fig. 1. Three class classification model

Now consider new data point star whose label need to be predicted. In order to find 'μ_{red}', 'μ_{green}' and 'μ_{yellow}' check how many objects of the particular class are similar to the test object, i.e. how closely it is related to different classes. This can be done by counting the number of data points which are present inside the circle. Then dividing this count by a total number of respective data points will give the value of respective 'μ'.

$$\mu_{red} = \frac{2}{6} = 0.33$$

$$\mu_{yellow} = \frac{4}{4} = 1.00$$

$$\mu_{green} = \frac{5}{5} = 1.00$$

According to the above value, the new data set is more likely to belong to the 'red' class. But now if the threshold is set lesser than 0.33 then it would be considered as concept drift; else it will be classified as it belongs to 'red' class.

Now to find the parameter 'μ' finding the Euclidean distance of the test data from every data point of the class makes the model's computation complexity high if there are a huge number of data points. To do the computations in the real-time, one more parameter 'aggr_const' (aggregator constant) is defined to create the group of data points within the class, can also be called as sub-centroids. These are created in such a way that they do not change the behaviour of the model. In other words, this can also be understood as *data compression*, i.e. treat some group of elements as a single element whose characteristic features are found by finding the centroid of that group.

5   Experiments

As it is already known that in the process of building the classification model there are two phases involved, a training phase and testing phase. In training phase of the model, training examples are required to develop the hypothesis for the model, and in this test scenario 80% of the data samples are selected as training examples and remaining 20% of the data samples are selected for testing phase to check the accuracy of the classification model. Both of these processes will be iteratively executed, and average accuracies will be compared.

Following are the datasets that are tested in the experiment:

5.1   SEA

It is an artificial dataset. It has three features ranging from 0 to 10 and two classes which are categorized as 0 or 1 [8].

For this dataset randomly:

- Chosen the value for the number of iterations as 100, and
- Threshold values as: [0.7, 0.4, 0.2].

Below Fig. 2 is the 2-Dimensional representation of the 3-Dimensional training data of SEA dataset with the help of dimensionality reduction technique called *Principal Component Analysis* [9] where x and y-axis represent the projected features of training data on a 2-D plane.

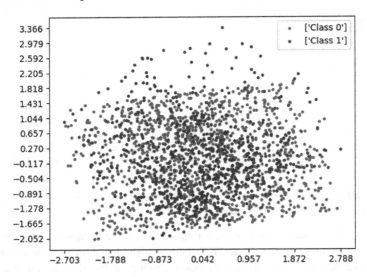

Fig. 2. 2D representation of SEA dataset using PCA

Observations. The following table shows the average accuracy of the model at different thresholds for 100 iterations (Table 1).

Table 1. SEA dataset Average accuracy vs Threshold

Average accuracy	Threshold		
	0.7	0.4	0.2
With CD Detection	82.01%	84.32%	98.13%
Without CD Detection	82.01%	82.10%	82.28%

Graphical Representation

- Here the graphical representation of the found average accuracies is shown.
- Below graphs show the accuracy of the model with respect to each iteration and comparison between the accuracy of the model with/without concept drift detection technique for each threshold.

Threshold 0.7: As shown in following graphical Fig. 3 accuracy came the same for both the cases and hence lines are overlapping. This means that there was no concept drift found.

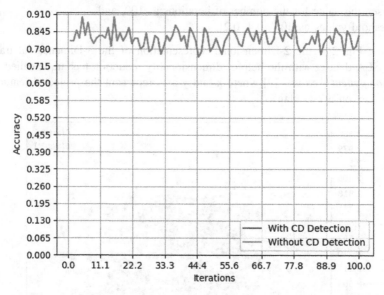

Fig. 3. Graph between accuracy and iterations consisting of two curves with CD Detection and without CD Detection for 0.7 Threshold

Threshold 0.4: As shown in graphical Fig. 4 it shows noticeable change (improvement) in the accuracy of the model. But there is a sudden drop in accuracy between iterations 22 and 33 which is due to noise coming in the data stream leading to misclassification of the data objects.

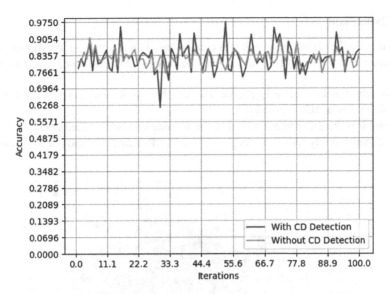

Fig. 4. Graph between accuracy and iterations consisting of two curves with CD Detection and without CD Detection for 0.4 Threshold

Threshold 0.2: As per the Fig. 5, this shows significant improvement in the accuracy of the model as it only allows the data elements which are very close to the centroid. At this value, it can be stated that the proposed approach becomes aggressive towards concept drift.

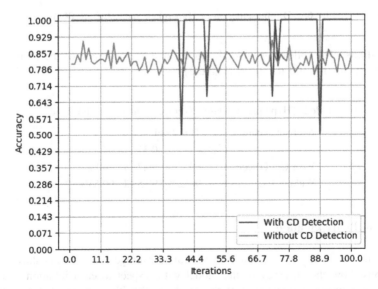

Fig. 5. Graph between accuracy and iterations consisting of two curves with CD Detection and without CD Detection for 0.2 Threshold

5.2 Hyperplane

It is also an artificial data set which was created based on the moving hyperplane. It has ten features ranging between 0 and 1 and two classes which are categorized as 0 or 1 [8].

Similarly for this dataset, choose the number of iterations equal to 100 and threshold values as 0.7, 0.3 and 0.2.

Figure 6 is the 2-D representation of the training data set reduced from 10-D using one of dimensionality reduction mechanism called Principal Component Analysis [9].

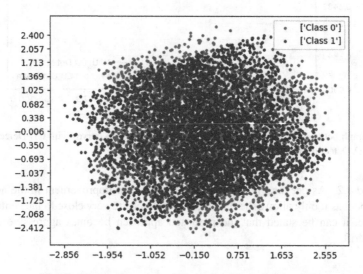

Fig. 6. 2D representation of Hyperplane dataset using PCA

Observations. The following table shows the average accuracy of the model at different thresholds for 100 iterations (Table 2).

Table 2. Hyperplane dataset Average accuracy vs Threshold

Average accuracy	Threshold		
	0.7	0.3	0.2
With CD Detection	90.25%	90.71%	100%
Without CD Detection	90.25%	90.54%	90.55%

Graphical Representation

- Here the graphical representation of the found average accuracies is shown.
- Graphs show the accuracy of the model with respect to each iteration and comparison between the accuracy of the model with/without concept drift detection technique for each threshold.

Threshold 0.7: As shown in graphical Fig. 7, accuracy is coming same for both the cases and hence lines are overlapping. This means that there was no concept drift found.

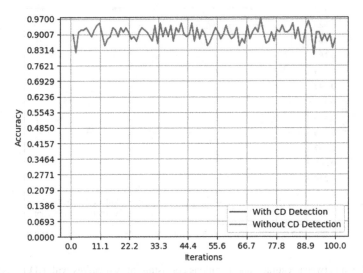

Fig. 7. Graph between accuracy and iterations consisting of two curves with CD Detection and without CD Detection for Threshold 0.7

Threshold 0.3: As shown in graphical Fig. 8, though average accuracies are same, concept drift was detected.

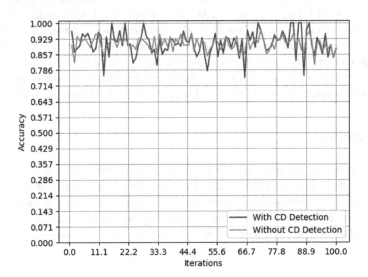

Fig. 8. Graph between accuracy and iterations consisting of two curves with CD Detection and without CD Detection for Threshold 0.3

Threshold 0.2: Here model shows 100% accuracy as shown in Fig. 9. At this value, it can be stated that the proposed approach becomes aggressive towards concept drift.

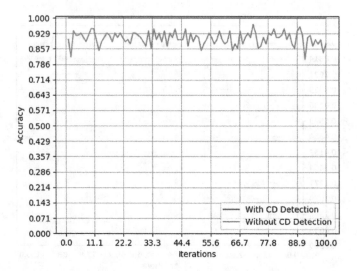

Fig. 9. Graph between accuracy and iterations consisting of two curves with CD Detection and without CD Detection for Threshold 0.2

6   Future Work

The proposed method has quite significant impact on the prediction quality of the classification model. However, this method can further be extended to detect the novel class and then retrain the classification model. Instead of not considering the data that resembles different characteristics than usual, a separate outlier buffer can be created to store such data and separate clustering algorithms e.g. density based clustering algorithms (DBSCAN: Density-based spatial clustering of applications with noise), etc. can be executed to detect the novel class and retrain the whole classification model.

7   Conclusion

The paper proposed the procedure to detect the concept drift and to test the same for maintaining the accuracy of the model. This procedure is quite flexible to work with both stationary and non-stationary environment. It uses some parameters which can be defined intuitively based on the amount of sensitivity needed to detect the concept drift. This procedure is quite simple to implement and easily understandable.

References

1. Zliobaitė, I.: Learning under concept drift: an overview. Technical report faculty of mathematics and informatics, Vilnius University, Vilnius, Lithuania (2009)
2. Khan, L.: Data stream mining: challenges and techniques. In: Proceedings of 22nd IEEE International Conference on Tools with Artificial Intelligence (2010)
3. Krempl, G., et al.: Open challenges for data stream mining research. SIGKDD Explor. Newsl. **16**(1), 1–10 (2014). https://doi.org/10.1145/2674026.2674028
4. Janardan, Mehta, S.: Concept drift in streaming data classification: algorithms, platforms, and issues. Procedia Comput. Sci. **122**, 804–811 (2017)
5. Wang, H., Abraham, Z.: Concept drift detection for streaming data. In: Proceedings of International Joint Conference of Neural Networks (IJCNN), Killarney, Ireland, pp. 1–9 (2015)
6. Kim, Y.I., Park, C.H.: Concept drift detection on streaming data under limited labeling. In: 2016 IEEE International Conference on Computer and Information Technology (CIT), pp. 273–280. IEEE (2016)
7. Nishida, K., Yamauchi, K.: Detecting concept drift using statistical testing. In: Corruble, V., Takeda, M., Suzuki, E. (eds.) DS 2007. LNCS (LNAI), vol. 4755, pp. 264–269. Springer, Heidelberg (2007). https://doi.org/10.1007/978-3-540-75488-6_27
8. Kadwe, Y., Suryawanshi, V.: A review on concept drift. IOSR J. Comput. Eng. **17**, 20–26 (2015). https://doi.org/10.9790/0661-17122026
9. Shlens, J.: A Tutorial on Principal Component Analysis, Systems Neurobiology Laboratory, Salk Institute for Biological StudiesLa Jolla, CA 92037 and Institute for Nonlinear Science, University of California, San Diego La Jolla, CA 92093-0402, 10 December 2005. Version 2

Homogenous Ensemble of Time-Series Models for Indian Stock Market

Sourabh Yadav[1(✉)] and Nonita Sharma[2]

[1] Gautam Buddha University, Greater Noida, Uttar Pradesh, India
sy9643391664@gmail.com
[2] Dr B R Ambedkar National Institute of Technology, Jalandhar, Punjab, India
nonita@nitj.ac.in

Abstract. In the present era, Stock Market has become the storyteller of all the financial activity of any country. Therefore, stock market has become the place of high risks, but even then it is attracting the mass because of its high return value. Stock market tells about the economy of any country and has become one of the biggest investment place for the general public. In this manuscript, we present the various forecasting approaches and linear regression algorithm to successfully predict the Bombay Stock Exchange (BSE) SENSEX value with high accuracy. Depending upon the analysis performed, it can be said successfully that Linear Regression in combination with different mathematical functions prepares the best model. This model gives the best output with BSE SENSEX values and Gross Domestic Product (GDP) values as it shows the least p-value as 5.382e−10 when compared with other model's p-values.

Keywords: Stock market · Forecasting · Time series · Univariate analysis
Multivariate analysis · Regression · Linear regression

1 Introduction

Stock market of any country is the key factor for determining the country's growth and economy. Stock market is a place where all the public listed companies trades their shares to raise their capital. Looking at the historical trends of the stock market, predicting the stock prices will not be an easily accomplished task. Therefore, predicting the stock market prices will definitely prove to be a great helping hand for those who invest in the stock market. It will help to determine the country's growth and economy for the future, which will assist the higher officials of any country for framing their policies for the development of the nation. Moreover, it will help the general public to understand the trends of the market, and when and how much one can invest for getting the maximum returns. There are several parameters of stock market, and BSE SENSEX is one of them. The BSE SENSEX, also called BSE 30 or SENSEX is a free floated market-weighted index of 30 well established and financially sound companies, listed on Bombay Stock Exchange.

Stock market trend analysis is one of the difficult tasks because of the daily ups and downs in prices of the stock. Hence it is important to build an accurate and precise prediction model for predicting the stock prices. Further, there are various approaches

© Springer Nature Switzerland AG 2018
A. Mondal et al. (Eds.): BDA 2018, LNCS 11297, pp. 100–114, 2018.
https://doi.org/10.1007/978-3-030-04780-1_7

to analyze the stock prices, but the statistical approach for analyzing the prices is one of the most widely used approach [11]. Statistical Analysis is collecting, exploring and presenting the data for understanding the patterns and trends (if any) present in the dataset. Furthermore, if time series approach is used, it will provide the better accuracy and precise prediction model [16]. Time Series for any dataset is an existence of data over a continuous time interval. Time series analysis is analyzing the time series data for the better understanding of the trends and patterns. There are various parameters of stock market, and BSE SENSEX is one of them. Moreover, there are many other additional factors which affect the BSE SENSEX, like Gross Domestic Product (GDP), Inflation, Exchange Rates like the value of US Dollar in Indian Rupee, and many other [1]. GDP for any country is the final value of the goods and services manufactured within the geographical boundaries of any country during a particular period of time. Inflation is the measure of the increase in prices of goods and services in a country annually. Exchange Rate is defined as the price of a country's currency in terms of another country's currency. If one sees these factors mathematically, these factors are directly proportional to the increase and decrease of the values of stock market prices.

This manuscript specifically targets for predicting the BSE SENSEX depending upon the historical values [17] and factors affecting the BSE SENSEX. Performing the univariate analysis or understanding the historical trends in the dataset, provides a model for predicting the stock prices depending upon past values. The historical data of the past 18 years was analyzed and best fit model is prepared depending upon the mean error of various forecasting models. Various forecasting models when applied in combination with each other and compared simultaneously, gives the best output [3]. Depending upon the results and errors of various forecasting model, error matrix is prepared for better understanding. To increase the accuracy of the results found in univariate analysis, the next target was of multivariate analysis. Hence, the next target is to determine the correlation values among the BSE SENSEX vector and all the factor affecting BSE SENSEX. Depending upon the correlation values, the correlation matrix is prepared to judge highly affecting factor. Moreover, multivariate analysis of the dataset provides a mathematical relation between highly affecting factors and prices of stock. Hence, the next target was to create a mathematical relation between BSE SENSEX values and additional factors affecting the BSE SENSEX. Further, one ensemble is also prepared, to improve the accuracy and precision of the model. Then, in the end, all the results are compared to find the best model for forecasting. The data used in the analysis is of 18 years for predicting forecasting model on the basis of univariate analysis and is of 15 years for performing multivariate analysis.

2 Problem Statement

The Main objective covered in this manuscript is to predict the BSE SENSEX value accurately and precisely. To achieve this objective, there are some sub-objectives. First sub-objective is to predict the BSE SENSEX value by univariate analysis or by analyzing the historical values and trends in the dataset and obtain a suitable forecasting model which give the least mean error. Second sub-objective is to improve the accuracy of the model by analyzing the factors affecting the BSE SENSEX and performing the multivariate analysis on most affecting factor in which mathematical equation comes out as an output. Third and last sub-objective targets to build an ensemble.

3   Proposed Method

Figure 1 depicts the proposed methodology for building the precise and accurate prediction model.

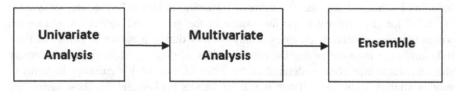

Fig. 1. Proposed methodology

3.1   Univariate Analysis

Step 1 for preparing the prediction model is to create a forecasting model depending upon the previous trends in the dataset or univariate analysis. Univariate analysis is one of the simplest forms, for analyzing the dataset in which previous values or historical values of the dataset is used for performing the analysis. 'Uni' means one, 'variate' here means variable, so one variable analysis is known as univariate analysis. For performing analysis on any dataset, univariate analysis acts as a basic step. Under this step, the first step is Data Collection. Data collection is the process of collection of data from all the relevant sources in a systematic fashion that enables one to answer the relevant questions and evaluate outcomes [7]. After collecting the data, data cleaning is the next step. Data Cleaning refers to the process of removing invalid data points from the dataset [14]. Cleaning is the process of removing the data points which are disconnected from the effect and assumption which are needed to be isolated. In this process, these particular data points are ignored, and analysis has been conducted on the remaining data. After data cleaning, the next step is an exploratory analysis of the dataset. For exploratory analysis, data is loaded in the statistical environment for performing the different statistical functions on the dataset. Further, the dataset is converted in time series. This means that data exists over a continuous time interval with equal spacing between every two consecutive measurements. Converting the dataset into time series always proves to be an effective method for the analysis of any dataset, especially in the stock analysis [2]. The next step involves plotting the time series object for analyzing the components of the time series data i.e. trend, seasonality, stationarity, and heteroskedasticity. Among these components, stationarity is most important. When the mean and variance are constant for a particular dataset, it is said that dataset holds the stationarity. (i.e. their joint distribution does not change over time).

The Plot of the time series will suggest that whether the data is stationary or not, which further suggests that data is volatile or not. If the data is not stationary, then it means there is a large deviation from the mean of the dataset. The data which not stationary, it will be quite unpredictable. Further, for testing the stationarity, different tests are performed like the Ljung-Box Test and Augmented Dickey-Fuller Test.

Next Step involves Testing for Stationarity under which two test are performed on the dataset i.e.

Ljung-Box Test: The Ljung–Box test is a type of statistical test of whether any of a group of autocorrelations of a time series are different from zero.

Augmented Dickey-Fuller (ADF) Test: The ADF test is unit root test for stationarity. Unit roots can cause unpredictable results in your time series analysis. The Augmented Dickey-Fuller test can be used with serial correlation. In this lag length is the parameter which is important in finding the meaningful results. In this lag length is the parameter which is important in finding the meaningful results [9].

Moreover, the Null Hypothesis states that large the p-value indicates non-stationarity and smaller p values indicate stationarity [8]. If in ADF test results are not in favor i.e. p-value comes out to be relatively high, then there is a need to do some further visual inspection, otherwise, next step i.e. Decomposition of the dataset can be skipped. So, if test depicts the high p-value then next step is decomposition of the dataset. This involves breaking down the dataset into parameters that are, Observed, Trend, Seasonal, Random [5]. When the Seasonal vector is plotted, it gives us indication towards stationarity or non-stationarity. If the data is stationary, the first phase of model estimation can be skipped. Basically, Model estimation comprises of two phases i.e. in the first phase, non-stationary data is transformed into stationary data and second phase, building a model. So the next step is the Model estimation. Firstly Auto Correlation Function (ACF) plot and Partial Auto Correlation Function (PACF) plot are prepared. These ACF and PACF plots tell about the Correlation factor of statistical analysis. Moreover, it helps to judge Co-variance of the dataset. When there is large autocorrelation within our lagged values, then, in that case, there is a need to take the difference of time series object in order to transform the series into a stationary series. The Difference of the series means calculating the differences between all consecutive values of a vector. This will helps to stabilize the mean, thereby making the time-series object stationary. Next step is to plot the transformed time series. This plot will suggest that whether the series is now stationary or not. To confirm the stationarity, ACF and PACF plots are again plotted for the differenced time series which clears the doubt about the stationarity. Further stationarity can be tested by different tests like the Ljung-Box Test, Augmented Dickey-Fuller Test, which will give the p-value very small in comparison to the previous p-values, which again proves the stationarity. Next job is the Building of the Model, which means deducing that which particular model applies best on our dataset depending upon our statistical results. Different models are:

Autoregressive Integrated Moving Average (ARIMA) Model: ARIMA is a forecasting technique that projects the future values of a series entirely based upon its own inertia. Its main application is in the area of short-term forecasting requiring at least 40 historical data points. It works best when the data exhibits a stable or consistent pattern over time with a minimum amount of outliers. It is the preferred choice because of its simplicity and wide acceptability. It offers great flexibility to work upon univariate time series [12].

BoxCox Transformation: BoxCox transformations are generally used to transform non-normally distributed data to become approximately normal.

Exponential Smoothing Forecast: This forecasting method relies on weighted averages of past observations where the most recent observations hold higher weight. This method is suitable for forecasting data with no trend or seasonal pattern.

Mean Forecast: This forecasting method relies on the mean of the historical data.

Naive Forecast: The naive forecasting method which gives an output as ARIMA (0, 1, 0) with a random walk model that is applied to time series object.

Seasonal Naive Forecast: This forecasting method works almost on the same principles as the naive method, but works better when the data is seasonal.

Neural Network: Neural networks are forecasting methods that are based on simple mathematical models of the brain. They allow complex nonlinear relationships between the response variable and its predictors. This model is very helpful when combined with the statistical computational approach for forecasting of stock market [15].

The model which have the least error or have the higher accuracy will be the best fit model for the dataset. Moreover, error analysis suggests the improvements that can be made in the results in the future [6].

3.2 Multivariate Analysis

Step 2 involves multivariate analysis for improving the results of step 1. Multivariate analysis is a statistical approach in which dataset is analyzed on the basis of different factors and the main objective is to prepare a combined model for better performance, analysis and, accuracy. Many times, univariate analysis is preferred because multivariate analysis results are difficult to interpret. For multivariate analysis, there are enough number of techniques, so depending upon the type of datasets, a particular technique is followed. Multivariate analysis can be performed by just analyzing the trend of all the factors which can have a great impact on the main dataset. Multivariate analysis is performed with the factors which have great influence on the dependent variable. The Dependent variable is a variable which needs to be detected after the analysis. Principle Component Regression (PCR) technique is the most commonly followed technique for the multivariate analysis. This technique is simply based upon Principle Component Analysis. PCR focuses to reduce the Dimensionality of the datasets. Moreover, it avoids the multicollinearity among the predictor variables. Results from Step 1 can be improved if a relationship analysis is carried out among the dataset vectors and factors affecting the dataset. For Relationship analysis, the statistical approach used is Regression. Regression is the statistical approach which is used to build a model in terms of mathematical equations for determining the relationship among the different factors with the main variable. In Regression one of the variables is known as Predictor variable whose value is to carried out by performing different experiments, and another variable is Response Variable. Response Variable is a variable whose value is procured by Predictor Variable. Generally, there are seven types of Regression available which are listed below:

Linear Regression: It is the most commonly know regression modeling technique. In this type of modeling technique, there can more than one independent variable which can be either discrete or continuous, but dependent variable must be continuous and the nature of the line of regression should be linear [13].

Logistic Regression: It is the type of regression which focuses to determine the probability of the event i.e. either success or failure [10]. Logistic Regression must be used or preferred when the dependent variable is in binary form i.e. 0 or 1, True or False, Yes or No.

Polynomial Regression: It is the type of regression model in which regression equation is of the form polynomial i.e. independent variable has the power more than 1. In this best-fit regression line is not particularly a straight line.

Stepwise Regression: It is the type of regression in which multiple independent variables are required. The selection of the predictor variable is done automatically. There is no intervention of humans. Its basic aim is to produce a best fit model with the highest possible accuracy.

Ridge Regression: It is the type of regression model which applied when independent variables a high absolute correlation value or have multiple collinearities. In this alpha value is set to be a 0.

Lasso Regression: It is the type of regression model which similar to ridge regression model but just uses absolute values instead of squares in penalty function. Moreover, the alpha value is set to be 1 in this model.

ElasticNet Regression Model: It is the hybrid model of ridge and lasso regression model. In this alpha value is set as 0.5.

The Linear Regression approach is preferred over other regression approaches as all other regression approaches are build by understanding the working of linear regression [21]. A Key requirement for linear regression is linearity among the variables. More-over, correlation values also help to judge the dependability of any response variable upon the predictor variable. The correlation values have the range of -1 to 1. So, larger the absolute value of the correlation coefficient, more the dependability of variables upon each other and more is the linearity among them. After determining the corre-lation value, the most influencing factor will be extracted. Furthermore, model fitting is done by applying different mathematical functions like logarithmic function or expo-nential function, on both response variable and predictor variable, for making models estimation simple and easier. Moreover, instead of passing a single factor i.e. most influencing factor, one can pass all the factors at the same time as an argument to the regression algorithm. Then whichever model performs better will be the best fit model. For determining the accuracy steps will be the same i.e. summarization of regression model.

3.3   Ensemble Technique

Step 3 involves the building of ensemble for the dataset. Ensemble, also known as Data Combiner, is a data mining approach that converges the strength of multiple models to achieve better accuracy in prediction. Basically Ensemble means combining multiple algorithms for improving the accuracy of the model. Ensemble method is one of the most influential developments in the field of data mining. They combine the multiple models into one, by extracting most accurate models from all those multiple models. Ensembling is performed depending upon the dataset. The necessary steps to perform the technique is outlined in below:

Algorithm: Ensemble of various Regression Methods
1. Perform univariate analysis and find the model that best fits the data.
2. Determine the predictor variables $L_1, L_2, \ldots \ldots L_k$ and response variable R_i
3. Evaluate the correlation coefficient $Cor(L_{j(j \in 1..k)}, R_i)$ between the all the predictor variables and response variables
4. Decrease the residuals by taking the logarithm.
5. If $Cor(L_{j(j \in 1..k)}, R_i) > 0.5$, the variable will be included in the ensemble otherwise it is not included in the ensemble.
6. Find the maximum value of the $\max(Cor(L_{j(j \in 1..k)}, R_i)$ and find the mathematical equation of response variable R_i, with the predictor variable depending upon maximum correlation value, $\max(Cor(L_{j(j \in 1..k)}, R_i)$, as E1.
7. Using the variables obtained in step 5, calculate the numeric sum of all the variables with each other and store it in another variable as $B = \sum L_{j(j \in 1..k)}$
8. Using the variable B, obtained in step 7, calculate the mathematical equation between variable B and R_i. model obtained in step 1, perform multivariate analysis between $L_{j(j \in 1..k)}$ and R_i as E2.
9. Compare the equations E1 and E2 obtained from step 6 and step 8.
10. If E1 fits to the data more accurately and precisely, then model obtained in step 6 is best fit else model obtained in step 8 best fit

All the prediction models can be analyzed depending on its accuracy and precision. The model with more accuracy and precision will be the best-fitted model for our dataset.

4   Results and Discussions

The tool which is used for forecasting is R. Various Packages related to the various functionalities described in Sect. 3, are included as: forecast package, tseries package. Datasets used in our analysis are BSE SENSEX, GDP of India, USD Prices in Rupee, and Inflation and the sources for these datasets are mentioned in Table 1.

BSE SENSEX dataset contains four different variables that are open, high, low, and close. The open variable represents the opening price of the stock market, high variable

Table 1. Sources of dataset

Dataset	Source
BSE SENSEX	Official website of BSE India [4]
GDP	Official website World Bank [20]
USD Prices in Rupee	Official website of Reserve Bank of India [18]
Inflation	Official website of European Union [19]

represents the highest price of the stock prices, low variable represents the lowest price of the stock prices, and close variable represents the closing price of stock prices. Results after applying the procedure mentioned are detailed below: In step 1 where univariate analysis been performed, after cleaning the data, which was the initial step, time series object was created for a vector of the dataset.

After the time series object was plotted in Fig. 2, it suggests to work upon different component like Trend, Seasonality, Stationarity, Heteroskedasticity. For testing the stationarity of the dataset, different tests were performed upon dataset like Augmented Dickey-Fuller Test and Ljung-Box Test.

Fig. 2. Plot of time series object

The values obtained from different stationarity test are mentioned in Tables 2 and 3 which further, suggests decomposing the dataset. After analyzing the plot of decomposition, nonstationarity of data is confirmed by analyzing the ACF and PACF plots. Analyzing the ACF and PACF plot suggests that, the difference of data points in the dataset is required to transform the series into stationary series. After taking the difference of the series again above quoted tests are applied on the series to cross-check its stationarity whose results are quoted in Tables 4 and 5.

From both the above Tables 4 and 5, it is clear that series has been transformed into a stationary series successfully as p-value is less 0.05. Once the series has become the stationary series, different model functions are applied on the series and depending upon their errors, best-fit model is selected. In the dataset, 4 vectors are separately analyzed by applying different model functions depending upon which error matrix is prepared.

Table 2. Augmented Dickey-Fuller Test

Parameters	Values
Data	open_ts
Dickey-Fuller	−1.265
Lag order	5
p-value	0.8841
Alternative hypothesis	Stationary

Table 3. Ljung-Box Test

Parameters	Values
Data	open_ts
X-Squared	2580.9
df	20
p-value	2.2e−16

Table 4. Augmented Dickey-Fuller Test

Data	z
Dickey-Fuller	−6.369
Lag order	5
p-value	0.01
Alternative hypothesis	Stationary

Table 5. Ljung-Box Test

Parameters	Values
Data	z
X-Squared	2580.9
df	20
p-value	<2.2e−16

As seen from the Tables 6, 7, 8, and 9 all the models have performed differently. Considering every model, best results are given by Exponential Smoothing Model as its Mean Error value is consistent for all 4 vectors i.e. Open, High, Low, Close. Moreover Neural Network model also works good but with the exception that the Low vector of dataset is showing relatively high Mean Error compared to other 3 vectors.

Further in step 2 multivariate analysis is performed to improve the accuracy. For performing multivariate analysis datasets like GDP dataset, Inflation dataset, Exchange Rates dataset (that can be the values of USD prices in terms of Indian Rupee) and open vector of BSE SENSEX are collected annually. Then the next step is to determine the linearity among the BSE SENSEX value and GDP values, Inflation, and Exchange

Table 6. Model estimation for open vector

	ME	RMSE	MAE	MASE	ACF1
ARIMA	−2.808	59.618	48.379	0.7	0.05
BoxCox	0.834	59.938	48.878	0.707	0.06
ETS	0.068	59.931	48.867	0.707	0.06
Meanf	0	59.928	48.865	0.707	0.06
Naive	−0.18	82.312	64.539	0.934	−0.457
Snaive	0.468	86.49	69.132	1	0.13
Neural network	−0.022	16.769	12.939	0.187	−0.034

Table 7. Model estimation for high vector

	ME	RMSE	MAE	MASE	ACF1
ARIMA	−2.212	56.043	44.505	0.675	0.132
BoxCox	−0.788	55.759	44.003	0.667	0.02
ETS	**0.027**	56.162	44.752	0.679	0.127
Meanf	0	56.16	44.751	0.679	0.127
Naive	−0.521	74.099	60.386	0.916	−0.43
Snaive	−0.641	83.238	65.926	1	0.204
Neural network	**−0.097**	13.049	13.765	0.209	−0.033

Table 8. Model estimation for low vector

	ME	RMSE	MAE	MASE	ACF1
ARIMA	−1.73	55.929	44.503	0.693	0.218
BoxCox	−0.053	53.792	42.937	0.669	−0.015
ETS	**0.022**	56.07	44.522	0.694	0.22
Meanf	0	56.067	44.52	0.694	0.22
Naive	−0.256	70.114	56.161	0.875	−0.309
Snaive	−0.391	79.903	64.19	1	0.272
Neural network	**0.27**	48.532	38.182	0.595	0.024

Table 9. Model estimation for close vector

	ME	RMSE	MAE	MASE	ACF1
ARIMA	−1.298	57.414	46.472	0.709	0.126
BoxCox	0.188	54.931	43.683	0.666	−0.002
ETS	**0.025**	57.547	46.314	0.706	0.129
Meanf	0	57.544	46.311	0.706	0.129
Naive	−0.307	76.075	58.987	0.9	−0.42
Snaive	−0.429	82.836	65.571	1	0.23
Neural network	**−0.031**	8.715	6.272	0.096	−0.167

Rates, respectively. So for that purpose, correlation coefficients are determined among the different vectors with the open vector of the dataset. Table 9 depicts the different correlation values of different vectors with the open vector:

After interpreting the Table 10, it is clearly observable that GDP vector is highly correlated with BSE SENSEX Open feature. So next step is to build a linear regression model between Open vector and GDP vector. Here predictor variable is the GDP vector and response variable is the open vector. As a result, there will be a linear equation between GDP and Open vector.

$$Open = 12793.6 * GDP - 2599 \qquad (1)$$

To check the accuracy of the equation, the regression object can be summarized. Now p-values will help to judge whether the model is accurately fitted or not. The accuracy will be high if each p-value in the summary is less than or equals to 0.05 approximately and R-squared values are also above 0.9 (Table 11).

Table 10. Correlation coefficients

Vector	Correlation coefficients
Inflation	0.4155946
GDP	0.9675431
Exchange rates	0.6650287

Table 11. Summary of linear regression model object with GDP vector

Parameters	Values
p-value for intercept	0.064
p-value for GDP coefficients	3.83e−09
Net p-value	3.828e−09
Multiple R-squared value	0.9361
Adjusted R-squared value	0.9312

In order to increase the accuracy of the linear regression model, some mathematical functions are implemented like logarithm is applied on both the vector i.e. GDP and Open vector. Applying mathematical function will decrease the residuals value and will also rectify the other problems which would be decreasing the accuracy of the model.

$$log(Open) = 1.35712 \, log(GDP) + 9.15080 \qquad (2)$$

Summary of Regression model is in Table 12.

Table 12. Summary of Linear Regression Model object with log (GDP) vector

Parameters	Values
p-value for intercept	<2e−16
p-value for logarithmic GDP coefficient	5.38e−10
Net p-value	5.382e−10
Multiple R-squared value	0.9527
Adjusted R-squared value	0.6650287

Then for better comparison among the models, next step is to build a combined model i.e. using all the factors that affect the BSE SENSEX i.e. Inflation, Exchange Rates, GDP value, as a predictor variable and response variable will remain same i.e. Open vector. As a result, the linear equation between open as a response variable and GDP, Inflation, USD Value as a predictor variable is constructed, which is quoted in Eq. 3.

$$Open = 13049.797 * GDP - 126.588 * Inflation + 9.607 * USD\ Value - 2497.615 \qquad (3)$$

For checking the accuracy regression object is summarized, and p-values and R-squared values can be extracted which help to interpret the accuracy. p-values and R-squared values are quoted in Table 13.

Table 13. Summary of combined linear regression model object

Parameters	Values
p-value for intercept	0.640
p-value for Inflation coefficient	0.567
p-value for GDP coefficient	5.18e−06
p-value for USD value coefficient	0.937
Net p-value	5.966e−07
Multiple R-Squared Value	0.9386
Adjusted R-Squared Value	0.9218

Further to increase the accuracy, logarithm of open and GDP, exponential of exponential of reciprocal of Inflation can be used for determining the equation.

$$\log(Open) = 1.318037 \log(GDP) - 0.208758 e^{e^{\frac{1}{(Inflation)}}} - 0.006184 USD\ Value \qquad (4)$$

For checking the accuracy, the regression model object is summarized, and Table 14 depicts the different p-values and R-squared values.

Next step is to build an ensemble. The Ensemble can be built by taking the numerical total of all the factors affecting BSE SENSEX i.e. GDP, Inflation, USD Value. In this firstly, the absolute value of the correlation coefficients need to be more than 0.5 for effective linearity. Correlation coefficients when calculated between open vector and total vector, it comes out to be 0.7751132. The correlation coefficient

Table 14. Summary of combined linear regression model with improvements

Parameters	Values
p-value for intercept	2.61e−08
p-value for exp (exp(Inflation)) coefficient	0.399
p-value for log (GDP) coefficient	4.11e−06
p-value for USD value coefficient	0.472
Net p-value	5.213e−08
Multiple R-squared value	0.9606
Adjusted R-squared value	0.9499

suggests that a linear regression model can be constructed as an experiment to improve the accuracy. For applying regression, open vector is used as the response variable and the total vector as predictor variable. As a result, there will be a linear equation between the open vector and the total vector which is represented by Eq. 5.

$$Open = 714.2 * Total - 27557.8 \tag{5}$$

For checking the accuracy of the result, summarization of regression object is required and Table 15 shows the summary of the model.

Table 15. Summary of ensemble

Parameters	Values
p-value for intercept	0.011661
p-value for total coefficient	0.000688
Net p-value	0.0006876
Multiple R-squared value	0.6008
Adjusted R-squared value	0.5701

The next and final step is to analyze the results of all the above linear regression model and find the best and suitable model for forecasting which can closely predict the value of BSE SENSEX. Table 16 show all the net p-values of all the above regression models, through which we can easily compare the results.

Table 16. Net p-values for all the above regression models

Parameters	Values
Net p-value for Open − GDP Model	3.828e−09
Net p-value for log(Open) − log(GDP) Model	5.382e−10
Net p-value for Open - GDP + Inflation + USD values Model	5.966e−07
Net p-value for log(Open) - log(GDP) + exp(exp(1/Inflation)) + USD values Model	5.213e−08
Net p-value for ensemble	0.0006876

It is clearly observable that, open-GDP model with mathematical function, gives the least p-value when compared with all other models.

5 Conclusion

In this manuscript, there is a research performed on the dataset of BSE SENSEX from January 1997 to January 2016, and the dataset of GDP of India (in Trillions), Inflation (in %), USD values (in Rupees), which is interpreted annually from the year 2001 to 2015. On applying different forecasting models in the beginning, then applying Linear regression techniques, it has been found that each and every model results differently and can be analyzed on the basis of mean error and the net p-values of the models. After analyzing the different mean error of forecasting model in the beginning, it has been concluded that Exponential Smoothing and Neural Network gives the consistently less mean error, with the small exception in the low vector where the mean error of neural network is comparatively high. Moreover, when linear regression algorithm is applied on the datasets for the improvements, it has been concluded that Linear Regression Model of logarithmic Open values of BSE SENSEX and logarithmic GDP values of India gives the best accuracy and precision, from all other quoted models.

References

1. Alam, P.: Factors affecting stock market in India. Splint Int. J. Prof. 3(9), 7 (2016)
2. Angadi, M.C., Kulkarni, A.P.: Time series data analysis for stock market prediction using data mining techniques with R. Int. J. Adv. Res. Comput. Sci. 6(6) (2015)
3. Armstrong, J.S.: Combining Forecasts. In: Armstrong, J.S. (ed.) Principles of Forecasting. International Series in Operations Research & Management Science, vol. 30. Springer, Boston (2001)
4. BSE Homepage. http://www.bseindia.com. Accessed 05 July 2018
5. Cleveland, W.P., Tiao, G.C.: Decomposition of seasonal time series: a model for the Census X-11 program. J. Am. Stat. Assoc. 71(355), 581–587 (1976)
6. Cole, R.: Data errors and forecasting accuracy. In: Economic forecasts and expectations: analysis of forecasting behavior and performance, pp. 47–82. NBER (1969)
7. Devers, K.J., Frankel, R.M.: Study design in qualitative research–2: sampling and data collection strategies. Educ. Health 13(2), 263 (2000)
8. Frick, R.W.: The appropriate use of null hypothesis testing. Psychol. Methods 1(4), 379 (1996)
9. Hall, A.: Testing for a unit root in time series with pretest data-based model selection. J. Bus. Econ. Stat. 12(4), 461–470 (1994)
10. Larsen, K., et al.: Interpreting parameters in the logistic regression model with random effects. Biometrics 56(3), 909–914 (2000)
11. Litterman, R.B.: A statistical approach to economic forecasting. J. Bus. Econ. Stat. 4(1), 1–4 (1986)
12. Mondal, P., Shit, L., Goswami, S.: Study of effectiveness of time series modeling (ARIMA) in forecasting stock prices. Int. J. Comput. Sci. Eng. Appl. 4(2), 13 (2014)
13. Montgomery, D.C., Peck, E.A., Vining, G.G.: Introduction to Linear Regression Analysis, vol. 821. Wiley, New York (2012)

14. Rahm, E., Do, H.H.: Data cleaning: problems and current approaches. IEEE Data Eng. Bull. **23**(4), 3–13 (2000)
15. Rao, A., et al. Survey: Stock Market Prediction Using Statistical Computational Methodologies and Artificial Neural Networks (2015)
16. Sapankevych, N.I., Sankar, R.: Time series prediction using support vector machines: a survey. IEEE Comput. Intell. Mag. **4**(2), 24–38 (2009)
17. Sharma, N., Juneja, A.: Combining of random forest estimates using LSboost for stock market index prediction. In: 2017 2nd International Conference for Convergence in Technology (I2CT). IEEE (2017)
18. The Reserve Bank of India Homepage. https://www.rbi.org.in. Accessed 10 July 2018
19. The Worldwide Inflation Data Homepage. http://www.inflation.eu. Accessed 10 July 2018
20. The World Bank Homepage. http://www.worldbank.org. Accessed 10 July 2018
21. Weisberg, S.: Applied Linear Regression, vol. 528. Wiley, New York (2005)

Improving Time Series Forecasting Using Mathematical and Deep Learning Models

Mohit Gupta[✉], Ayushi Asthana, Nishant Joshi,
and Pulkit Mehndiratta

Department of Computer Science, Jaypee Institute of Information Technology,
Noida, India
mohitatjammu@gmail.com, 3145ayushi@gmail.com,
nishantjoshi10996@gmail.com,
pulkit.mehndiratta@jiit.ac.in

Abstract. With the increase in the number of internet users, there is a deluge of traffic over the web, handling Internet traffic with much more optimized and efficient approach is the need of the hour. In this work, we have tried to forecast Internet traffic on TCP/IP network using web traffic data of Wikipedia articles, provided by Kaggle (https://www.kaggle.com/). We work on the stationarity of time series and use mathematical concepts of log transformation, differencing and decomposition in order to make the time series stationary. Our research presents an approach for forecasting web traffic for these articles using different statistical time series models such as Auto-Regressive (AR) model, Moving Average (MA) model, Auto-Regressive Integrated Moving Average (ARIMA) Model and a deep learning model - Long Short-Term Memory (LSTM). This research work opens the possibility of efficient traffic handling thus, leading to improved performance for an organization as well as better experience for the users on the internet

Keywords: Time series · Forecasting · Trend · Seasonality · Deep learning
Statistics · Stationarity

1 Introduction

Today websites are much concerned about handling the increasing web traffic. Web traffic is the amount of data received or sent by users on the internet. It is determined by the number of users, the number of pages they visit and the number of visits a page receives. The continuous growth in web traffic makes TCP/IP traffic forecasting a critical task and an active area of research to explore. Forecasting web traffic allows organizations to plan the allocation of computer sources, estimate the growing user demand, set deadlines for the ongoing projects and thus helping them to be more resourceful.

The basic idea of a time series is, it's a sequence of successive data points taken with respect to time. Time series forecasting makes use of the previous observed values to forecast future values. Time series does not hold the basic assumptions of linear regression; thus, a very basic property of time series makes it special and challenging to

A. Mondal et al. (Eds.): BDA 2018, LNCS 11297, pp. 115–125, 2018.
https://doi.org/10.1007/978-3-030-04780-1_8

work on. Currently, time series forecasting has vast applications in areas such as stock prices [1, 2], weather forecasting [3], economy [4] and even health sectors [5] thus a very important and interesting area to work upon.

1.1 Time Series and Stationarity

A time series is a collection of observations $[y_1, y_2, y_3 \ldots y_t]$ taken with respect to time. It is a sequence of points typically consisting of successive measurements made over a time interval. Time series forecasting is done with the help of special mathematical and statistical models because basic machine learning models do not account for the dependence of time, where a data point at a particular time may be dependent on previous values. The components of a time series which needs to be taken into consideration before forecasting it are: seasonality, trend, cycles and other irregular variations. Seasonality [6] is a component of time series where data has regular and predictable changes that occur at regular interval. These are relatively short repetitive fluctuations. Trend [7] in time series refers to long-term increase or decrease in the data, which may be linear or nonlinear. A series is said to have cycles if it exhibits rises and falls that are not of the fixed period. These components of any time series lead to non-stationarity, which makes it very challenging to use such times series for forecasting purpose.

A time series is said to be strict stationary if the joint probability distribution of a function is independent of time (t). The concept of stationarity [8] implies that the statistical properties like mean, autocorrelation, and variance do not vary with time.

1.2 Techniques to Make Time Series Stationary

Differencing
Differencing [9] is a technique for transforming time series data. Trend in a time series is removed by subtracting the trend line with our time series plot, this type of time series is called trend-stationary and the method used is called de-trending. But in many cases, this method is not sufficient for achieving stationarity. So, another approach to transforming a series using differencing as explained in Sect. 3.4.

Decomposition
A time series is made up of various components combined together and each of them has its own significance and importance with respect to stationarity. Decomposition [10] provides a better way to understand the role of these components during series analysis and forecasting. We have used this approach as explained in Sect. 3.5.

2 Related Work

Forecasting has always been an area of active research and practical usage. Earlier traditional models like Extrapolation, Trending and Curve Fitting Methods were used to forecast. But they are limited in their approach as they can work only on limited and short time frames and that too, if inflections points do not exist. In an article [11], researchers have forecasted electricity consumption using hybrid model combining Trend Extrapolation and Least Squares Support Vector Machine (LSSVM) and found hybrid models to be better. In [12] authors show that de-trending approaches can be used to make time series stationary and has a uniform effect on computational intelligence models.

Junbo et al. in [13] used AR model to forecast future short-term speed of any vehicle in traffic and predicting multiple acceleration states, they are mapping acceleration measurements to the Markov states using the fuzzy encoding which can generate the better result. In paper [5], the ARIMA model is used to predict blood glucose concentration by using CGM (Continuous Glucose Monitoring) data. The results are obtained by first verifying the non-stationary time series with Augmented Dickey-Fuller Test. In [14], authors have proposed unit root test for heterogeneous panels based on the mean of individual root statistics. An article forecasted the stock prices and value of ANTM share from November 2007 to December 2007 is used [15]. The authors have also conducted the research where the Root Mean Square Error (RMSE) value for ARIMA is more than ANN.

The article is divided into various sections as follows - Sect. 3 discusses methodology which explains how the data is used and the implementation of our work. Section 4 presents the results we achieved using our implemented approach. Section 5 presents the discussion about our approach and the work done in this field. Finally, Sect. 6 contains the conclusion and future scope of our experimental design.

3 Methodology: Case Study – Wikipedia Web Traffic

3.1 Dataset

The dataset that has been used for this research consists of views of Wikipedia articles. The training data has approximately 145 k time series observation values. Each of these time series values represent the static views of different Wikipedia articles starting from July 1st 2015 up until December 31st 2016 i.e. corresponding to each article name, we have day-wise total traffic received by it. This dataset has been taken from Kaggle. com. We have separated articles as per languages and presented our results on the time series of the Wikipedia article of English Language (Fig. 1).

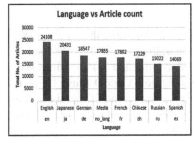
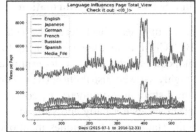

Fig. 1. Depicts that language vs article count of the dataset and distribution of total views of articles vs days (as per language) respectively.

3.2   Implementation

Initially, using the Wikipedia web traffic dataset we derive our time series for the first English language article to work on. Then the most important step to make our time series stationary is performed by using mathematical models – Log Transformation, Differencing, and Decomposition. We use first-order differencing with the additive approach to remove the random variations in the time series. Decomposition proves out to be the best of all these mathematical models as it clearly decomposes our time series into trend, seasonality and residual components. The stationarity of our time series is tested using Dickey-Fuller Test (Fig. 2).

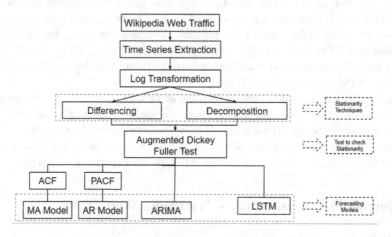

Fig. 2. Depicts the flow chart our work

We use forecasting models - AR (Auto Regressive Model), MA (Moving Average Model), ARIMA (Auto Regressive Integrated Moving Average) and Deep Neural Network – Long Short-Term Memory (LSTM) to forecast the resultant series. To set hyperparameters i.e. orders of model for AR and MA, we use PACF and ACF respectively. Finally, we use LSTM in a very efficient manner to forecast our values with 0.03 learning rate along with Adam optimizer, a second order gradient descent.

3.3   Time Series Log Transformation

We prepare a time series to forecast the TCP/IP traffic by using the views of Wikipedia article. By TCP/IP traffic, we are referring to the traffic generated by the total views of each Wikipedia article. We use logarithmic transformation for our time series - to unskew highly skewed distribution (which does not hold Central Limit Theorem) thus, making the distribution symmetric by considering assumption of Central Limit theorem i.e. the mean varies normally and variance stabilizes. We feed in this transformed time series to the stationarity techniques mentioned above.

3.4   Differencing

We use first-order differencing which brings out the change from one period to the next one.

$$Z_{t_{new}} = Z_{t_{initial}} - Z_{t-m} \tag{1}$$

Where, $Z_{t\ new}$ means observation value after differencing $Z_{t\ initial}$ (initial observation value) from the previous value, Z_{t-m} at time period t. We use m = 1 as we use first order differencing method (Fig. 3).

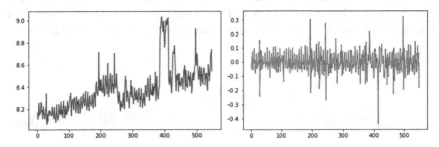

Fig. 3. Depicts the log-transformed series and resulted time series after differencing respectively.

3.5   Decomposition

Decomposition is a technique used to break down and visualize time series into its various components and then de-trend the log transformed series. The below equation explains the decomposition technique.

$$z(t) = T_t + S_t + R_t \tag{2}$$

Where Z(t) is an original time series. T_t, S_t, R_t are trend, seasonality and residuals respectively. Thus, finally achieving stationarity time series for applying forecasting techniques (Fig.4).

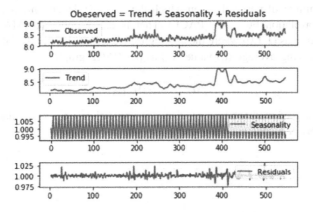

Fig. 4. Depicts the resultant decomposed time series

3.6 Augmented Dickey Fuller Test – Stationarity Test

After applying stationarity techniques, we are testing the series with Augmented Dickey fuller test [16]. It is applied to test the null hypothesis. Our time series when tested will result in a p-value and ADF statistical value. If p-value > 0.05, it means series accepts the null hypothesis (H0) and the data has a unit root and is non-stationary. Otherwise, it rejects the null hypothesis (H0), the data is stationary. ADF value is compared with the critical values. If the ADF value is less than the critical values then we can reject the null hypothesis, and hence can assume that our time series is stationary. If ADF value is less than all the three percentages (1%, 5% and 10%) then we say that there is less than 1% chance of the presence of these random trends in our time series and hence we are 90% confident that the series is stationary (Fig. 5).

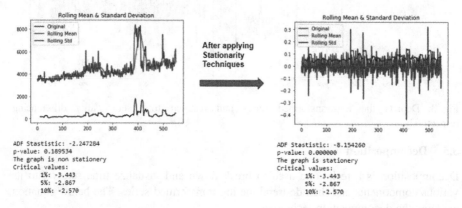

Fig. 5. Depicts the results of Augmented Dickey Fuller test before and after applying stationary technique.

3.7 Auto Regressive Model

Autoregressive model [17, 18] is a model used for time series forecasting where current values are dependent on past values known as the order of the autoregressive model. This order (number of past values on which the series is dependent) is found out by Partial Autocorrelation Function (PACF). Therefore, we apply ARIMA model with parameters p, d, q as (1,1,0) where the autoregressive term (p) is 1, level of differencing (d) is 1 and moving average (q) value is 0 as per PACF graph.

$$y_t = c + \varphi_1 y_{t-1} \tag{3}$$

where y_t is the predicted total no of views of any article on a particular day, c is a constant and ϕ_1 is a constant parameter.

3.8   Moving Average Model

In case of moving average model [14, 19], the series is dependent on the random error terms rather than on the past values. The number of past error terms is the order of MA model, found by using Autocorrelation Function (AF). Therefore, as per AF, an ARIMA model with parameters p, d, q as (0,1,1) where moving average term(q) is 1.

$$y_t = c + e_t + \theta_1 e_{t-1} \tag{4}$$

where e_t is white noise and e_{t-1} is the past error term (Fig.6).

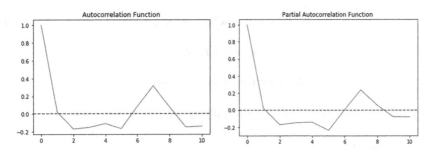

Fig. 6. Depicts that results of ACF and PACF respectively.

3.9   Auto Regressive Integrated Moving Average Model

ARIMA model [20] is a combination of autoregressive and moving average model along with differencing which corresponds to the integrated term in the name of the model. The total number of views for Wikipedia articles are forecasted with the help of ARIMA model with parameters p, d, q as (1,1,1).

For our Wikipedia dataset, the equation for our approach works –

$$y_t = c + e_t + \theta_1 e_{t-1} + \varphi_1 y_{t-1} \tag{5}$$

where y_t is the predicted total no of views of any article on a particular day, c is a constant e_t is white noise, θ_1 is a constant parameter and e_{t-1} is the past error term. Thus, with the help of this model, the total number of views of any article on a particular day can be forecasted.

3.10   Long Short-Term Memory (LSTM) Network

LSTM [21] is one of the RNN's special variants. It is introduced by Schmidhuber and Hochreiter. LSTM's can handle long-term dependencies. These are much more advanced than RNN as it resolves the problem of vanishing or boosting gradient Descent. It works using three gates – input, output and forget gates.

For our work of forecasting web traffic with LSTM, the first step is to select which information to take in, using input gate and then to control the information being thrown out of the cell state, using forget gate. Then, which new updated information will get stored in the cell state, C_t and which information will pass on (using output gate and activation function). We have Adam optimizer, a second order gradient method.

$$\hat{c}_t = \tanh\left(\mathbf{w}_c^x\mathbf{x}_t + \mathbf{w}_c^a\mathbf{a}_{t-1} + \mathbf{b}_c\right) \tag{6}$$

$$input\ gate, \mathbf{i} = \sigma\left(\mathbf{w}_i^x\mathbf{x}_t + \mathbf{w}_i^a\mathbf{a}_{t-1} + \mathbf{b}_i\right) \tag{7}$$

$$forget\ gate, \mathbf{f} = \sigma\left(\mathbf{w}_f^x\mathbf{x}_t + \mathbf{w}_f^a\mathbf{a}_{t-1} + \mathbf{b}_f\right) \tag{8}$$

$$outut\ gate, \mathbf{o} = \sigma\left(\mathbf{w}_o^x\mathbf{x}_t + \mathbf{w}_o^a\mathbf{a}_{t-1} + \mathbf{b}_o\right) \tag{9}$$

$$c_t = \mathbf{i} * \hat{c}_t + f * c_{t-1} \tag{10}$$

$$activation, \mathbf{a}_t = \mathbf{output} * \mathbf{A}(c_t) \tag{11}$$

Where, x_t is the information and w_j^i: applying weights on i to calculate j.

3.11   Testing of Data

From Wikipedia dataset, we have used last 100 days views of English article for our testing purposes. We have used our trained models - AR, MA, ARIMA and LSTM Model to forecast the values of future 100 days. Thus, by comparing these 100 forecasted values with the original values we find the Root Mean Square Error (RMSE) value for each model (Fig. 7).

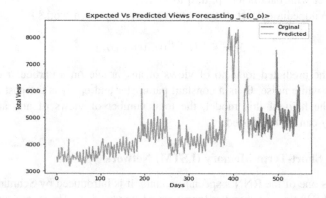

Fig. 7. Shows the separation of time series of the English article as training and testing data.

4   Result

We have predicted views of Wikipedia articles using various machine learning and deep learning models. We have used AR, MA, ARIMA and LSTM models for forecasting views of Wikipedia articles. ARIMA model has the least error among all the models because of its characteristic of using both past lags and past errors. The most important task to optimally forecast using these models was to work on the stationarity of the series. Log transformation was unable to make the series stationary but helped in removing skewness of the data while differencing and decomposition were successful in achieving stationarity. We also determined the orders of AR, MA and ARIMA model using Auto Correlation and Partial Autocorrelation function so that we get the best results from these models.

Here we can see that the deep learning model LSTM has higher RMSE as compared to other models because the size of the dataset is not very large. For this dataset ARIMA model is a good method to forecast web traffic. Therefore, the ARIMA model achieved high accuracy and can be very useful for forecasting web traffic but the advantage of LSTM over ARIMA is that no explicit parameters are required for LSTM (Table 1).

Table 1. Models along with their root mean square error

Model name	RMSE value
Auto Regressive Model (AR)	0.0233276
Moving Average Model (MA)	0.0288327
Auto Regressive Integrated Moving Average Model (ARIMA)	0.0232821
Long Short-Term Memory Cells (LSTM)	0.0354931

5   Discussion

In the last decade analysis of web traffic has become an active area of research in various subfields of computer networks and machine learning. The models of time series forecasting have evolved over the years and these concepts of machine learning, deep learning, data analysis, computer networks are being used together to further improve these models. This paper will surely be a useful work for sites to manage their traffic and for researchers to dive down further into time series forecasting. Our work helps in achieving state of the art results for any application of time series forecasting. We have used techniques to improve our models rather than to apply the models directly. In the paper [5], Augmented Dickey-Fuller test is applied to check for the stationarity of the series thus telling us that this test can be adopted for our approach as well. In the paper [15] stock prices were predicted using the ARIMA model where ARIMA turned out to be the fast model for predicting stock prices therefore, we also used this model for our problem set. In our work, we extend this approach by using mathematical techniques like differencing, decomposition and log transformation to remove trend and seasonality from the series thus making these models more efficient.

Internet traffic forecasting has been done using various concepts of computer networks and machine learning in the past but this approach of using concepts like stationarity, decomposition, differencing on models will surely help sites to handle traffic optimally.

6 Conclusion/Future Scope

We have successfully forecasted future web traffic for Wikipedia articles using various statistical, mathematical and deep learning models. Thus, we were able to achieve state of the art results using transformations like Log transformation, Differencing and Decomposition. The use of AR, MA, ARIMA and LSTM resulted in better forecasting. We can also perform these techniques on dynamic internet traffic of Wikipedia articles, so as to make our models more robust in real-time forecasting. This work can be further extended to know which particular domain of people in which part of the world, of what age group are most likely to visit which kinds of articles and how machine learning can be used to tempt them to open these sites thus helping websites in increasing their views and help them in diverting traffic towards their websites and be ahead among competitors. So, the work can further extend to optimize business solutions.

References

1. Chen, J., Hong, H., Stein, J.C.: Forecasting crashes: trading volume, past returns, and conditional skewness in stock prices. J. Financ. Econ. **61**(3), 345–381 (2001)
2. Liu, H., Song, B.: Stock trends forecasting by multi-layer stochastic ANN bagging. In: 2017 IEEE 29th International Conference on Tools with Artificial Intelligence (ICTAI), pp. 322–329. IEEE, November 2017
3. Tektas, M.: Weather forecasting using ANFIS and ARIMA mod els. Environ. Res. Eng. Manag. **51**(1), 5–10 (2010)
4. Harvey, A.C., Todd, P.H.J.: Forecasting economic time series with structural and Box-Jenkins models: a case study. J. Bus. Econ. Stat. **1**(4), 299–307 (1983)
5. Yang, J., Li, L., Shi, Y., Xie, X.: An ARIMA model with adaptive orders for predicting blood glucose concentrations and hypoglycemia. IEEE J. Biomed. Health Inf. (2018)
6. Franses, P.H.: Seasonality, nonstationarity and the forecasting of monthly time series. Int. J. Forecast. **7**, 199–208 (1991)
7. Nelson, C.R., Plosser, C.R.: Trends and random walks in macroeconomic time series: some evidence and implications. J. Monet. Econ. **10**(2), 139–162 (1982)
8. Witt, A., Kurths, J., Pikovsky, A.: Testing stationarity in time series. Phys. Rev. E **58**(2), 1800 (1998)
9. Granger, C.W., Joyeux, R.: An introduction to long-memory time series models and fractional differencing. J. Time Ser. Anal. **1**(1), 15–29 (1980)
10. Beveridge, S., Nelson, C.R.: A new approach to decomposition of economic time series into permanent and transitory components with particular attention to measurement of the 'business cycle'. J. Monet. Econ. **7**(2), 151–174 (1981)

11. Wu, Y., Pan, Z., Luo, X., Gao, J., Zhang, Y.: A hybrid forecasting method of electricity consumption based on trend extrapolation theory and LSSVM. In: 2016 IEEE PES Asia-Pacific Power and Energy Engineering Conference (APPEEC), pp. 2333–2337. IEEE, October 2016

12. Pouzols, F.M., Lendasse, A.: Effect of different detrending approaches on computational intelligence models of time series. In: The 2010 International Joint Conference on Neural Networks (IJCNN), pp. 1–8. IEEE, July 2010

13. Jing, J., Filev, D., Kurt, A., Özatay, E., Michelini, J., Özgüner, Ü.: Vehicle speed prediction using a cooperative method of fuzzy Markov model and auto-regressive model. In: 2017 IEEE Intelligent Vehicles Symposium (IV), pp. 881–886. IEEE, June 2017

14. Said, S.E., Dickey, D.A.: Testing for unit roots in autoregressive-moving average models of unknown order. Biometrika 71(3), 599–607 (1984)

15. Adebiyi, A.A., Adewumi, A.O., Ayo, C.K.: Comparison of ARIMA and artificial neural networks models for stock price prediction. J. Appl. Math. 2014, 7 (2014)

16. Engle, R.F., Yoo, B.S.: Forecasting and testing in co-integrated systems. J. Econ. 35(1), 143–159 (1987)

17. Akaike, H.: Fitting autoregressive models for prediction. Ann. Inst. Stat. Math. 21(1), 243–247 (1969)

18. Deodatis, G., Shinozuka, M.: Auto-regressive model for nonstationary stochastic processes. J. Eng. Mech. 114(11), 1995–2012 (1988)

19. Haining, R.P.: The moving average model for spatial interaction. Trans. Inst. Br. Geogr. 3, 202–225 (1978)

20. Arima, A., Iachello, F.: Ann. Phys. (NY) 99, 253 (1976)

21. Hochreiter, S., Schmidhuber, J.: Long short-term memory. Neural Comput. 9(8), 1735–1780 (1997)

Emerging Technologies and Opportunities for Innovation in Financial Data Analytics: A Perspective

Anirban Mondal[1] and Atul Singh[2(✉)]

[1] Department of Computer Science, Ashoka University, Sonepat, Haryana, India
anirban.mondal@ashoka.edu.in
[2] BAI7, IIM Bangalore, Bengaluru, India
atul.singh16@iimb.ernet.in

Abstract. Several key transformations in the macro-environment coupled with recent advances in technology have opened up tremendous opportunities for innovation in the financial services industry. We discuss the implications and ramifications of these macro-environmental trends for data science research. Moreover, we describe novel and innovative IT-enabled applications, use-cases and techniques in retail financial services as well as in financial investment services. Furthermore, this paper identifies the research challenges that need to be addressed for realizing the full potential of innovation in financial services. Examples of such research challenges include context-aware analytics over uncertain and imprecise data, data reasoning and semantics, cognitive and behavioural analytics, design of user-friendly interfaces for improved expressiveness in querying financial service providers, personalization based on fine-grained user preferences and financial Big Data processing on Cloud-based infrastructure. Additionally, we discuss new and exciting opportunities for innovation in financial services by leveraging the new and emerging financial technologies as well as Big Data technologies.

Keywords: Financial analytics · Big data analytics
Machine learning · Artificial intelligence · Deep learning
Natural Language Processing (NLP)

1 Introduction

Over the past decade, there have been several key transformations across many aspects of the macro-environment in which the financial services industry operates [17]. We observe that an all-encompassing trend across the entire gamut of the financial industry has been *democratization*, which has been happening at an unprecedented scale throughout this decade. Furthermore, nowadays financial data is generated not only by governments and firms, but also by the users of social media (e.g., websites, blogs and twitter feeds).

© Springer Nature Switzerland AG 2018
A. Mondal et al. (Eds.): BDA 2018, LNCS 11297, pp. 126–136, 2018.
https://doi.org/10.1007/978-3-030-04780-1_9

Such a transformation in the macro-environment, coupled with the rapidly increasing amounts of data about such aspects (which was previously challenging to obtain), has opened up new and exciting avenues for creating a major paradigm shift in terms of innovation in financial services as a key value proposition to end-users (both individuals as well as firms). However, to realize innovation in financial services, the challenge is to build capabilities for effectively analyzing the huge amounts of data (often from disparate and heterogeneous sources) and to obtain actionable insights for improving financial data management.

In particular, we observe that the changes in the macro-environment of the financial industry have resulted in the emergence of technology startup firms, which were previously outside of the financial sector. Such technology startup firms are designated as "Financial Technology" firms and are often simply referred to by the abbreviation of *fintech* firms [14,30,31]. Thus, fintech essentially concerns any technology-based innovation in the financial industry. Innovations associated with fintech are happening across a wide range of financial services. Interestingly, even large well-established banks and financial firms have started funding fintech firms in an effort to leverage the cross-boundary disruption innovation ushered in by fintech firms. In fact, global investments in the fintech space during 2010–2017 has been US $97.7 billion with an impressive compounded annual growth rate in fintech funding of 47% [4]. Understandably, the emergence of fintech firms with novel enabling technologies carry huge implications for data science research.

The remainder of this tutorial paper is organized as follows. Section 2 discusses the macro-environmental trends in the financial industry as well as the ramifications of these trends. Section 3 presents some of the novel and innovative IT-enabled applications, use-cases and techniques in retail financial services as well as in financial investment services. Section 4 discusses the research challenges and indicates how addressing these research challenges can enable a class of novel and innovative applications for the financial industry.

2 Macro-environmental Trends in Financial Services

In this section, we will set the context by briefly discussing the macro-environmental trends in financial services based on our previous work in [17]. Interested readers may refer to [17] for more details about these macro-environment trends.

2.1 Political/Legal Trends

Governments have been adopting increasingly protectionist measures (e.g., increased import duties and more trade barriers) in order to protect the economic sustainability and growth of their respective countries. Consequently, cost-cutting by means of outsourcing financial services to developing countries has been on a decreasing trend. Hence, for reducing operational costs, the financial industry has been focusing on innovations that support increased automation.

Moreover, the implication of the shift in trend from defined benefit pension plans towards defined contribution pension plans [16,28] is that the retirement benefits of users would essentially be highly uncertain and volatile due to the dependency on market conditions. Understandably, predicting the market conditions 20–40 years into the future requires the design of new financial risk modelling, risk management and long-term investment planning approaches, which need to be highly robust as well as cognizant of the macro-environmental factors. Since financial services for retirement planning need to be provided across a wide gamut of society, these services should typically be low-cost, thereby once again highlighting the need for automation.

Furthermore, financial accounting frauds (e.g., Enron [10], WorldCom [8] and Madoff [18]) have resulted in increasingly stringent legal and compliance requirements such as the Sarbanes-Oxley Act (SOX) [32]. Ensuring compliance and generating periodic reports (e.g., 10 K reports) *manually* to satisfy compliance and audit criteria is a typically expensive and error-prone process (primarily due to human errors) [22]. Hence, for purposes of cost reduction and accuracy in fraud detection and report generation, automated technologies are becoming imperative.

2.2 Economic Trends

Alternative investments typically fall outside of the standard asset classes (e.g., stocks, bonds and cash), hence they could be useful for portfolio diversification. Examples of alternative investments include P2P lending, crowdfunding, real estate, futures, derivative contracts and hedge funds [1]. In recent years, the growth of alternative investments has been largely fueled due to the relatively low returns as well as lack of trust in the more traditional avenues for investing. Thus, the alternative investment industry has been growing at a fast pace. Globally, alternative assets were US$6.5trillion as on July 2017 [2]. Moreover, the size of the global P2P lending is expected to grow at a CAGR of 48.2% between 2016 and 2024 and the market will be $897.85 billion by the year 2024 [9].

Crowdfunding has also been growing at a tremendous pace as an avenue for users to invest in entrepreneurship and innovation [21]. Given that P2P loans and crowdfunding are essentially unsecured and concern absolute strangers, it has become imperative to build more accurate and robust loan default prediction models based on the data. Furthermore, the emergence of alternative currencies (e.g., Bitcoin [26]) has also been becoming a major trend for the financial services industry. We perceive the alternative investment industry as a manifestation of the broader trend towards the *democratization* of financial services.

2.3 Socio-Cultural/Demographic Trends

In emerging markets, the trends of micro-financing [29] and correspondent banking models [15] have been becoming prominent primarily to improve financial inclusion (including loan facilities) for a large percentage of the population, which is excluded from any banking system. Moreover, technological developments, such as the availability of relatively low-cost smartphones, have also contributed significantly to the growth of financial inclusion e.g., by enabling mobile payments. Additionally, the correspondent banking model is largely enabled by mobile devices and technologies since the correspondents use mobile devices to connect to banking services requested by low-income customers living in remote areas, which do not have any brick-and-mortar bank in their vicinity.

Furthermore, the dramatic pace of economic growth in emerging markets has contributed to the middle class expanding at a rapid pace, thereby resulting in significantly increased demand for home loans and car loans. Given that the number of people, who would be targeted for financial inclusion as well as home/car loan opportunities, have been increasing dramatically over the years, there is a clear need for scalable information technologies, which would be adequately robust to satisfy the requirements of such financial services to a much larger percentage of the population. Such technologies should also provide adequate security and privacy and they should be largely automated in order to be able to provide *personalized* services to a tremendously large number of customers.

Interestingly, Millennials will inherit $30 trillion in wealth [33], thereby implying a massive paradigm shift in investment banking services. The ramification is that social media data and investment services leveraging social media data would become more important for investment decision-making as opposed to the services provided by investment specialists. In essence, this would make it imperative to build effective predictive models for actionable analytics on social media data, which often comes from disparate, heterogeneous and unreliable sources.

3   Applications and Use-Cases

In this section, we discuss innovative IT-enabled applications, use-cases and techniques in retail financial services as well as in financial investment services.

3.1   Blockchain in Financial Services

Blockchain technology [37, 38] and its applications have been revolutionizing the financial sector. The key reasons for the popularity and adoption of Blockchain by the financial services industry concern scalability (dramatically increasing number of users), security (including the provenance of financial data) as well as cost-efficiency and speed of financial transactions. In particular, observe that the workflows of the financial services industry had been designed for a time when the number of customers using financial services was relatively low with respect to the number of customers today and the number of transactions used to be considerably lower as well. Moreover, given the popularity of the Internet and mobile devices, a large percentage of financial transactions are nowadays occurring online. This has resulted in new kinds of security-related threats and vulnerabilities, which cannot be effectively handled by the largely centralized security mechanisms that exist in banks today. Notably, a decentralized paradigm, such as Blockchain, would intuitively be more effective in handling such threats and vulnerabilities primarily due to the absence of a single point of failure.

In this regard, *smart contracts* and *smart assets* are becoming increasingly popular [7,13]. Financial services typically entail a significant amount of legal contracts (e.g., futures contracts, contracts associated with M&A activities and so on). These contracts require a significant amount of paperwork and often require third-party service providers for ensuring that the clauses in the contracts are honoured by all concerned parties. In a similar vein, trade finance (e.g., lending, insurance, international trade) [6,20] concerns several partners in the supply chain and requires a significant amount of asset management, which also includes paperwork, fraud detection and third-party service providers. Given the global nature of supply chains nowadays coupled with the impact of increasing globalization, the number of trading partners can be reasonably expected to increase, thereby increasing the overall complexity of asset management.

Observe that the data provenance capabilities and the robustness (due to decentralization) associated with Blockchain technology inherently supports non-repudiation. For example, if a contract or an asset movement has been recorded on the Blockchain, the concerned parties can neither deny nor manipulate it in any way. Interestingly, a significant amount of analytics can also be done on the Blockchain for detecting frauds and for detection of money laundering activities because every asset movement and/or every contract gets recorded permanently on the Blockchain with a corresponding time-stamp without the requirement of any middlemen or third-party service providers. Moreover, Blockchain technology is also beginning to see applications concerning loyalty-based rewards programs [5,36]. From a broader perspective, observe how all of these applications of Blockchain also align with the trend of democratization that we have been witnessing in the financial services industry during this decade.

3.2 AI in Financial Services

AI has been becoming increasingly important in the financial services industry [11]. The reasons for the popularity of AI for providing financial services include time and cost savings, data-driven decision-making and insights for improved investment risk planning, more effective fraud detection and more accurate predictions concerning customer churn. For example, let us consider the use-case of loan approval, wherein AI can facilitate improved analytics for more accurate loan default predictions as well as significantly reduce the time of loan processing. Observe that due to the automation provided by AI, the cost of the loan approval process is also significantly reduced.

Furthermore, consider the case of a firm, which is trying to predict its sales performance for the next few months. Given the prevalence of the Internet and social media, a rich amount of highly contextual information can be used to analyze and understand the brand sentiment of the firm and its products (or services) by using AI and sentiment analysis algorithms on the social media data. Moreover, observe that data about brand sentiment typically comes from a wide gamut of heterogeneous and disparate data sources e.g., customer reviews on websites, call-center voice data about customer complaints, user blogs, videos on various websites, twitter feeds, news websites and so on. Linking such disparate data and obtaining actionable analytics from such data can be done efficiently with the use of AI algorithms.

Incidentally, chatbots enabled by AI have also been gaining popularity in the financial services industry primarily for facilitating improved customer service [24,27]. The key reasons for the adoption of chatbots by financial services firms are as follows. First, employing humans to answer queries issued by customers is often time-consuming and entails high costs, especially given the huge population of customers. Second, customers may ask a wide range of queries from appropriate form-filling to home-loan planning to insurance planning to investment planning all the way to retirement planning. Understandably, only a relatively small number of human experts would be capable of adequately answering these queries, given the significant breadth of the financial services industry. Observe that hiring such experts would be prohibitively expensive, thereby making it necessary for financial services firms to recruit and train humans in only specific areas of finance. This often results in accuracy issues with the suggestions provided by human experts, who may not be adequately trained for the job.

Third, the queries asked by customers in the financial services industry are often too complex for humans to handle because the data for answering the queries are spread across a wide gamut of databases. Thus, AI, predictive analytics and cognitive technologies become imperative to address such complex customer queries. Fourth, the need for *personalization* has been increasing. Providing personalized customer services for a huge population of ever-increasing customers with increasingly complex queries becomes practically infeasible even if a huge number of human customer service specialists were to be hired by financial services firms (even if we suspend reality for a moment and ignore the cost issues). Interestingly, AI and machine learning algorithms with their capability to learn and improve over time are ideally poised to effectively handle such chatbot-related customer services.

3.3   Machine Learning in Financial Services

Machine learning has been becoming increasingly important for providing improved financial services [3]. Machine learning aims at using historical data for training purposes, and then using the learning to create software agents with cognitive decision-making capabilities that are similar to those of humans. In particular, we can classify machine learning into the following three categories, namely *supervised machine learning, unsupervised machine learning* and *reinforcement learning*. Now we shall discuss applications of these three categories of machine learning in financial services domains such as retail banking, corporate banking and asset management.

Retail banking (also known as consumer banking) essentially concerns banking for the masses. Typical retail banking services include personal bank account management (e.g., savings accounts, fixed deposit accounts etc.) and loan/mortgage management. Interestingly, given the onset of democratization and the principle of financial inclusion in financial services, the nature of retail banking has been changing over the years. For example, many of the services traditionally provided by retail banks in brick-and-mortar settings are nowadays also being provided online or by means of mobile devices. Notably, this also brings in additional challenges around security, privacy and authentication of customer data.

Furthermore, the goal of financial inclusion (in part due to government regulations) poses interesting challenges for machine learning in emerging markets. In emerging markets, a significantly large population of customers have typically low-incomes and as such, limited data is available about their past interactions with banks and other financial firms. The implication is that in such cases, machine learning algorithms would need to predict the probability of loan default in the absence of adequate data for determining the creditworthiness of such customers. Additionally, such loan default predictions need to be performed at scale in the case of emerging markets, thereby further reinforcing the need for machine learning. Incidentally, given the scalability requirements of retail banking nowadays, analytics of various business processes and workflows has become a necessity for facilitating the detection of bottlenecks in various processes as well as for process improvement. Moreover, the use of image processing techniques for identifying forgery in documents also constitutes a major opportunity for doing analytics in retail banking.

Now let us examine the applications of financial services in investment banking. Investment banking typically requires a significant amount of predictive analytics across a wide gamut of investment application domains such as structuring an M&A deal, underwriting activities, wealth management services (both for individuals as well as for corporations), asset valuation and the valuation of a given firm. Investment banks use supervised and unsupervised machine learning algorithms in these investment application domains. As an example, consider the case of a firm X, which is interested in the acquisition of another firm Y. This is a common scenario in investment banking. Here, firm X has to perform a valuation of firm Y (including assets, liabilities, equity) for determining whether it would be a good business decision to acquire firm Y. For this purpose, the investment banker needs to look at several types of data associated with firm Y. Such types of data include business/legal documents and company filings (e.g., 10 K reports) over the past 5–6 years, social media data about the reputation of the firm Y in the market environment as well as in the non-market environment, recent news about firm Y, online videos concerning firm Y and so on.

Observe that the data is highly heterogeneous, multi-modal (e.g., text, video etc.) and largely unstructured. Linking all this data poses serious research challenges and requires a significant amount of domain knowledge in finance. Notably, nowadays the valuation of any given firm does not depend solely on quantitative factors, but also on the factors influencing the non-market environment e.g., reputation of the firm. Furthermore, given the complexity of the data, it is practically too challenging for even a human expert to link together all this data and derive actionable insights from the data. Herein lies the role of machine learning in evaluating, classifying the risk and understanding the cost versus benefit analysis associated with firm X's acquisition of firm Y.

In a similar vein, portfolio management also requires investment bankers to understand the *effective* valuation of several firms e.g., for determining which stocks to invest in. Such valuation necessitates an in-depth analysis (including identification of correlations) of data ranging from 10 K reports to social media data. Moreover, portfolio managers need to consider the investment requirements of their respective clients, and such requirements typically vary widely among the clients (both individuals and firms) depending upon factors such as risk appetite, mid-to-long-term growth objectives and so on. Furthermore, portfolio managers need to diversify the portfolio of their clients for mitigating risk.

To put things into perspective regarding the complexity as well as the scalability requirements of portfolio management, it is important to note that there is a huge number of listed securities all over the world. Additionally, nowadays there are several alternative investment options as well such as crowdfunding. This makes it practically infeasible for a human expert to link all kinds of data regarding each of the securities and/or alternative investment options, and analyze and obtain a reasonable understanding of the investment options, while incorporating portfolio diversification as well as fraud detection in the investment planning process. Machine learning can be very useful during the different parts of the investment planning process. For example, one could use natural language processing (NLP) on the textual data concerning securities and then create knowledge graphs from the information associated with the textual data. Such knowledge graphs could be queried for additional insights and could even be valuable for triggering alerts e.g., an alert could be triggered when the net value of a given portfolio goes below a certain threshold [34].

Incidentally, wealth and portfolio management services have been around for a long time, but the cost of such services have narrowed their applicability to only the relatively rich segments of the population. Thus, machine learning can play a significant role in providing wealth and portfolio management services at scale for a much larger percentage of the population than ever before. Such wealth and portfolio management services are beginning to be offered through software solutions called robo-advisors [35], thereby further democratizing financial services. Observe that the role of machine learning in such scenarios goes well beyond the mere automation of financial services. In essence, machine learning is used in such scenarios for facilitating decision-making and gaining insights about the financial future of various kinds of firms.

Additionally, we would also like to point out that Reinforcement Learning, which is a type of machine learning technique, has been steadily gaining importance in the financial services industry. This is primarily due to the inherently dynamic nature of processes and workflows in financial services. Observe that supervised machine learning techniques are generally not adequate to address dynamically changing contexts because the decisions made based on the training data may no longer be valid during deployment because of significant contextual changes. In this regard, Reinforcement Learning solutions (e.g., actor-critic model [19]) prove to be valuable since they are capable of addressing contextual changes, which typically occur during the delivery/deployment of financial services. As a single instance, JP Morgan Chase [25] uses Deep Reinforcement Learning (DRL) for facilitating decision-making during trading operations by considering contextual changes in the market conditions. Furthermore, the work in [12] presents a DRL approach for creating a risk-conscious portfolio, while the work in [23] uses a DRL approach for portfolio management.

4   Research Challenges and the Way Forward

The sheer scale and importance of the financial services industry coupled with the changes in technology as well as the changes in the macro-environment pose significant challenges. Some of the key research issues for realizing innovations in the financial services industry include modelling and integrating large-scale and complex data that is generated from disparate and heterogeneous sources, putting a context around the data for context-aware reasoning and analytics purposes, cognitive analytics on the financial data, semantic understanding and interpretation of the financial data. Moreover, more research is required on effective heuristics for detecting fraud, intelligent approaches for identifying cases of creative accounting and techniques for working with huge amounts of uncertain data from different modalities and identifying cross-connections and correlations.

Furthermore, doing effective analytics on financial data requires a significant amount of domain knowledge not only in finance, but also in business. Hence, there are practical challenges associated with the analysis, representation and modelling of financial business requirements in tandem with the data, incorporating financial domain knowledge and addressing specific legal and regulatory constraints that are imposed by the financial domain. Thus, innovations concerning a wide gamut of document technologies and analytics techniques are increasingly becoming critical in the financial domain.

In this work, we have provided an overview of the financial services industry and pointed out a wide gamut of applications and use-cases, many of which are expected to be opportunities for innovation in the near future. Notably, addressing the research challenges as well as the practical challenges would be key to realizing the next-generation innovations in the financial services industry. We hope that academicians and industry practitioners from Computer Science, Finance and other related disciplines will collaboratively work together to realize significant innovations for the financial services industry.

References

1. Alternative Investment. http://www.investopedia.com/terms/a/alternative_invest ment.asp. Accessed 30 Sept 2010
2. Global shift into alternative assets gathers pace. https://www.ft.com/content/ 1167a4b8-6653-11e7-8526-7b38dcaef614. Accessed 30 Sept 2010

3. How AI and automation will shape finance in the future, November 2017. https://www.forbes.com/sites/workday/2017/11/03/how-ai-and-automation-will-shape-finance-in-the-future. Accessed 30 Sept 2010

4. Fintech funding sets new records in 2017 (2018). https://www.finextra.com/pressarticle/72819/fintech-funding-sets-new-records-in-2017. Accessed 30 Sept 2010

5. Making blockchain real for customer loyalty rewards programs (2018). https://www2.deloitte.com/us/en/pages/financial-services/articles/making-blockchain-real-customer-loyalty-rewards-programs.html. Accessed 30 Sept 2010

6. Ahn, M.J.: A theory of domestic and international trade finance. No. 11–262, International Monetary Fund (2011)

7. Alharby, M., van Moorsel, A.: Blockchain-Based Smart Contracts: A Systematic Mapping Study (2017). CoRR abs/1710.06372. http://arxiv.org/abs/1710.06372

8. Asthana, S., Balsam, S., Kim, S.: The effect of Enron, Andersen, and Sarbanes-Oxley on the US market for audit services. Account. Res. J. 22(1), 4–26 (2009)

9. Bajpai, P.: The Rise of Peer-to-Peer (P2P) Lending (2016). http://www.nasdaq.com/article/the-rise-of-peertopeer-p2p-lending-cm685513. Accessed 30 Sept 2010

10. Brody, R.G., Melendy, S.R., Perri, F.S.: Commentary from the american accounting association's 2011 annual meeting panel on emerging issues in fraud research. Account. Horiz. 26(3), 513–531 (2012)

11. Danielsson, J., Macrae, R., Uthemann, A.: Artificial intelligence, financial risk management and systemic risk. Tech. rep., University of Zurich, Department of Informatics, November 2017. Accessed 30 Sept 2010

12. Gao, X., Chan, L.: An algorithm for trading and portfolio management using q-learning and sharpe ratio maximization. In: Proceedings of the International Conference on Neural Information Processing, pp. 832–837 (2000)

13. Gatteschi, V., Lamberti, F., Demartini, C., Pranteda, C., Santamaría, V.: Blockchain and smart contracts for insurance: Is the technology mature enough? Future Internet 10(2), 20 (2018)

14. Gomber, P., Koch, J.A., Siering, M.: Digital finance and fintech: current research and future research directions. J. Bus. Econ. 87(5), 537–580 (2017)

15. Kishore, A.: Business correspondent model boosts financial inclusion in India (2012). https://www.minneapolisfed.org/publications/community-dividend/business-correspondent-model-boosts-financial-inclusion-in-india. Accessed 31 Jan 2015

16. Krishnan, V.S., Cumbie, J., Ice, R.: Defined benefit plans vs. defined contribution plans: An evaluation framework using random returns. https://papers.ssrn.com/sol3/papers.cfm?abstract_id=2954196

17. Krishnapuram, R., Mondal, A.: Upcoming research challenges in the financial services industry: a technology perspective. J. Inst. Dev. Res. Bank. Technol. (IDRBT), 66 (2017)

18. Kuhn, J.R., Sutton, S.G.: Learning from Worldcom: implications for fraud detection through continuous assurance. J. Emerg. Technol. Account. 3(1), 61–80 (2006)

19. Lillicrap, T.P., Hunt, J.J., Pritzel, A., Heess, N., Erez, T., Tassa, Y., Silver, D., Wierstra, D.: Continuous control with deep reinforcement learning (2015). CoRR abs/1509.02971. http://arxiv.org/abs/1509.02971

20. Marc, A.: Boosting trade finance in developing countries: What link with the WTO. World Trade Organization Staff Working Paper Ersd-2007-04. http://ssrn.com/abstract 1086278 (2007)

21. Mollick, E.: The dynamics of crowdfunding: an exploratory study. J. Bus. Ventur. 29(1), 1–16 (2014)

22. Mondal, A., Sandor, A., Popa, D.N., Stavrianou, A., Proux, D.: System and method for facilitating interpretation of financial statements in 10 K reports by linking numbers to their context, US Patent App. 14/715,998. 24 November 2016

23. Moody, J., Saffell, M.: Learning to trade via direct reinforcement. IEEE Trans. Neural Netw. **12**(4), 875–889 (2001)

24. Morgan, B.: 5 ways chatbots can improve customer experience in banking, August 2017. https://www.forbes.com/sites/blakemorgan/2017/08/06/5-ways-chatbots-can-improve-customer-experience-in-banking/. Accessed 30 Sept 2010

25. Mosic, R.: Deep Reinforcement Learning Based Trading Application at JP Morgan Chase, July 2017. https://medium.com/@ranko.mosic/reinforcement-learning-based-trading-application-at-jp-morgan-chase-f829b8ec54f2. Accessed 30 Sept 2010

26. Nakamoto, S.: Bitcoin: A peer-to-peer electronic cash system. https://bitcoin.org/bitcoin.pdf (2008). Accessed 30 Sept 2010

27. Nguyen, M.H.: How chatbots and artificial intelligence will save banks and the finance industry billions, October 2018. https://www.businessinsider.com/chatbots-banking-ai-robots-finance-2017-10?IR=T. Accessed 30 Sept 2010

28. Orlova, N.S., Rutledge, M.S., Wu, A.Y.: The transition from defined benefit to defined contribution pensions: Does it influence elderly poverty?. Tech. rep, Center for Retirement Research (2015)

29. Peachy, S., Roe, A.: Access to finance: What does it mean and how do savings banks foster access. WSBI, World Savings Banks Institute (2006)

30. Puschmann, T.: Fintech. Bus. Inf. Syst. Eng. **59**(1), 69–76 (2017). https://doi.org/10.1007/s12599-017-0464-6

31. Schueffel, P.: Taming the beast: a scientific definition of fintech. J. Innov. Manag. **4**, 32–54 (2016)

32. Securities, U., Commission, E., et al.: The laws that govern the securities industry (2012). http://www.sec.gov/about/laws.shtml#sox2002. Accessed 31 Jan 2015

33. Shin, L.: How millennials' money habits could shake up the financial services industry (2015). http://www.forbes.com/sites/laurashin/2015/05/07/how-millennials-money-habits-could-shake-up-the-financial-services-industry/. Accessed 30 Sept 2010

34. Singh, A.: Relationships matter: Mining relationships using deep learning, September 2018. https://confengine.com/odsc-india-2018/proposal/7264/relationships-matter-mining-relationships-using-deep-learning. Accessed 30 Sept 2010

35. Sironi, P.: FinTech innovation: from robo-advisors to goal based investing and gamification. Wiley, Chichester (2016)

36. Swaminathan, K.: How blockchain can crack the holy grail of loyalty programs, July 2018. https://www.finextra.com/blogposting/15430/how-blockchain-can-crack-the-holy-grail-of-loyalty-programs. Accessed 30 Sept 2010

37. Yli-Huumo, J., Ko, D., Choi, S., Park, S., Smolander, K.: Where is current research on blockchain technology?–a systematic review. PloS One **11**(10), e0163477 (2016)

38. Zhao, J.L., Fan, S., Yan, J.: Overview of business innovations and research opportunities in blockchain and introduction to the special issue. Financ. Innov. **2**(1), 28 (2016). https://doi.org/10.1186/s40854-016-0049-2

Web and Social Media Data

Design of the Cogno Web Observatory for Characterizing Online Social Cognition

Srinath Srinivasa[✉] and Raksha Pavagada Subbanarasimha

International Institute of Information Technology, Bangalore, India
sri@iiitb.ac.in, raksha.p.s@iiitb.org

Abstract. It is important to occasionally remember that the World Wide Web (WWW) is the largest information network the world has ever seen. Just about every sphere of human activity has been altered in some way, due to the web. Our understanding of the web has been evolving over the past few decades ever since it was born. In its early days, the web was seen just as an unstructured hypertext document collection. However, over time, we have come to model the web as a global, participatory, socio-cognitive *space*. One of the consequences of modeling the web as a space rather than as a tool, is the emergence of the concept of *Web observatories*. These are application programs that are meant to observe and curate data about online phenomena. This paper details the design of a Web observatory called *Cogno*, that is meant to observe online social cognition. Social cognition refers to the way social discourses lead to the formation of collective worldviews. As part of the design of Cogno, we also propose a computational model for characterizing social cognition. Social media is modeled as a "marketplace of opinions" where different opinions come together to form "narratives" that not only drive the discourse, but may also bring some form of returns to the opinion holders. The problem of characterizing social cognition is defined as breaking down a social discourse into its constituent narratives, and for each narrative, its key opinions, and the key people driving the narrative.

Keywords: Web observatory · Social cognition
Opinion marketplace · Abstraction · Expression · Social media

1 Introduction

The World Wide Web is the largest, global information network that humanity has ever seen. The overall structure of the web is neither engineered into its present shape, nor is it a naturally occurring structure that appeared without any human intervention. The web and its dynamics are emergent consequences of a large number of independent human decisions. Researchers have tried to model and understand the web and its dynamics, ever since its inception.

In the early days, the web was considered to be a large, unstructured database. Concepts from Relational Database systems were projected onto the

© Springer Nature Switzerland AG 2018
A. Mondal et al. (Eds.): BDA 2018, LNCS 11297, pp. 139–154, 2018.
https://doi.org/10.1007/978-3-030-04780-1_10

web (for example, WebSQL [20]) in an attempt to utilize the largest data store the world had ever seen. Later on, strict database models gave way to more loosely-defined models that viewed the web as a digital library. This resulted in concepts from Information Retrieval (IR) and Library Sciences, being applied to the web [5, 7, 16].

A major inflection point in our understanding of the web came with models that focused on hyperlink structures rather than the documents themselves [8, 15, 24]. Unlike in a conventional application program, the structure of hyperlinks on the web is not determined a priori. The author of a web page can link to any other web page on the web, at any time. The structure of hyperlinks represent largely independent judgments and reveal latent intentions of a large number of humans who have created these links. Hyperlinks can be interpreted in different ways. The fact that the author of a document has placed a hyperlink to another document, makes the hyperlink as a "relevance indicator." Similarly, when a user clicks on a hyperlink, her attention flows from one document to another – making hyperlinks the "attention pathways" of the web.

Yet another major inflection point in our understanding of the web, came with Web2.0 at the turn of the new century, that emphasized on the *social* nature of web activity. Web2.0 applications were designed to encourage user participation and feedback, which were in turn used for discerning a variety of underlying semantics. The social nature of the web, has only grown stronger over the years. Current-day proliferation of several web enabled micro-platforms like smart phones and IoT devices have greatly emphasized and leveraged on the social nature of web activity. Earlier, users needed to explicitly connect to the Internet and start a web browser to be connected to the web. But in the current day, users are connected to the web implicitly through several smart devices. In addition, advances in AI and information integration have ensured that a user's *identity* can be maintained on the web with high precision. Hence for instance, a user may be disconnected from all devices, but his location may still be discerned by web applications through IoT enabled cameras and face recognition technology. This causes the person to be "participating" in some web application logic, even when completely disconnected from the web.

The above line of argumentation have led us to move away from viewing the web as a tool, or as an extension of some human faulty like the mind. Instead, we now see the web as a logical socio-cognitive *space* that subsumes us, rather than as a tool that extends our capabilities. Figure 1 schematically depicts this distinction. While a tool is something that we explicitly and intentionally decide to use in order to extend our capabilities, a space is something that subsumes us and may affect us even without our explicit, intentional engagement with it.

In this model of the web, humans are *participants*, rather than users. Users have a dominant relationship with tools, unlike participants in a space. Participants in a space are more like *components* that affect one another and which may be put to use, to characterize the space. This is true of the web, since applications on the web uses humans, as much as humans use the web.

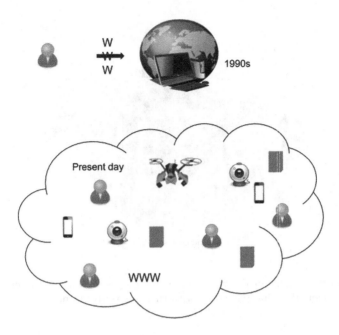

Fig. 1. Contrasting web as a tool versus as a space

This change in perspective, has also resulted in the emergence of a new concept called "Social Machine" [14,28]. A Social Machine is an ensemble of automation and human components, whose emergent dynamics is a result of decisions taken by both algorithmic and human decision-making processes. For instance, the dynamics of a social media platform like Facebook is a result of algorithmic decisions like Timeline algorithms, as well as human decisions about what to post and what to comment on.

Rather than viewing the web as a network formed of documents and hyperlinks, we now view the web as a network of social machines which constantly affect human populations and the contents of the web. Figure 2 depicts our overall model of the web, as classified into three "realms" based on their social activity. The "social" realm shown as the inner-most circle, is where much of the social activity takes place. This is the realm comprising of social machines like Facebook, Twitter, Stackoverflow, etc. The "trigger" realm indicated by the middle circle, refers to publishing sites like news portals, media houses and publishing platforms (for example CNN, BBC, Medium, Wordpress, etc.), which may not host a lot of social activity themselves, but whose activity often *triggers* social activity in the social realm. Finally, there is the "inert" realm comprising of mostly information content (like online encyclopedias, research sites, etc.) that have very little social activity, but whose contents may be used in conversations in the social realm.

This paper presents the design of a Web Observatory called Cogno, to observe the social realm of the web. The concept of a web observatory [18,30–32] is a

Fig. 2. The three realms of the web. Commercial logos shown are for representative purpose only. Copyright for logos rests with their respective owners.

natural consequence of modeling the web as a space, rather than as a tool. Analogous to physical observatories, web observatories are application programs that observe some aspect of the web to collect and curate datasets for further research.

An important element of web observatory design is development of a mathematical model for the phenomena of interest. The Cogno web observatory focuses on *social cognition*. Social cognition refers to the way collective worldviews are formed in a population due to the exchange of views. This paper presents a model for characterizing social cognition that is implemented as part of the Cogno web observatory.

2   Related Literature

World wide web has become the largest source for information and it has become a platform of expression and communication for many. It is very hard if not impossible for a person or a organization to understand the dynamics of the web. This gave rise to a global project called Web Observatory [31] aimed to build an ensemble of multiple web science research groups to collectively address the problem of understanding the web. The primary task of web observatories across the world is to archive and share various datasets and analysis tools. A global grid of web observatories has the potential to unleash real-time analytics and even social intervention, to shape the future of the web in a constructive way.

We see web observatories fundamentally as infrastructures for Big Data Analytics. Web observatories address all the Vs of Big Data. This has resulted in the emergence of several distributed frameworks and design patterns for web observatories [18,32]. There are many common data sets that the big data enterprises use and many of these are replicated for different installations. This cannot be implemented for web observatories considering the vast and rapidly growing data and the high cost of replication and curation.

As of October 2018, 15 web observatories exist worldwide, focusing on various aspects of the web. For instance, South Australian Government Web Observatory [34] is focused on enhancing the power of open data portals to provide access and share open government data-sets, encourage others to develop analytic applications to address issues related to ageing population. Another web observatory named *Zooniverse* [26] is a citizen science platform that invites the public to participate and contribute to the analysis of various research projects of various topics like space, nature, biology, medicine, climate science and the humanities. Web observatory allows researchers to scale up the process of analysis my many folds by the public contribution. Zooniverse also plans to further use tools available with other web observatories to perform analysis on their data. Web observatories should have a standard and well maintained structure to store and analyze data, so that inter-operability between web observatories is effortless. The Southampton University Web Observatory(SUWO) [13] provides support for harvesting, storage and access to various web related datasets. Since, data can be vast on the web, SUWO provides harvesting of data either through data-source-centric mode or topic-centric mode. They also provide a portal on which a number of analytic tools can be used to understand and analyze visual representations of different datasets.

The social media observatory [19] at Indiana University provides an interactive platform called *Truthy*, which collects and analyzes discourses on Twitter. The observatory focus on understanding diffusion of memes across the social network. The observatory provides features such as (1) various computed statistics for memes and users, (2) an interactive diffusion network and (3) API access to data and computed statistics. They propose that social media observatories should have six design goals namely: reliability, reproducibility, topic filtering, visualization, open access and legal compliance. Although the social media observatory provides an access to study opinion diffusion across the network, their primary focus is not social cognition. To the best of our knowledge, there is a dearth of platforms like *cogno*, that specifically focus on understanding online social cognition.

With regards to the question of modeling social cognition, many recent research focus on understanding the impact of social media on human behaviour [1,9,35] by looking at opinion formation and diffusion process. Most of these models look at either sentiment diffusion or meme diffusion across the network. In our work, we propose a specific model for an opinion, based on two components called *abstractions* and *expressions*.

Various topic modeling techniques like LDA, Embedding Based Topic Modeling etc, have been proposed to model user topic distribution on social media [11,25,29]. Many of these techniques are based on associating entities in social media to a encyclopedic knowledge base like DBpedia or human generated topics. Even though multiple topics are generated by these models, they do not model the narratives emerging from the discourse. As shown later in this paper, augmenting semantic clustering with sentiment scores depicts clusters formed due to underlying social dynamics, rather than due to the contents of the posts.

Understanding the influence of a user has elicited a lot of research interest [2,6] based on quantifying retweets, followers, mentions or the cascade effect of the user on the network. Many research efforts also focus on expert identification [12,23,33] in a network by applying topic modeling techniques based on topic consistency of users. A machine learning approach has been used for identifying campaigns [10,17] by extracting various features of the vocabulary used or the URLs used in the network. Dynamics of communication on social media has been explored by analyzing the time series analysis of information flow in the network [4].

While these are relevant to identifying opinion drivers (Described in Sec. 4), most of these focus on the network aspect or position of the users and seldom focus on intention or the content of the users on social media. An expert of an influencer may or may not be an opinion driver.

3   Modeling Social Cognition

The objective of the Cogno web observatory is to observe *online social cognition*. Social cognition refers to the way opinions are exchanged in a social setting and the way such exchanges result in the formation of collective worldviews. In this section, we present a computational model to characterize social cognition.

Discussions happening around a topic of interest, is called a *social discourse*. A social discourse D_t, where a set of people are discussing about topic t is modeled as follows:

$$D_t = (U, N, \delta, O, \gamma) \tag{1}$$

Here U is the set of people participating in D_t, and N is a set of "narratives" that are driving the discourse. A narrative is a consistent storyline that is formed by several compatible opinions expressed on the topic. A narrative may or may not be completely grounded in facts. But the internal consistency of opinions gives a narrative, both persuasive power and resilience. A strong narrative is one that is resilient against contrary opinions, and has the tendency to prevail over the population.

Each narrative may have one or more "drivers" – people who play key roles in building and propagating the narrative – either voluntarily or involuntarily. The term $\delta : U \to N$ represents a partial, surjective function that identifies potential users for each narrative, who could be driving the narrative.

The building blocks of a narrative are *opinions* expressed by the participants as part of the discourse. The set of all opinions expressed in the discourse is represented by the term O in Eq. 1 above. Each narrative $n \in N$ can be seen as a subset of "compatible" opinions from O, that is: $n \in 2^O$. A small set of opinions in a narrative can be seen as "key arguments" or perspectives that characterize the narrative. The term $\gamma : O \rightarrow N$ represents a partial, surjective function that assigns one or more key opinions to each narrative. Typically, a given opinion cannot be a key opinion for more than one narrative.

Opinions are in turn, modeled based on two constituent elements [27], namely: *abstractions* and *expressions*. Abstraction refers to the perspective of the opinion-holder about the topic of discourse, and expression refers to the opinion-holder's sentiment about the issue. Formally, an opinion $o \in O$ is represented as $o = (a, e)$, where a is the abstractive component of the opinion, and e is the expressive component of the opinion.

Two people having the same perspective about an issue, but displaying different sentiments, are said to have different opinions about the issue. Hence for instance, suppose that two people may hold the perspective "Demonetization was a failure", but one seems happy about it, while the other person seems sad about it. We say then that the two opinions are different (and in this case, contrary) from one another.

When several opinions are expressed as part of a social discourse, "compatible" opinions often come together to form a "narrative". To extend the earlier example, a negative sentiment stating demonetization was a failure, may be compatible with another opinion that says "Demonetization was necessary" with a positive sentiment. The combination of these two opinions forms a small narrative that "justifies" both the opinions. The narrative may go somewhat like this: "Demonetization was necessary, but the way it was implemented was a failure. The failure hence, was not with the idea itself but with the way it was implemented."

A narrative is a logical framework comprising of a consistent set of opinions, and often "filled-in" by other default ground truths, or implications, that builds a semantic worldview.

Although we cannot observe the complete semantic framework of a narrative on social media posts, we can still observe its formation by computing semantic compatibility of opinions and finding opinion clusters.

Narratives form the primary drivers of a social discourse. Some narratives end up having enormous resilience and persuasion power in a population. Some narratives may live for several centuries, and may even become historical theories that are widely accepted by scholarly communities. Narratives using which governments form policies may even affect the course of history.

A *characterization of social cognition* is defined as the breaking down of a social discourse into its dominant narratives, identifying the key arguments of each narrative, and the key participants who are driving the narrative.

Figure 3 represents the overall flow of the proposed model for breaking a social discourse into its constituent narratives.

The process starts with fetching all the related posts for a given trending topic on social media. All posts are then pre-processed to remove punctuation marks, numbers, url, stopwords, whitespaces, searched keyword and lowering the text.

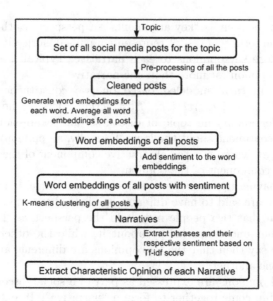

Fig. 3. Process overview of Characterizing Narrative formation

The set of all remaining words in the post are considered to contribute to the abstraction. Semantic characterization of each word is done by computing their word embeddings. Word embeddings are vector representations of the words, which capture the context of each word. We use the Continuous Skipgram model based on predicting the context given a word. This was first proposed by Mikolov et al. [22] using a 3-layer feed-forward neural network model, and implemented in word2vec [21].

For all words in a post, respective word embeddings are generated, where each embedding is a vector of m dimensions. Here $m \ll n$, where n is the total number of words in the vocabulary of terms encountered in the dataset. The dimensionality of the word vector is also the number of neurons in the hidden layer of the shallow feed-forward network that is used to compute word embeddings.

Semantic embedding of a post is then calculated by averaging word embeddings of all constituent words in the post. Semantic embedding of a post now captures the abstraction part of the post.

To add the expression element of the opinion expressed, sentiment analysis is performed on the post, by looking at terms describing sentiment. The dominant sentiment expressed in the post is mapped to a range $[-1, +1]$ indicating both arity and intensity. This sentiment is added as the $(m+1)^{th}$ dimension to the post embedding. For the current implementation on Twitter, we used the Syuzhet sentiment analysis package for R^1 for this purpose.

---

[1] https://cran.r-project.org/web/packages/syuzhet/index.html.

Table 1. Twitter datasets selected for experimentation

Twitter search keyword	Description	Number of Tweets	Tweet dates
#Inauguration[a]	Donald Trump Presidential Inauguration	10000	22 Nov 2016 – 21 Apr 2017
#Womensmarch[b]	Womens March Campaign	10000	02 Feb 2017 – 08 Feb 2017
#Demonetization[c]	Demonetization of 500 and 100 Rupee notes in India	10000	07 Feb 2017 – 08 Feb 2017

[a] https://www.kaggle.com/adhok93/inauguration-and-womensmarch-tweets
[b] https://www.kaggle.com/adhok93/inauguration-and-womensmarch-tweets
[c] https://www.kaggle.com/arathee2/demonetization-in-india-twitter-data

After this step, every post in the dataset will have a $(m+1)$-dimensional vector representation encapsulating both abstraction and expression components. These post embeddings are then clustered using k-means clustering technique. Each cluster thus obtained represents a narrative.

For each narrative, we also compute its set of one or more *characteristic opinions*. This is a set of keyphrases along with their sentiment, that can be used to label a narrative. Keyphrases are extracted based on the following patterns:

1. Named Entity (Ex: Donald Trump, Albert Einstein)
2. Number followed by a noun (Ex: 30 states, 45th President)
3. Adjective followed by a noun (Ex: scintillating display, beautiful inauguration)
4. Foreign word (Ex: IAMWithHer, Metoo, Womensmarch)
5. Noun followed by verb (Ex: Terrorists arrested, students celebrated)

The most relevant phrases are chosen from the candidate set of phrases obtained from the above patterns, based on their tf-idf score, and their respective sentiment intensity.

Results and Evaluation: The above process of identifying narratives was tested on three topics on Twitter. Table 1 describes the various datasets selected for experimentation. Table 2 shows one of the narratives identified for the topic #Demonetization. It shows the characteristic opinion extracted for the narratives and top few tweets of the narrative.

Results of the model were evaluated with help of human evaluators in two steps. The first step involved evaluation of coherency of narratives. Users were asked to rate the coherency of each narrative of each topic. Second step involved evaluation of relevance of characteristic abstraction and expression generated by the model. The model was found to be 89% accurate. Average precision and recall of the model were 0.59 and 0.75 respectively. A score called Novelty

Table 2. An example narrative from the topic #Demonetization

Characteristic Opinion of the Narrative	
Abstraction	**Expression**
Fund shortage	Negative
Man ends life	Negative
Daughter's wedding	Negative
Gujarat demonetization	Negative
Top Tweets	**Tweet count**
Man ends life over fund shortage ahead of daughter's wedding in Gujarat ! #Demonetization	140
A man shaved his head at JantarMantar in protest against #Demonetization on 13th day of the move	20
Demolition Man vs Demonetization Man India Has Suffered Under Both	13
Sudden declaration of demonetization has affected common man's life	1

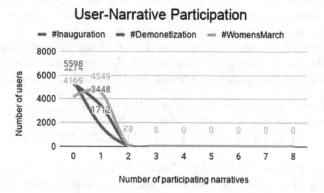

Fig. 4. User-Narrative Participation

represented the proportion of relevant characteristic opinion extracted which, evaluators couldn't identify themselves. Novelty score for the model was 0.34.

Figure 4 shows the participation of users in narratives. It can be noted that around 61% of the users didn't participate in any narratives and among the users who participated, 97% of the users participated in only one narrative. This is used to argue that the narratives identified represent underlying *social* characteristics, rather than just characteristics of the tweet contents.

4   Identifying Opinion Drivers on Social Media

The second part of the characterization problem is identifying key drivers of a narrative. A computational model for this may be found in [3]. We present an overview of the model in this section.

An opinion driver is a user who displays an *intent* to drive the topic into some particular direction. Since the intent to drive an opinion is a latent variable, we model this based on the following two observables:

1. High amount of proactive participation in the topic and,
2. Narrow focus in the vocabulary used.

For a given social discourse D_t for topic t, we calculate the "driver score" for each user $u \in U$ participating in the discourse. Broadly, the likelihood of u being an opinion driver, is computed as follows:

$$L(u) = Evidence(u) \cdot prior(u) \tag{2}$$

Evidence score for each user refers to the narrow focus in the vocabulary of terms used by the user. This is calculated by first computing the entropy of her vocabulary:

$$H_u = - \sum_{i=1}^{m} p_u(t_i) \log_2 p_u(t_i) \tag{3}$$

Here, $p_u(t_i)$ is the probability of user u using term t_i as part of her vocabulary. The evidence score for a user is calculated by comparing this entropy with the maximum entropy obtained across all users in the discourse:

$$Evidence(u) = \frac{(\hat{H} + 1) - H_u}{\max_u((\hat{H} + 1) - H_u)} \tag{4}$$

The prior probability mentioned in Eq. 2 is calculated by modeling intent to drive opinions as a Dirichlet process over a base distribution computed by user activity over the observed social discourse.

After the likelihood score is computed, all users in the discourse are arranged in descending order of their Likelihood score. Users are divided at a point where the difference between any two consecutive users is greater than the expected value of the difference.

All users with likelihood score greater than the dividing point are considered as potential opinion drivers for the topic t.

Results and Evaluation: Proposed model of identifying opinion drivers were implemented and evaluated on 6 different trending topics on Twitter. Two of them, (#Ignis and #SpiceJetBigOrder) were previously known to be promoted topics on Twitter while the other four (#Demonetization,#Jallikattu, #Kansas and #Dangal) were organically trending.

Figures 5 and 6 shows the distribution of opinion driver likelihood score for all user of the topic #Kansas and #SpiceJetBigOrder respectively. The red dot

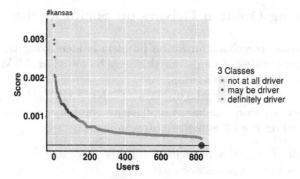

Fig. 5. Likelihood score distribution for users of topic '#Kansas' divided in to 3 classes along with null user (Color figure online)

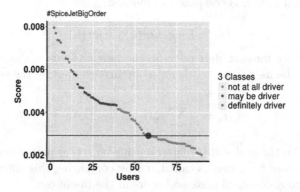

Fig. 6. Likelihood score distribution for users of topic '#SpiceJetBigOrder' divided in to 3 classes along with null user (Color figure online)

in the figures represent a hypothetical "null" user whose levels of activity and range of vocabulary were computed based on statistical expectation measures. We see that, for organically trending topics like #Kansas, the null user score is way below all the other scores. This is because, for organically trending topics, the overall vocabulary would be very rich and diverse. However, for promoted campaigns, the null user score is much higher, and well inside the rest of the scores. This is because, in a campaign, the overall vocabulary is already controlled.

Table 3 shows the top 3 set of words used by the top 3 drivers for three different topics. Of these #Ignis was a campaign, while the other two were organically trending topics. We see that in campaigns all drivers had a large overlap in their most used terms. However, in organically trending topics, drivers were pushing the discourse in different directions.

Table 3. Vocabulary of top 3 opinion drivers

Topic	Drivers	Top-1st word	Top-2nd word	Top-3rd word
#Ignis	Driver 1	{@nexaexperience}	{launch}	{electronation, ignis}
	Driver 2	{@nexaexperience}	{electronation}	{launch, looking, forward}
	Driver 3	{@nexaexperience}	{electronation, watch}	{launch, live}
ABVP	Driver 1	{#delhiuniversity}	{ramjas, college}	{umar, khalid}
	Driver 2	{#modiministry}	{ramjas, college}	{khalid, protest, clash}
	Driver 3	{well, done}	{traitors}	{medicine, long, due}
#Kansas	Driver 1	{feb, pm}	{#shooting}	{man}
	Driver 2	{#espn2, #tcu}	{#aclu}	{#doj}
	Driver 3	{kansas, #noplacelikeks}	{#ksbucketlist}	{city}

It was also found that the vocabulary distribution of opinion drivers represented the overall vocabulary distribution of the discourse more accurately, as compared to users with high impact or activity.

5   Conclusions and Future Work

This paper described the overall principles behind the design of the Cogno web observatory. Web observatories observing social phenomena, and understanding the various narratives driving social discourse, are likely to become more important in the future, given the central roles played by social media in just about every aspect of human activity.

At the time of this writing, Cogno is mostly a theoretical framework, with specific parts of it implemented and prototyped separately. We are yet to weave the disparate elements mentioned in this work, into a single web application, and make it scalable over large datasets in interactive response time.

In the future, we plan to also incorporate several more models of social cognition into Cogno. Prominent among these, is a model for understanding the *formation* of narratives over time, and correlating it with specific other features. For example, one of the questions we are asking is whether a user's network position correlates with their tendency to form and drive narratives.

We envisage that a web observatory like Cogno would be of interest to strategic heads of several stakeholders like governments, law-enforcement agencies, non-governmental organizations, universities and corporate bodies.

References

1. Adamic, L.A., Lento, T.M., Adar, E., Ng, P.C.: Information evolution in social networks. In: Proceedings of the Ninth ACM International Conference on Web Search and Data Mining, WSDM 2016, pp. 473–482. ACM, New York (2016). https://doi.org/10.1145/2835776.2835827
2. Bakshy, E., Hofman, J.M., Mason, W.A., Watts, D.J.: Everyone's an influencer: quantifying influence on twitter. In: Proceedings of the Fourth ACM International Conference on Web Search and Data Mining, WSDM 2011, pp. 65–74. ACM, New York (2011). https://doi.org/10.1145/1935826.1935845
3. Bhanushali, A., Subbanarasimha, R.P., Srinivasa, S.: Identifying opinion drivers on social media. OTM 2017. LNCS, vol. 10574, pp. 242–253. Springer, Cham (2017). https://doi.org/10.1007/978-3-319-69459-7_17
4. Borge-Holthoefer, J., Perra, N., Gonçalves, B., González-Bailón, S., Arenas, A., Moreno, Y., Vespignani, A.: The dynamics of information-driven coordination phenomena: a transfer entropy analysis. Sci. Adv. 2(4), e1501158 (2016). https://doi.org/10.1126/sciadv.1501158. http://advances.sciencemag.org/content/2/4/e1501158
5. Ceri, S., Bozzon, A., Brambilla, M., Della Valle, E., Fraternali, P., Quarteroni, S.: Web information retrieval. Springer, Heidelberg (2013). https://doi.org/10.1007/978-3-642-39314-3
6. Cha, M., Haddadi, H., Benevenuto, F., Gummadi, K.P.: Measuring user influence in twitter: the million follower fallacy. In Proceedings of International AAAI Conference on Weblogs and Social, ICWSM 2010 (2010)
7. Craswell, N., Hawking, D.: Web information retrieval. Information Retrieval: Searching in the 21st Century, pp. 85–101 (2009)
8. Desikan, P., Srivastava, J., Kumar, V., Tan, P.N.: Hyperlink analysis-techniques & applications. Army High Performance Computing Center Technical Report (2002)
9. Ferrara, E., JafariAsbagh, M., Varol, O., Qazvinian, V., Menczer, F., Flammini, A.: Clustering memes in social media (2013). CoRR abs/1310.2665. http://arxiv.org/abs/1310.2665
10. Ferrara, E., Varol, O., Menczer, F., Flammini, A.: Detection of promoted social media campaigns. In: Tenth International AAAI Conference on Web and Social Media, pp. 563–566 (2016)
11. Muñoz García, O., García-Silva, A., Corcho, O., de la Higuera Hernández, M., Navarro, C.: Identifying topics in social media posts using DBpedia. In: Jean-Dominique, M., Hrasnica, H., Genoux, F. (eds.) Proceedings of the NEM Summit, NEM Initiative, Eurescom - The European Institute for Research and Strategic Studies in Telecommunications - GmbH, Heidelberg, Germany, pp. 81–86, September 2011
12. Ghosh, S., Sharma, N., Benevenuto, F., Ganguly, N., Gummadi, K.: Cognos: Crowdsourcing search for topic experts in microblogs. In: Proceedings of the 35th International ACM SIGIR Conference on Research and Development in Information Retrieval, SIGIR 2012, pp. 575–590. ACM, New York (2012). https://doi.org/10.1145/2348283.2348361
13. Hall, W., et al.: The southampton university web observatory. In: 1st International workshop on Building Web Observatories, ACM Web Science 2013, 1–3 May 2013, April 2013. https://eprints.soton.ac.uk/352287/
14. Hendler, J., Berners-Lee, T.: From the semantic web to social machines: a research challenge for ai on the world wide web. Artif. Intell. 174(2), 156–161 (2010)

15. Henzinger, M.R.: Hyperlink analysis for the web. IEEE Internet Comput. **1**, 45–50 (2001)
16. Lewandowski, D.: Web information retrieval. Inf. Wissenschaft Praxis **56**(1), 5–12 (2005)
17. Li, H., Mukherjee, A., Liu, B., Kornfield, R., Emery, S.: Detecting campaign promoters on twitter using markov random fields. In: Proceedings of the 2014 IEEE International Conference on Data Mining, ICDM 2014, pp. 290–299. IEEE Computer Society, Washington, DC (2014). https://doi.org/10.1109/ICDM.2014.59
18. Madaan, A., Tiropanis, T., Srinivasa, S., Hall, W.: Observlets: empowering analytical observations on web observatory. In: Proceedings of the 25th International Conference Companion on World Wide Web, WWW 2016 Companion, International World Wide Web Conferences Steering Committee, Republic and Canton of Geneva, Switzerland, pp. 775–780 (2016). https://doi.org/10.1145/2872518.2890593
19. McKelvey, K., Menczer, F.: Design and prototyping of a social media observatory. In: Proceedings of the 22nd International Conference on World Wide Web. WWW '13 Companion, pp. 1351–1358. ACM, New York (2013). https://doi.org/10.1145/2487788.2488174
20. Mihaila, G.A.: WebSQL: an SQL-like query language for the World Wide Web. University of Toronto (1996)
21. Mikolov, T., Chen, K., Corrado, G., Dean, J.: Efficient estimation of word representations in vector space (2013). CoRR abs/1301.3781. http://dblp.uni-trier.de/db/journals/corr/corr1301.html#abs-1301-3781
22. Mikolov, T., Sutskever, I., Chen, K., Corrado, G.S., Dean, J.: Distributed representations of words and phrases and their compositionality. In: Burges, C.J.C., Bottou, L., Welling, M., Ghahramani, Z., Weinberger, K.Q. (eds.) Advances in Neural Information Processing Systems, vol. 26, pp. 3111–3119. Curran Associates, Inc., New York (2013)
23. Pal, A., Counts, S.: Identifying topical authorities in microblogs. In: Proceedings of the Fourth ACM International Conference on Web Search and Data Mining, WSDM 2011, pp. 45–54. ACM, New York (2011). https://doi.org/10.1145/1935826.1935843
24. Park, H.W.: Hyperlink network analysis: a new method for the study of social structure on the web. Connections **25**(1), 49–61 (2003)
25. Qiang, J., Chen, P., Wang, T., Wu, X.: Topic modeling over short texts by incorporating word embeddings (2016). CoRR abs/1609.08496. http://arxiv.org/abs/1609.08496
26. Simpson, R., Page, K.R., De Roure, D.: Zooniverse: Observing the world's largest citizen science platform. In: Proceedings of the 23rd International Conference on World Wide Web, WWW 2014 Companion, pp. 1049–1054. ACM, New York (2014). https://doi.org/10.1145/2567948.2579215
27. Sivaraman, N.K., Srinivasa, S.: Abstractions, expressions and online collectives. In: Proceedings of the ACM Web Science Conference, WebSci 2015, pp. 58:1–58:2. ACM, New York (2015). https://doi.org/10.1145/2786451.2786499
28. Smart, P.R., Shadbolt, N.R.: Social machines. In: Encyclopedia of Information Science and Technology, 3rd Edn., pp. 6855–6862. IGI Global, Hershey (2015)
29. Steinskog, A., Therkelsen, J., Gambäck, B.: Twitter topic modeling by tweet aggregation. In: NODALIDA (2017)
30. Tinati, R., Wang, X., Tiropanis, T., Hall, W.: Building a real-time web observatory. IEEE Internet Comput. **19**(6), 36–45 (2015)

31. Tiropanis, T., Hall, T., Shadbolt, W., De Roure, D., Contractor, N., Hendler, J.: The web science observatory. IEEE Intell. Syst. **28**(2), 100–104 (2013). https://doi.org/10.1109/MIS.2013.50
32. Tiropanis, T., Hall, W., Hendler, J., de Larrinaga, C.: The web observatory: a middle layer for broad data. Big Data **2**(3), 129–133 (2014). https://doi.org/10.1089/big.2014.0035
33. Wagner, C., Liao, V., Pirolli, P., Nelson, L., Strohmaier, M.: It's not in their tweets: modeling topical expertise of twitter users. In: Proceedings of the 2012 ASE/IEEE International Conference on Social Computing and 2012 ASE/IEEE International Conference on Privacy, Security, Risk and Trust, SOCIALCOM-PASSAT 2012, pp. 91–100. IEEE Computer Society, Washington, DC (2012). https://doi.org/10.1109/SocialCom-PASSAT.2012.30
34. Wang, X., et al.: Building a web observatory for south Australian government: Supporting an age friendly population. Web Science Conference, June 2015
35. Xiong, F., Liu, Y.: Opinion formation on social media: an empirical approach. Chaos Interdiscip. J. Nonlinear Sci. **24**(1), 013130 (2014). https://doi.org/10.1063/1.4866011

Automated Credibility Assessment of Web Page Based on Genre

Shriyansh Agrawal, S. Lalit Mohan, and Y. Raghu Reddy$^{(\boxtimes)}$

International Institute of Information Technology, Gachibowli, Hyderabad, India
{shriyansh.agrawal,lalit.mohan}@research.iiit.ac.in,
raghu.reddy@iiit.ac.in

Abstract. With more than a billion web sites, volume and variety of content available for consumption is huge. However, credibility, an important quality characteristic of web pages is questionable in many cases and tends to be non-uniform. Credibility can increase or reduce the importance of web page leading to potential gain or loss of user base. Credibility without factoring genre of content (for example, Help, Article, Discussion, etc.) can lead to incorrect assessment. Depending on the genre, the importance of features such as web page date time modified, grammar, image to text ratio, in and out links, and other web page features differ. We propose a genre credibility assessment based on web page surface features and their importance in a genre. Further, we built a *WEBCred* framework to assess *GCS* (Genre based Credibility Score) with flexibility to add/modify genres, its features and their importance. We validated our approach on 10,429 'Information Security' related web pages; the assessed score correlated 35% with crowd sourced Web Of Trust (WOT) score and 39% with Alexa ranking.

Keywords: Quality · Credibility · Web page · Genre
Information security

1 Introduction

With more than a billion websites, Internet has become a primary source to cater to the content requirements of nearly 4 billion users[1]. Most of the time, users are bombarded with more information than needed, information that can be conflicting, ambiguous, inconsistent, irrelevant, etc. This can lead to mistrust between the consumer of the information and its source. If the quality of web page from a particular source site is perceived to be bad or not credible, users will abandon it and seek other ways or sites that meet their needs. On the other hand, a credible web page [20] acts as a catalyst for human cooperation and well explained in the seminal work of Hovland [15] on information source importance. In the past couple of decades, researchers have studied and analysed various

---

[1] http://www.internetlivestats.com/.

© Springer Nature Switzerland AG 2018
A. Mondal et al. (Eds.): BDA 2018, LNCS 11297, pp. 155–169, 2018.
https://doi.org/10.1007/978-3-030-04780-1_11

aspects of credibility, and developed frameworks [35,37] for measuring credibility on the web. Some of the early literature on web credibility assessment [12,22,36] was based on aesthetic/surface features. With the resurgence of AI, the research focus has translated to the cognitive aspects of credibility features and led to the development of browser plugins or APIs (Application Program Interface) such as Web of Trust (WOT) [24], Alexa [9] and fact-checking sites (Snopes.com, FactCheck.org and others). However, the available work on frameworks and APIs does not provide flexibility to the user to decide the features for measuring credibility. The need for user intervention on relevance, results and credibility is reiterated in Karen Sparck Jone's speech during acceptance of ACM SIGIR Gerard Salton Award [19]. Most of the research focus on web search has been on making results more relevant and is mostly based on user search history. Search engine optimization techniques and page rank [28] algorithms built on graphs are based on location, past search history, site popularity and not necessarily on user understanding of credibility. Novice users using search engines do not know how to start and lack sufficient knowledge for finding the best possible results. For a novice, surface features such as fonts, colour, images and other layouts of the web page create the first impression on credibility. While for most of the regular users of the internet, content relevance, information source, evolution of content and other fine-grained features of credibility decide web page usage. To enhance the quality of search results, we focus on credibility assessment based on web page genre. Crowston et al. [21] studied classification of web pages and listed possible genres of the web. The importance of genre on effectiveness of communication for web pages is also highlighted by various researchers [30,32]. In 2004, Meyer et al. [11] classified web pages into eight genres: Help, Article, Discussion, Shop, Portrayals of companies and institutions, Private portrayal, Link collection and Downloads. Based on the available literature, a web page genre is identified based on content type and other features such as links, language grammar, spellings, text vis-a-vis image, etc.

There is literature on web page genre classification and usage of surface features for credibility assessment separately. However to the best of our knowledge, there isn't any work done by researchers on genre based credibility assessment. Here are some examples that necessitate genre based credibility: (a) a political article on news website has a different credibility as compared to the same/similar article on a personal portrayal web page, (b) the importance of 'content grammar' on a discussion/forum site or features like 'real world presence' on a shopping site creates a different impact on a web page, (c) lack of web page modified date time on a security related article could lead to unnecessary panic. Given the various types of information available on web pages, there is a strong need understand the various web page elements that contribute to that information and assessment of the page itself.

We conducted surveys to validate the need for genre based assessment and the appropriate web page elements for credibility assessment. The survey results supported our argument about the need for Genre Credibility Score (*GCS*) of a web page. Further, we built *WEBCred* framework to automate the assessment of credibility score of a web page. The framework allows the addition of new genres, features and their weightages providing flexibility for user intervention. Using *WEBCred*, we assessed the credibility of web pages and validated results with the ranking of Alexa (popularity based algorithm) and the score of WOT (crowdsourced/review based). As the number of websites is in billions and contains content of all possible domains, providing a credibility score for cross-domain sites needs to be considered, but it's beyond the scope of this paper. In this paper, we restrict the work to the evaluation of the credibility score of web pages belonging to an 'Information Security' domain. However, it needs to be noted that the proposed approach can be extrapolated to other domains. The remainder of the paper is organized as follows: In Sect. 2, we detail the proposed approach for credibility assessment and present the survey results and its analysis. In Sect. 3, we validate the approach with Alexa and WOT rankings and finally, we conclude and provide an overview of the future work in Sect. 4.

2 Approach

A web page is a collection of elements with content, links and meta-data that serve a purpose. Web pages have evolved from containing static content to machine-generated dynamic (scripts based) content. However, the purpose of a web page, i.e. information dissemination, is still intact. The content of the web page is in the form of text, image, video or any other binary format. The text may have misspells or grammatical issues, contact details and others apart from semantics and or discourse. Links can be broken or it may refer to outer domain web pages or can also be cited by other web pages. Meta-data of a web page contains last modified date time, domain, language count, title and others encoded into the page but not rendered on the page. The characteristics of content, links and meta-data are referred to as features in this paper. The organization and the usage of these features differ from page to page and the content type in the page.

In 2008, Chen et al. [10] used web page features as the basis to automate an approach for genre based classification. Presence or absence of certain web page features in a particular genre instance has an impact on the credibility of the page. For example, (i) Shopping web page available in multiple languages or regional articles published in regional language often increase usage count of web pages; (ii) Responsiveness of web page is usually appreciated. However, the importance of the same differs across the various genre of web pages; (iii) Web page consisting of more text is considered to have more knowledge base, thereby the importance of text2image ratio vary for News vs Link-collection web pages; (iv) Banner advertisements may reduce the credibility of a web page content especially when the advertising is unexpected. However, its importance on Shopping web pages differs as compared to Help genre web pages; (v) Conventionally, the response time for a request in Help page is expected to be less compared to other genres of web page due to its need for faster assistance. There are also several other features such as - authority/expertise of owner, social hits, the polarity of content, discourse across varied sources, etc. [14,23] that have varying degree of impact on credibility of a web page. We used the surface features listed in Table 1 for credibility assessment. These surface features are identified based on citations in the literature on genre identification and credibility assessment. We also performed a survey to validate their applicability and importance in a genre.

Table 1. Genre credibility assessment features

Feature name	Citation
Advertisements	[13]
Internet Domain	[27]
Real World Presence	[26]
Misspell	[16]
Text2Image Ratio	[18]
Modification DateTime	[26]
In Links, Out Links, Broken Links	[18,36]
Internationalization	[25]
PageLoadTime	[13]
Responsive Design	[1]

2.1 Survey

An online survey [4] was conducted to validate the need for Genre based credibility and also the importance of web page features. To obtain unbiased data, we conducted the survey on a wide sample set. We leveraged paid crowdsource platform provided by CrowdFlower (now known as Figure Eight Inc.), unpaid personnel via professional mailing lists, delegates in academic conferences of universities, social media channels such as Facebook and LinkedIn groups. The number of possible genres in web is potentially an incomplete list. In the initial study, we realized that some of the genres may not be obvious to the subjects of the survey and may lead to an inconclusive study. As a result, we scoped the survey to 4 common genres - {Help, Article, Shop and Portrayal - Organization} that are easily identifiable by novice or regular internet users.

Six questions (Q1–Q6) captured survey participants demographics - Age, Gender, Location, Qualification, Name and Email. Name and Email of the participant were captured as conventionally anonymous surveys can lead to quality issues. The survey contained a question (Q7) pertinent to the importance of each feature: Advertisements, Misspell, Text2Image Ratio, Modification Date Time, Page Load Time, Responsive Design, Real World Presence, Internationalization, Internet Domain, Links - In, Out and Broken. The Q7 survey question is listed as a $i * j$ matrix of 12 features across 4 genres, where each cell represented the importance of ith feature (f) with respect to j^{th} genre (g). The subject was mandated to mark the importance of 12 listed features (from Table 1) across all four genres on a Likert scale of 1 to 4 with one being least important and four being most important. An explanation and examples of web pages were provided to avoid misunderstanding/confusion on genres. To elicit survey participant input, Question (Q8) was provided to seek inputs/comments. One of the questions (Q9) had a multi-select option to select various genres that users have seen on internet web pages; this question also has an option for capturing 'Other Genres' and 'None' to identify and remove bots/casual participants.

2.2 Survey Results and Analysis

A total of 374 [5] participants responded to the survey from 43 different countries. The demographics of the participants are shown in Fig. 1. We invalidated 11% (count - 41) of responses that selected 'None' as web page genres (Q9), indicating that the participants were casual or did not have clarity on what was being requested. 64% of the responses were from CrowdFlower, and the remaining were from posts on social media and mailing lists. Academic (students, researchers and faculty) participants accounted for 51% of total responses while the other 49% participants were associated with industry - IT and other profiles such as administration. However, there is no published data on internet usage patterns based on user profile, and we did not analyze further as their participation was almost equal. 40% of survey participants were females, and 60% of the participants were males; the usage is similar to worldwide internet stats on gender [34] age. The age group of participants ranged from 19–78 with 67% of them $<= 35$ years, the measure is not very different from internet usage by age group [33].

The survey had participation from 43 countries is broadly classified into India, North America (NA), South America (SA), Africa (AF), Europe (EU) and Rest of Asia (RA). 36% of survey participants were from India, which is similar to the participation of Indians on crowdsourcing platforms. However, the participation from North American countries including the US was very low [17]. About 35% of the participants provided comments, and most of them corroborated on the need for genre based credibility assessment. Some of the interesting comments to Q8 were (i) 'never trusts web page content'; (ii) 'in link advertisements on a shopping website are helpful'; (iii) 'Level of language is not same across, localisation of web page provides comfort to the user'. The other themes that emerged from participants comments were (i) upset with bombardment of advertisements; (ii) design of web pages is important; (iii) need on source/authority of information; (iv) importance varies from job to job, so need a mechanism to alter the relevance parameters based on user intervention. 90% of the participants stated that Articles, Discussion Forums, Shop, Help and Portrayal as most common genres in response to Q9, there were also comments to include 'Social Network' as a genre that is already included as part of discussion/forum in the genre classification. From the survey results, the importance (rank) of a feature (f) for a genre (g) was established (shown in Fig. 2) applying a rank aggregation

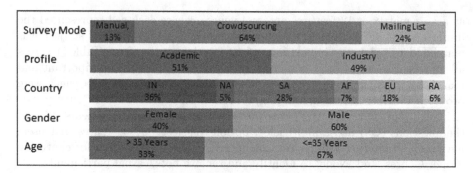

Fig. 1. Survey results - demographics

[29] on crowdsourced responses, non-crowdsourced responses and a combination. The method was also used to break a tie instance on 'inlinks' feature between Shop and Portrayal-Organization, the rank was given in favour of Portrayal-Organization. The results implied that the importance of feature varies across genres though in few cases (Articles and Portrayal - Organization) the variation margin was low. Most of the participants opined that web page features for Help genre are least important except for 'Internationalization' and 'Outlinks'. Domain ($gTLD$) importance is more for Portrayal - Organization and Articles indicating that participants are particular when they follow/ready/browse information on topics or organizations. Interestingly, the importance of 'Misspell' is least important in Help genre indicating that participants are not worried about language when they are seeking support related information. Higher importance of responsive design for Articles genre indicates that participants are following most of the updates on smart phones and/or tabs. While there is a growing presence of e-commerce/shopping on mobile platforms, survey participants did not feel responsive design as very important for Shop genre. This correlates to some of the e-commerce websites backtracking from 'mobile only' strategy to improved desktop website[2]. Based on the survey results on feature-genre importance, we propose a credibility assessment approach of the web page (detailed in the next section) for a given web genre.

2.3 Credibility Assessment

Genre based credibility score is based on the feature importance/weightages calculated from survey responses. The experimentation is performed on 'Information Security' domain. With increasing internet access and IT dependency, security[3] vulnerabilities, threats and incidents have increased manifold. Credibility assessment of security-related internet content could provide improved confidence to citizens. We started with 157,000 information security web page

---

[2] https://tinyurl.com/MyntraDesk.
[3] https://tinyurl.com/PWCSecurity.

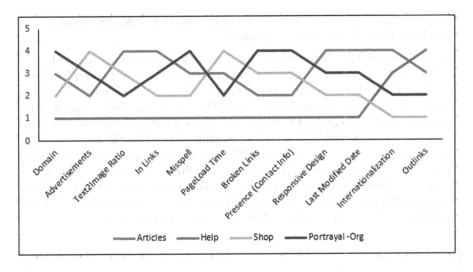

Fig. 2. Survey results - feature-genre importance

corpus [31] and selected a random sample of about 10% of web pages for extracting the feature instance values using an automated tool *APIBox*. The *APIBox* is a tool built by us for easing the extraction of feature values for crawled web pages. From the 10% sample, we shortlisted 10,429 web pages, as the remaining pages had *HTTP* errors and incomplete values. Following is the brief description of *APIBox* that consumes other open APIs and utilities -

- **Crawling** - For the given URL, the web page content is parsed to extract text and hyperlinks after removal of HTML tags and Javascript code. On the extracted content, *APIBox* initiates threads for features assessment, broadly on content, links, presence and others.
- **Content** - The English text in the web page is parsed to obtain unique words and evaluated for spell errors. The count of advertisements, banners and unwanted frames; ratio of the viewport (viewing region) of text and non-text (images, video); and the last modified datetime are extracted to measure the freshness of the web page.
- **Links** - Links can be inside or outside the domain. The count of broken-links (HTTP status code between 400–500) and the count of links referring to other websites (outlinks) in the webpage are identified using a URL regular expression. There could be other websites that have links pointing/citing the webpage (known as inlinks).
- **Presence** - The text in the header and the footer of the page are parsed for real world presence (contact address, telephone number, email and Social Media presence) of the site. The *gTLD* (Generic Top-level Domain), the domain of the URL is also identified.
- **Others** - The non-functional aspects such as pageload time, responsive design and internationalization (varied language support) are measured considering

162   S. Agrawal et al.

the growing usage of internet in developing countries (different languages) on smaller screens (including smartphones).

Some of the open APIs that are used for obtaining feature values in a web page are - NLTK[4] and Beautiful Soup[5] for text extraction and evaluation including spell error; Easylist[6] service to identify advertisements on web pages, including unwanted frames and images; YSlow[7] with PhantomJS to check response time of the web page; Mercenie[8] API service to validate a web page responsive design behaviour; Web Archives[9] APIs if last modified DateTime is not present in metadata of web page; and Google API[10] to extract inlinks of the web page. These APIs were selected based on the available literature on their wider usage, ease of code integration and the licensing terms. A snapshot of the sample set of pages with feature value instances is shown in Fig. 3. Source code of *APIBox* is available at [3] and the dataset of web pages is available at [2].

URL	Genre	Misspell	Ads	Text2Img	Last Modified	Real World	Domain	PageLoadtime	Language	Responsive	Broken Links	Inlinks	Outlinks
https://noscript.net/chang elog	Portrayal-O	41	11	0.9948	10/09/2017 10:39:31	{"contact": 1, "email": 0, "sitemap": 0, "Help": 0}	net	26	1	FALSE	14	256000	10
http://www.milincorporat ed.com/	Shop	5	5	0.20617	01/01/2016 14:42:54	{"contact": 1, "email": 0, "sitemap": 0, "Help": 0}	com	251	1	FALSE	0	102	3
https://twitter.com/CTIN_ Global/status/7906706406	Others	162	14	0.76681	21/08/2017 10:41:13	{"contact": 0, "email": 0, "sitemap": 0, "Help": 0}	com	53	44	FALSE	24	2	10
http://www.ncbi.nlm.nih. gov/pubmed/25204641	Articles	2	37	0.24987	03/09/2017 17:17:56	{"contact": 1, "email": 0, "sitemap": 0, "Help": 1}	gov	53	1	TRUE	7	0	32
http://howtodoinjava.com /spring/spring-aop/spring-	Help	7	64	0.68717	31/08/2017 18:31:25	{"contact": 0, "email": 0, "sitemap": 0, "Help": 0}	com	103	1	TRUE	1	0	17

Fig. 3. Snapshot - *APIBox* data

A correlation heatmap shown in Fig. 4 was prepared for the feature instance values extracted from the 10,429 web pages. The analyzed data states that only pairs of (advertisements, outlinks) and (misspelled, langcount) are correlated, most of the other feature pairs showed 'marginal' to 'no correlation'. The possible reasons for correlation are (i) most of the advertisements take the user to an outer domain that is considered as outlinks (ii) Internationalized words are identified as misspell as the focus was only on the English text. Hence, based on the evaluated data, we suggest that the given set of features (f) (Table 1) are orthogonal and/or not related. Further, we propose a linear equation (which is detailed below) to assess Genre Credibility Score (GCS) of web pages.

3   Scoring, Results and Validation

Features' value (f_i) provided by *APIBox* are on different scales for web pages. Value of Modified DateTime may not be always in few milliseconds for web pages.

---

[4] http://www.nltk.org/.
[5] https://pypi.org/project/BeautifulSoup/.
[6] https://easylist.to/easylist/easylist.txt.
[7] http://yslow.org/phantomjs/.
[8] http://tools.mercenie.com/responsive-check/.
[9] https://archive.org/.
[10] https://developers.google.com/custom-search/json-api/v1/overview.

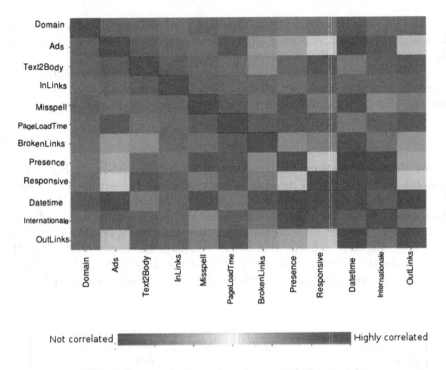

Fig. 4. Features heatmap based on correlation matrix

Some web pages may have more spelling mistakes or advertisements compared to the count of broken links. To normalize features' (f_i) value to a measurable scale, the same sample set of 10,429 information security web pages (wp) were used to calculate the mean (μ) and the standard deviation $(\sigma$ - Eq. 1) of the features (Advertisements, Real World Presence, Text2Image Ratio, In and Out Links, Broken Links and Page Load Time). Each feature value (f_i) for a given sample is normalized (v_i) to $\{-1, 0, 1\}$ across genres based on Eq. 3.

$$\sigma_i = \sqrt{\frac{\sum (f_i - \mu_i)^2}{N_{w_p}}} \tag{1}$$

$$F = (f_1, f_2, f_3, .., f_n) \tag{2}$$

$$v_i = \left\{ \begin{array}{ll} 0, & f_i \in [\mu - \sigma, \mu + \sigma] \\ \pm 1, & \text{otherwise} \end{array} \right\}$$

$$\text{where } i \in [1, n] \tag{3}$$

$$V = (v_1, v_2, v_3, .., v_n) \tag{4}$$

Other features such as Internationalization, Misspell and Responsive Design are normalized to 0 or 1 based on it's presence. Modified DateTime value is normalized to 1 if the web page is updated within a month ($<30days$, configurable) otherwise 0. We considered 6 Top Level Domain ($gTLD$) with scores as: 1 -

.gov, 1 - .edu, 0 - .org, 0 - .com, 0 - .net and -1 to all others. While there is no documented guideline, it is generally acknowledged by the user community that the content in .gov and .edu has more credibility.

Finally, a normalized value (v) of each (n) feature (f) with weightages (w) based on web page genre (g) are identified. These values are used in a linear equation (Eq. 5) to calculate Genre Credibility Score (GCS) of web page (p).

$$GCS_p = \sum_{i=1}^{n} (w_g^{f_i} * v_p^{f_i}) \tag{5}$$

Weightage of a feature (f) for a genre (g) is calculated (shown in Fig. 5) from the rank and its weighted mean in a genre based on the survey results data. Reciprocal of calculated GCS is ordered monotonically to obtain $WEBCred$ Rank (WR) among set of 10,429 web pages. URLs which have higher GCS are ranked lower. However, for some instances, there can be a tie between URLs with same GCS. In such scenario, $WEBCred$ compares the feature value of URLs to break the tie. Selection of features to break the tie is advocated by ordering feature weightages. The feature which has the highest weightage is used first and similarly more are selected if required.

Category	Articles	Help	Shop	Portrayal -Org
Domain	0.079	0.059	0.069	0.111
Advertisements	0.053	0.059	0.138	0.083
Text2Body Ratio	0.105	0.059	0.103	0.056
In Links	0.105	0.059	0.069	0.083
Misspell	0.079	0.059	0.069	0.111
PageLoad Time	0.079	0.059	0.138	0.056
Broken Links	0.053	0.059	0.103	0.111
Presence (Contact Info)	0.053	0.059	0.103	0.111
Responsive Design	0.105	0.059	0.069	0.083
Last Modified Date	0.105	0.059	0.069	0.083
Internationalization	0.105	0.176	0.034	0.056
Outlinks	0.079	0.235	0.034	0.056

Fig. 5. Feature weightage for genres

3.1 *WEBCred*

To identify the most credible web page in a genre, we built a framework (*WEBCred*) for automated assessment extending *APIBox* along with the formulated equations. The process diagram for *WEBCred* is shown in Fig. 6. Since classification is not automated in the current version of the framework, the user enters URL and pre-selects applicable genre. For the selected genre, the relevant features are displayed along with their pre-calculated weights. The list of features identified based on the available literature and survey results may not be complete for a genre; hence, $WEBCred$ has the flexibility to select/add other features (web page elements). Users are given the flexibility to alter the calculated weights of a feature. After features and their corresponding weights are selected, the user submits the request for credibility assessment. The input URL and the features with weights are provided to the $APIBox$ (step 6). If the prior assessment of feature values of the URL does not exist in the persistent storage (DB) or if the page is modified (i.e. last modified datetime of a page is altered), then $APIBox$ evaluates the feature value as explained earlier and stores it in DB. The data in the persistent database is refreshed on a periodic basis (currently, configured for every 4 h) to handle feature value changes in a webpage. Further, the collected feature values are normalized, and a final credibility score is calculated based on the attached weightages. The framework provides users with the flexibility to include new features (such as semantics), modify or remove existing features, add, modify or remove genres for credibility assessment. As the feature values are either normalized to $(-1, 0$ and $1)$ or $(0$ and $1)$, the GCS score ranges from -1 to $+1$. The source code of $WEBCred$ is available on Github at [8]. All related data used for analysis is available at [7]. Figure 7 shows the interface of the deployed $WEBCred$ [6].

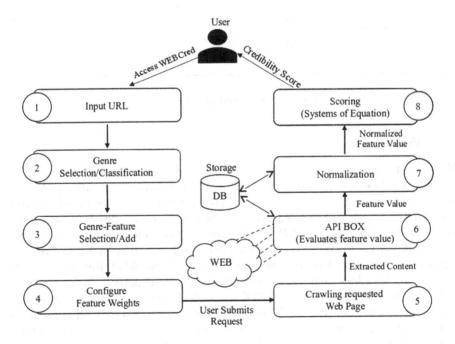

Fig. 6. WEBCred process diagram

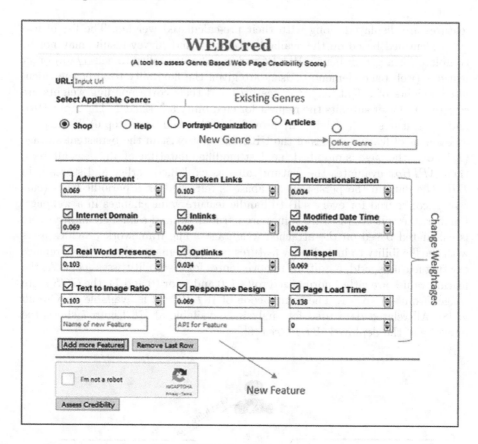

Fig. 7. *WEBCred* user interface

3.2 Validation

For given focused set of websites, we limit manual classification of web pages to 4 widely used genres - {Help, Article, Shop and Portrayl-Organization}. Assessment requires pre-calculated weightages (W) for selected genre (g) along with normalized value (V) of each feature (f). The calculated weightage of each genre is included as a default in $WEBCred$ framework. For validation of the approach, we calculated the correlation of $WEBCred$ Rank (WR) with Alexa Ranking (popularity based, widely used by search engines) and WOT Ranking (crowd-sourced/reviews based) using 10-fold cross-validation method. WR correlates 39% with the Alexa ranking and 35% with WOT ranking. Between Alexa and WOT, the correlation is 36%. The lower correlation states that all approaches work on different ideology. $WEBCred$ encompasses genre features that are not being used by page ranking algorithms as the credibility of the site is not their primary motive but popularity. However, Shop genre has a higher correlation with Alexa score, and Portrayal-Organization has a higher correlation with WOT score. It is due to the similarity in priority of features, shop genre web pages and Alexa give high priority to "inlinks/popularity", while portrayal genre and WOT ranking give more importance to "Contact info".

4   Conclusions and Future Work

The proposed genre based approach factors web page elements/features for credibility assessment. The framework is flexible to add newer genres, features and their weights including modifying existing APIs being used for measuring credibility score. The validation with Alexa ranking and WOT score demonstrated that the study with 10,429 URLs in the information security domain could be extended to other domains as well. Currently, *WEBCred* requires one to pre-select the Genre. We plan to work on the following extensions in the future work:

- Validate the ranked web pages by domain experts for agreement on the genre based approach.
- Extend *WEBCred* to include automated genre classification so that weightages can be calculated dynamically based on web pages data.
- Include *WEBCred* as a browser plugin.
- Validate the approach for more URLs in information security and other domains. Also, we plan to study the same for web pages that may fall into multiple domains. This may lead to reworking the existing credibility evaluation to include domain-specific parameters.
- As the feature values are normalized and numeric, we also plan to experiment the credibility assessment using supervised learning algorithms such as SVM and Naive Bayes.
- We plan to extend the validation to subject matter experts, use the approach for possibly enriching existing domain ontologies using credibile web pages.

As the experiment is done on information security web pages, we can also use the information to identify credible sources of vulnerabilities, threats, incidents and controls. The approach can be combined with existing ranking algorithms of search engines to display quality relevance results.

References

1. Aggarwal, S., Herre, V.O., Reddy, Y.R., Indurkhya, B.: Providing web credibility assessment support. In: Proceedings of the 2014 European Conference on Cognitive Ergonomics (2014)
2. Agrawal, S., Sanagavarapu, L.M., Reddy, Y.R.: API Box Collected Data. http://tinyurl.com/ybqpjtne. Accessed 19 July 2018
3. Agrawal, S., Sanagavarapu, L.M., Reddy, Y.R.: API Box Source Code. https://tinyurl.com/APIBoxSource. Accessed 19 July 2018
4. Agrawal, S., Sanagavarapu, L.M., Reddy, Y.R.: Genre Web Credibility Survey. https://form.jotform.me/80636000478453. Accessed 19 July 2018
5. Agrawal, S., Sanagavarapu, L.M., Reddy, Y.R.: Survey Data. https://tinyurl.com/WEBCredSurvey. Accessed 19 July 2018

6. Agrawal, S., Sanagavarapu, L.M., Reddy, Y.R.: Web Credibility Website. https://tinyurl.com/WEBCredFramwork. Accessed 19 July 2018
7. Agrawal, S., Sanagavarapu, L.M., Reddy, Y.R.: WEBCred collected Data. https://tinyurl.com/WebCredDataRaw. Accessed 19 July 2018
8. Agrawal, S., Sanagavarapu, L.M., Reddy, Y.R.: WEBCred Source Code. https://tinyurl.com/WebCredFramework. Accessed 19 July 2018
9. Alexa: Alexa Page Rank. https://www.alexa.com. Accessed 19 July 2018
10. Chen, G., Choi, B.: Web page genre classification. In: Proceedings of the Symposium on Applied Computing (2008)
11. Meyer zu Eissen, S., Stein, B.: Genre classification of web pages. In: Biundo, S., Frühwirth, T., Palm, G. (eds.) KI 2004. LNCS, vol. 3238, pp. 256–269. Springer, Heidelberg (2004). https://doi.org/10.1007/978-3-540-30221-6_20
12. Fogg, B.J., Tseng, H.: The elements of computer credibility. In: Proceedings of the SIGCHI Conference on Human Factors in Computing Systems (1999)
13. Fogg, B.J., Soohoo, C., Danielson, D.R., Marable, L., Stanford, J., Tauber, E.R.: How do users evaluate the credibility of web sites? A study with over 2,500 participants. In: Proceedings of the Conference on Designing for User Experiences (2003)
14. Hilligoss, B., Rieh, S.Y.: Developing a unifying framework of credibility assessment: construct, heuristics, and interaction in context. Inf. Process. Manage. **44**, 1467–1484 (2008)
15. Hovland, C.I., Weiss, W.: The influence of source credibility on communication effectiveness. Public Opin. Q. **15**, 635–650 (1951)
16. Iding, M.K., Crosby, M.E., Auernheimer, B., Barbara Klemm, E.: Web site credibility: why do people believe what they believe? J. Instr. Sci. **37**, 43–63 (2009)
17. Ipeirotis, P.G.: Demographics of Mechanical Turk (2010)
18. Ivory, M.Y., Hearst, M.A.: Statistical profiles of highly-rated web sites. In: Proceedings of the SIGCHI Conference on Human Factors in Computing Systems (2002)
19. Jones, K.S.: A look back and a look forward. In: Proceedings of the 11th Annual International Conference on Research and Development in Information Retrieval (1988)
20. Jøsang, A., Keser, C., Dimitrakos, T.: Can we manage trust? In: Herrmann, P., Issarny, V., Shiu, S. (eds.) iTrust 2005. LNCS, vol. 3477, pp. 93–107. Springer, Heidelberg (2005). https://doi.org/10.1007/11429760_7
21. Crowston, K.: Reproduced and emergent genres of communication on the world wide web. In: Proceedings of the Thirtieth Hawaii International Conference on System Sciences (1997)
22. Lazar, J., Meiselwitz, G., Feng, J.: Understanding web credibility: a synthesis of the research literature. Found. Trends Hum. Comput. Interact. **1**, 139–202 (2007)
23. Metzger, M.J., Flanagin, A.J.: Credibility and trust of information in online environments: the use of cognitive heuristics. J. Pragmat. **59**, 210–220 (2013)
24. MyWOT: Web of Trust. https://www.mywot.com. Accessed 19 July 2018
25. Oakleaf, M.: Writing information literacy assessment plans: a guide to best practice. Commun. Inf. Lit. **3**, 4 (2010)
26. O'Grady, L.: Future directions for depicting credibility in health care web sites. Int. J. Med. Inform. **75**, 58–65 (2006)
27. Olteanu, A., Peshterliev, S., Liu, X., Aberer, K.: Web credibility: features exploration and credibility prediction. In: Serdyukov, P., et al. (eds.) ECIR 2013. LNCS, vol. 7814, pp. 557–568. Springer, Heidelberg (2013). https://doi.org/10.1007/978-3-642-36973-5_47

28. Page, L., Brin, S., Motwani, R., Winograd, T.: The pagerank citation ranking: bringing order to the web. Technical report (1999)
29. Pihur, V., Datta, S., Datta, S.: RankAggreg, an R package for weighted rank aggregation. J. BMC Bioinform. **10**, 62 (2009)
30. Pollach, I.: Electronic word of mouth: a genre analysis of product reviews on consumer opinion web sites. In: Proceedings of the 39th Annual Hawaii International Conference on System Sciences (2006)
31. Sanagavarapu, L.M., Sarangi, S., Reddy, Y.R., Varma, V.: Fine grained approach for domain specific seed URL extraction. In: Proceedings of the 38th Annual Hawaii International Conference on System Sciences (2018)
32. Santini, M., Power, R., Evans, R.: Implementing a characterization of genre for automatic genre identification of web pages. In: 21st International Conference on Computational Linguistics and 44th Annual Meeting of the Association for Computational Linguistics (2006)
33. Statista: Internet Usage by Age. https://www.statista.com/statistics/751005/india-share-of-internet-users-by-age-group/. Accessed 19 July 2018
34. Statista: Internet Usage by Gender. https://www.statista.com/statistics/491387/gender-distribution-of-internet-users-region/. Accessed 19 July 2018
35. Sundar, S.S.: The MAIN model: a heuristic approach to understanding technology effects on credibility. J. Digit. Media Youth Credibility **73100**, 78–92 (2007)
36. Wathen, C.N., Burkell, J.: Believe it or not: factors influencing credibility on the web. J. Am. Soc. Inf. Sci. Technol. **53**, 134–144 (2002)
37. Yamamoto, Y., Tanaka, K.: Enhancing credibility judgment of web search results. In: Proceedings of the SIGCHI Conference on Human Factors in Computing Systems (2011)

CbI: Improving Credibility of User-Generated Content on Facebook

Sonu Gupta[1(✉)], Shelly Sachdeva[2], Prateek Dewan[3],
and Ponnurangam Kumaraguru[3]

[1] Jaypee Institute of Information Technology, Noida, India
gupta.sonu1607@gmail.com
[2] National Institute of Technology, Delhi, New Delhi, India
shellysachdeva@nitdelhi.ac.in
[3] Indraprastha Institute of Information Technology, Delhi, New Delhi, India
{prateekd,pk}@iiitd.ac.in

Abstract. Online Social Networks (OSNs) have become a popular platform to share information with each other. Fake news often spread
rapidly in OSNs especially during news-making events, e.g. Earthquake
in Chile (2010) and Hurricane Sandy in the USA (2012). A potential
solution is to use machine learning techniques to assess the credibility
of a post automatically, i.e. whether a person would consider the post
believable or trustworthy. In this paper, we provide a fine-grained definition of credibility. We call a post to be credible if it is *accurate, clear,
and timely*. Hence, we propose a system which calculates the *Accuracy,
Clarity, and Timeliness* (A-C-T) of a Facebook post which in turn are
used to rank the post for its credibility. We experiment with 1,056 posts
created by 107 *pages* that claim to belong to news-category. We use a
set of 152 features to train classification models each for A-C-T using
supervised algorithms. We use the best performing features and models
to develop a RESTful API and a Chrome browser extension to rank posts
for its credibility in real-time. The random forest algorithm performed
the best and achieved ROC AUC of 0.916, 0.875, and 0.851 for A-C-T
respectively.

Keywords: Online social media · Facebook · Credibility

1 Introduction

With the advent of time, OSNs have replaced traditional media like print media
and television as a source of information about the latest happenings around
the globe. Instant updates and easy sharing nature of OSNs have contributed to
this shift. They have also become a go-to resource for journalists during news
gathering. Facebook is the most popular social networking site (SNS) with 2.2
billion monthly active users on average as of January 2018.[1] Given the increasing

---

[1] https://en.wikipedia.org/wiki/Facebook.

© Springer Nature Switzerland AG 2018
A. Mondal et al. (Eds.): BDA 2018, LNCS 11297, pp. 170–187, 2018.
https://doi.org/10.1007/978-3-030-04780-1_12

popularity, it has emerged as a news source and as a medium to disseminate information. Therefore, OSNs witness an upsurge in user activity whenever a high impact event takes place. Users log-on to Facebook and other SNSs to check for updates, to share posts and their opinions on these events. Albeit a vast volume of content is posted on OSNs, not all of the information is accurate and reliable. Some users intentionally post fake news while other share such posts without verifying its content. The effect of such rumors can be highly misleading and can cause panic among people. Credibility on an OSN is a matter of great concern as information spreads quickly here. Figure 1 shows an example of a fake Facebook post.[2] During the 2016 US presidential election, a satirical news website asserted that Francis endorsed Trump for president. The story was almost entirely fabricated but picked over 960,000 Facebook engagements.

Fig. 1. An example of a fake Facebook post.

On Facebook, pages are more popular than user profiles. Generally, celebrities, businesses, and organizations create Facebook pages to connect with everyone. User profiles and pages follow other pages of their interests. Various news channels have pages and keep updating latest news. There are no restrictions on the number of followers a page can have while a user profile can have a maximum 5000 friends. So, a page enjoys a broader audience than a user profile. Thus, pages are the best medium to spread any information quickly. Also, it has been observed that user-profile owners often post their opinions only and to post about news-related events they tend to share posts created by pages instead of writing one by themselves. So, for the purpose of this study, we focus on the posts created by pages.

---

[2] https://www.cnbc.com/2016/12/30/read-all-about-it-the-biggest-fake-news-stories-of-2016.html.

Many researchers have studied the credibility of information on Twitter. There are a few real-time systems to detect misinformation in Twitter. But there is only a little research on the credibility of user-generated content on Facebook. Detecting misinformation on the Facebook faces more challenges than Twitter. Unlike Twitter which provides both streaming API and search API, Facebook provides only Graph API.[3] Using Graph API, we cannot search for posts using a keyword. We can fetch a post directly via Graph API only if it is a public post and its post ID is known. In Twitter, we can get at most 3200 tweets for a particular user using the Twitter API, but Facebook does not provide any way to fetch posts for a specific user.[4] Facebook also restricts the information access to great extends due to its various privacy policies. Facebook has a more complex structure due to its diverse features like pages, events etc. which are not present on Twitter. Thus, the techniques used for credibility assessment on Twitter cannot be directly applied on Facebook.

Haralabopoulos et al. [10] assert that if an OSN user has a strong position in the network, it is expected for an inaccurate and poorly timed post to have a modest impact in the network. Therefore, the authors claim that *Accuracy, Clarity, and Timeliness* are the three critical parameters to define information quality on OSN. Accuracy is the condition of validating the information in the sense of being true, correct, reliable, precise, and free of errors. Clarity implies the clearness, ease of consumption, readability, and absence of ambiguity. Timeliness is used to describe up-to-date, seasonable, or well-timed information, which is a crucial factor in most online breaking news services and social networks. Following up [10], we define credibility as a function of *Accuracy, Clarity, and Timeliness*. The *pages* have a strong influence in the network given the huge number of followers. Hence, we call a post made by a *page* credible if it is accurate, clear and timely. In this paper, we propose a novel technique which ranks the posts created by *pages* for its credibility. It classifies each post on the basis of A-C-T. These classification results are used to rank the posts from 1–4. To the best of our knowledge, this is the first work which ranks the Facebook posts for its credibility in terms of A-C-T. Our major contributions of this work are:

- a ranking model to assess the credibility of posts in terms of A-C-T.
- an extensive set of 152 features.
- a RESTful API and a browser extension to assess the credibility of posts in real-time.

The rest of the paper is organized as follows: Sect. 2 describes the related work. In Sect. 3, we explain the proposed technique and discuss the methodology used for data collection and labeling in Sect. 4. In Sect. 5, we discuss various features and present the classification results. In Sect. 6, we conclude by highlighting the limitations and future work directions.

---

[3] https://developers.facebook.com/docs/graph-api.
[4] https://dev.twitter.com/overview/api.

2   Related Work

Many researchers have studied the credibility of information on OSNs. There are many solutions based on both computational science and social science. The credibility score of a message on online social media can be computed by using (a) web-page dependent features like share count, likes etc., (b) or by comparing the messages with those of trusted news sources.

2.1   Credibility Assessment of Content on Twitter

Out of all OSNs, the credibility of Twitter messages (tweets) is studied the most by the research community. Mendoza et al. characterized Twitter data generated during the 2010 earthquake in Chile [12]. They studied the response of Twitterers in this emergency situation. They observed that fake news spread rapidly and thus create chaos in the absence of real information from traditional sources. Gupta et al. distinguish between fake and real images propagated during Hurricane Sandy on Twitter using decision tree classifier [9]. Castillo et al. showed that automated classification techniques could be used to detect news topics from conversational topics and computed their credibility based on various features [2]. They were able to achieve a precision and recall of 70–80% using J48 decision-tree algorithm. Gupta and Kumaraguru applied machine learning algorithms (SVM-rank) and information retrieval techniques (relevance feedback) to compute the credibility of tweets using message based and source based features [7]. They observed that the dispersion of information differs during crisis and non-crisis events. In [1], Alrubaian et al. proposed a system to compute credibility score on Twitter using five procedures; tweet collecting and repository, credibility scoring technique, reputation scoring technique, user experience measuring technique, and trustworthiness value, the last of which is an output of the preceding three procedures. Due to the absence of network-based and entity-based features on Facebook, the above techniques cannot be applied directly to calculate the credibility of Facebook posts.

Li and Sakamoto showed that displaying both retweet count and collective truthfulness rating minimizes the spread of inaccurate health-related messages on Twitter [11]. They suggested that collecting and displaying the truthfulness rating of crowds in addition to their forwarding decisions can reduce the false information on social media. However, we believe that computing truthfulness rating on the basis of crowd's response can be a victim of a collusive attack by malicious entities and can have adverse effects. In [10], Haralabopoulos et al. proposed three solutions to address the credibility challenge which includes community-based evaluation and labeling of user-generated content in terms of A-C-T along with real-time data mining techniques. The above solution entirely relies on community-based evaluation. Thus, it fails to generate a reliable credibility score if no/few users evaluated the post. It also depends on the critical thinking of the crowd as mentioned in [15]. So, instead of relying on the crowd, in this work, we use supervised machine learning algorithms to identify A-C-T of a post.

Credibility Assessment Tools: Researchers developed and deployed tools to compute the credibility score on Twitter in real-time. Ratkiewicz et al. created a web service called Truthy that helps in tracking political memes on Twitter [13]. It detected astroturfing, smear campaigns, and misinformation in the context of U.S. political elections. In [8], Gupta et al. presented a semi-supervised ranking model using SVM-rank for calculating credibility score. They developed TweetCred as a browser extension, web application, and a REST API, to calculate real-time credibility scores. Inspired by the TweetCred, we also developed a user-friendly browser extension to rank Facebook posts in real-time.

2.2   Credibility Assessment of Content on Facebook

Saikaew and Noyunsan studied the credibility of information on Facebook [14]. They developed two chrome browser extensions. The first extension was used to measure the credibility of each post by asking users to give a score from 1(the lowest value) to 10 (the highest value). These post evaluations were used to train an SVM model. The second extension was used to automatically evaluate the real-time credibility of each post using the SVM model. Also, the model was trained using only 8 features such as likes, comment counts etc. In order to solve this issue, we train a model with our proposed extensive feature set to rank the Facebook posts for its credibility in real-time using a Google chrome-browser extension. To the best of our knowledge, our work is the first research work which ranks the Facebook posts for its credibility in terms of A-C-T.

3   Credibility Assessment of User-Generated Content on Facebook

We propose a supervised machine learning technique to improve the credibility of user-generated content on Facebook. For this, we obtain ground truth and train classification models on it. The resulting models are used to develop a RESTful API and a browser extension which ranks the Facebook post for its credibility on the basis of A-C-T. Each step is explained in detail in the following sections.

3.1   Proposed Credibility Assessment Model

We define credibility as a function of A-C-T. Therefore, we require three independent classification models, i.e. each for A-C-T. Figure 2 describes the technique proposed in this paper to assess the credibility of a Facebook post. The first step is to collect data using Facebook's Graph API. It returns data in a semi-structured form, and wherefore we store it in a NoSQL database. For our work we use MongoDB. Each document stores data of a post. From all the posts we collected, we randomly sample η posts to curate the ground truth dataset. We host an annotation portal on AWS and with the help of human annotators, we collect ground truth for A-C-T. With the help of previous work and our analysis, we curate a feature set. We use it to train various supervised binary classification models each for A-C-T. We rank each post on the basis of classification result for A-C-T. If a post is accurate, clear and timely, we rank it as *1*. If a post has positive results for any two of A-C-T, it is ranked as *2*, whereas if it has a positive result for anyone, then it is ranked as *3*. A post would be ranked as *4* if it is neither accurate, clear nor timely, i.e. it has negative results for A-C-T.

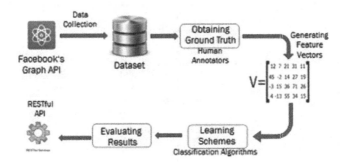

Fig. 2. Describes our proposed technique to assess credibility of user-generated content on Facebook.

It is worth mentioning that we experimented with the standard ranking models like SVM-rank[5], but couldn't achieve decent results due to the lack of ground truth data. For the same reason, we were unable to use unsupervised learning models. Therefore, we preferred to use standard supervised machine learning algorithms.

3.2   Credibility Assessment Tools

Our aim in this research work is to provide a user-friendly tool which can compute credibility scores for Facebook posts in real-time. Using the best performing models, we develop a RESTful API which receives the input from the trained classification models. On top of the API, we develop a Google Chrome browser extension that displays the results in real-time on user's Facebook news-feed in the form of an alert symbol. In this section, we describe the working of the tools.[6]

Credibility Investigator API: Figure 3 shows the data flow step of the CbI extension and the API. Credibility Investigator API is a RESTful API written in python using the Flask framework.[7] Due to Facebook's API restrictions, CbI API works only on public posts which are accessible through Facebook's Graph API. The API provides a POST method to submit public post's ID from a user's news

---

[5] https://www.cs.cornell.edu/people/tj/svm_light/svm_rank.html.

[6] Both the tools are in the development stage; hence, they are not available online.

[7] http://flask.pocoo.org.

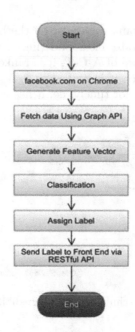

Fig. 3. Data flow steps of the CbI extension and API.

feed for analysis. Once a post is submitted to the API, it generates feature vectors which are given as input to our A-C-T pre-trained models. Since the objective is to provide a real-time alert to the Facebook user, we need to minimize the time taken in feature extraction and classification. In order to achieve this goal, we implemented multiprocessing such that features are extracted simultaneously, which helps in saving a lot of processing time. Hence, the output of the models is the rank of the post.

Browser Extension: A large no. of people use browsers to access Facebook. Therefore, we developed a Google Chrome browser extension, named, Credibility Investigator (CbI). Once installed and enabled, it seamlessly integrates credibility ranks with the user's news feed. CbI loads every time a user logs in to Facebook. It extracts post IDs of all the public posts from user's news feed. The post IDs are sent to the RESTful API. If the API returns rank *4*, a *red* alert symbol is displayed with the post, which indicates that the post is *not credible*. If the rank is *3*, an *orange* alert symbol is displayed, which indicates that the post is *maybe not credible* and for rank *2*, a *yellow* alert symbol is displayed on the user's news feed along with the post. The yellow alert symbol is to show that the post *maybe credible*. This helps the user to take an informed decision that whether to trust the post or not. In order to minimize the change in user's news feed, we prefer not to display any alert symbol if the post is ranked *1*. Figure 4 shows a screen-shot of the news feed of a Facebook user when the extension is enabled. An orange alert symbol appears next to the post's time-stamp. When

Fig. 4. (Best viewed in color print) Screenshot of the News Feed of a Facebook user when the extension is enabled. An *orange* alert symbol appears next to the post times-tamp which indicates that the post is *maybe not credible*. (Color figure online)

we hover a mouse pointer on the *orange* alert symbol, it displays a message - *maybe not credible*.

4   Data Collection and Labeled Dataset Creation

In this section, we describe how we collect data from Facebook for analysis and to build a true dataset of credible and non-credible Facebook posts.

4.1   Data Collection

We collected data using Facebook's Graph API search endpoint. The API returns only the public posts. Using a search endpoint, we cannot directly query posts for a specific event. But we can query pages using a keyword. The response consists of all the pages related to the given keyword. Figure 5 shows our data collection procedure. Our dataset consists of posts created by various news-related pages. We used the keyword 'news' to collect page IDs of various news-related pages. Using page ID, we collected most recent 100 posts and their details from each page. For the purpose of this study, we considered only those pages that have more than 5000 likes on them. Also, there are several news-related pages which are marked as verified by Facebook, for instance, @thehindu, @washingtonpost. We assumed all the posts created by verified pages to be credible. Hence, we have not included such posts in our dataset. We also excluded all the posts that were not in the English language. Thus, we collected an initial dataset of 10,416 public posts published by 107 unique pages on Facebook.

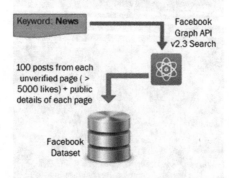

Fig. 5. Describes the data collection procedure used.

4.2 Labeled Dataset Creation

To train our model, we require a labeled dataset. To label the Facebook posts, we followed a similar approach as used by authors in [2,7,8] to label tweets from Twitter for credibility. Unlike [8], we curated ground truth for A-C-T instead of credibility. We took help from human annotators to obtain the ground truth regarding the A-C-T of information present in the posts. As shown in Fig. 6, we developed a web interface for labeling the dataset. We hosted the annotation portal on Amazon Web Services (AWS) EC2 instance.[8] Annotators were asked to sign-up on the portal and then sign in using the same credentials. All the annotators were the frequent users of Facebook. The average age of an annotator was 22. They were given a set of instructions in which the definitions of A-C-T was mentioned. We provided them with the links to Facebook posts. For each post, they had to choose from 6 options for A-C-T individually.

We asked them to select one of the following options for each post: C1. Definitely accurate/clear/timely
C2. Maybe accurate/clear/timely
C3. Neutral accurate/clear/timely
C4. Maybe not accurate/clear/timely
C5. Definitely not accurate/clear/timely
C6. Skip

We also provided an option to skip the entire post if they were not sure of their response. We obtained labels for 1,056 Facebook posts selected randomly from 10,416 posts. Each post was evaluated by three annotators to maintain the

---

[8] https://aws.amazon.com/ec2/.

Fig. 6. Screenshot of the web interface used by human annotators to label Facebook posts for A-C-T.

confidence in the labels, and the mode was calculated to give the label to that post. If all the users had different answers, we calculated median and mean for all such posts. During experiments we found median values give better results than mean values. So, we choose median values over mean values to give the label to that post. To input this data to the binary classifiers, we created two classes with this annotated data, each for A-C-T. Class 1 contains all the posts with the score greater than 3. And remaining posts constitutes class 2. Thus, it becomes a binary classification problem. Table 1 shows the description of our final dataset.

Table 1. Descriptive statistics of Facebook dataset.

1.	No. of Facebook Pages	500
2.	No. of Verified Facebook Pages	223
3.	No. of pages with likes > 5000	483
4.	No. of pages with likes > 5000 + verified = true	223
5.	No. of pages with likes > 5000 + verified = false and language = English	107
6.	No. of posts which are posted by pages with likes > 5000 + verified = false and language = English	10,416

5   Automatic Credibility Assessment

Our goal is to develop a system for ranking Facebook posts on the basis of credibility. Our system classifies each post on the basis of A-C-T and then ranks the posts for its credibility. We adopt supervised machine learning algorithms for classification. First, we perform feature extraction from the posts. Second, we train multiple classification models for A-C-T using several feature sets. Lastly, we compare the accuracy of different machine learning algorithms, using the training labels obtained in Sect. 4. In this section, we also describe Facebook's current technique to identify fake news.

5.1   Facebook's Current Techniques to Identify Fake News

Facebook relies on the community to identify fake news. As shown in Fig. 7, users can report fake news with the help of a report button. If many people report a story, then Facebook sends it to third-party fact-checking organizations. If the fact checkers agree that the story is fake, users see a flag on the story indicating that it has been disputed, and that story may be less likely to show up in News Feed. Users can still read and share the story, but now there a flag which indicates that the post is fake. The drawback of this technique is that this it is slow and the post would be flagged only after going viral.

Fig. 7. Facebook's current technique to take community feedback to combat fake news.

5.2   Credibility Assessment Features

The data is collected using Facebook's Graph API. It includes posts, user's reactions to that post and the source page details. Generating feature vectors from the posts is an important step that affects the accuracy of the statistical model trained from this data. Here, we use a collection of features from previous work on user-generated contents on Facebook [4–6]. Along with that, we curate some new features to enhance the model. The features can be broadly divided into three groups (G1) Page-based, (G2) Post-based, and (G3) Page history-based. These three groups of features are used to classify posts on the basis of A-C-T which in turn are used to rank the posts from *1–4*. If a post is *accurate, clear, and timely,* the post is ranked as *1.* If a post is *either not-accurate, not-clear or not-timely,* it is ranked as *2.* If a post is classified as either of two from following, *not-accurate, not-clear, and not-timely,* it is ranked as *3.* And if a post is *neither accurate, clear nor timely,* it is ranked as *4.*

Page-Based Features (G1): Source of the post is an important factor to measure the trustworthiness of the post. With the larger audience, pages play an important role in the dissemination of information. Users share posts created by the pages, and it thus accelerates the diffusion of information in the network. Our system focuses on posts that are either created by pages or are the shared posts with a page as a source. Table 2 presents the page based features. We have three kinds of features, (a) Boolean, (b) Numeric, and (c) Nominal.

Table 2. Page-based features.

Feature set	Features
Boolean (20)	Affiliation, birthday, can post, cover picture, current location, working hours, description present, location, city, street, state, zip, country, latitude, longitude, personal interests, phone number, public transit, website field, founded
Numeric (36)	Average sentence length for description, average word length for description, parking capacity, category list length, check-ins, no. of email IDs in description, fraction of HTTP URLs in description, description length, fraction of URLs shortened, fraction of URLs active, likes, page name length, no. of subdomains in URLs, path length of URLs, no. of redirects in URLs, no. of parameters in URLs, [no. of exclamation marks, no. of question marks, no. of alphabets, no. of emoticons, no. of English stop words, no. of English words, no. of lower case characters, no. of upper case characters, no. of newline characters, no. of words, no. of unique words, no. of sentences, no. of total characters, no. of digits, no. of URLs] in description, description repetition factor, talking-about count, were-here count
Nominal (2)	Category, description language

Post-Based Features (G2): Some researchers have shown that post based features are useful to access information trustworthiness [2,6,8]. We have used features from these previous works along with some new features. Table 3 presents post-based features. The post based features include text-based features and message link-based features.

Page-History Based Features (G3): On Facebook, pages are often used to disseminate the information. Every page is associated with a category. It is

Table 3. Post-based features.

Feature set	Features
Numeric (38)	audience engaged, [average no. of upper case characters, average length, average word length, no. of English words, no. of English stop words] for description, message, and name fields, description, message, name, no. of comments, no. of reactions (love/like/sad/wow/anger/haha), no. of shares, no. of URLs, total no. of unique domains, has http, the no. of Para, no. of redirects, the number of subdomains, is a shortened URL, are emoticons present, no. of urls, positive, negative and neutral sentiment of user's 100 comments in [chronological, ranked, reverse-chronological] order
Nominal (2)	status type [added photos, added video, created event, created note, mobile status update, published story, shared story, wall post], type [event, link, music, note, offer, photo, video, status]

assumed that the topic of posts created by the page is in accordance with the page category. In our preliminary analysis, we found that this is not the case every time. There are several pages with *News* in their category but they do not or rarely post anything related to news. Thus, page history plays an important role in assessing the credibility of the post especially when it comes to news. It's a background check for the post before believing in its content. To calculate the page-history of a page, we considered the last 100 posts or posts from last 7 days, whichever is lesser. Table 4 presents the page history-based features.

In [1], researchers found that the messages with the least credibility are associated with negative social events. Such messages also contain strong negative sentiment words and opinions. So, it can be concluded that sentiment history is a good indicator of the trustworthiness of the user. We can calculate sentiment history of a page also. To do that, our system finds the sentiment of all the posts collected for the source page. We estimate the frequency of posts with negative, positive and neutral sentiment and use them as features.

5.3   Classification Algorithms

The dataset which we obtained after the labeling was unbalanced. In the real-world scenario, the datasets are always imbalanced. We experimented with both over-sampling and down-sampling techniques. For over-sampling, we used a well-known technique called Synthetic Minority Over-sampling Technique (SMOTE) [3]. But it further decreased the performance of the classifier. We also experimented with down-sampling, but it resulted in over-fitting of the model. Therefore, we adjusted the weights by changing the weight parameter of the classifier to balanced. We went ahead with this and performed all the experiments.

We tested and evaluated various classification algorithms to classify the data on the basis of A-C-T: Naive Bayesian, K-nearest neighbors, decision tree, random forest, gradient boosting classifiers, artificial neural networks and support

Table 4. Page-history based features.

Feature set	Features
Numeric (54)	Daily activity ratio, audience engaged, [average no. of upper case characters, average length, average word length, no. of English words, no. of English stop words] for description, message, and name fields, no. of posts containing the field [description, message, name], no. of comments, total reactions. No. of reactions (haha/like/love/wow/sad/angry), no. of shares, no. of posts with status type [added photos, added video, created event, created note, mobile status update, published story, shared story, wall post], no. of posts with type [event, link, music, note, offer, photo, video, status], total no. of URLs, total no. of unique domain, no. of posts with positive, negative, neutral sentiment

Table 5. Results for supervised learning experiments for *Accuracy* for four classifiers over four different feature sets. Random Forest performed the best with Page-Based features.

Classifier	Feature set	Acc (%)	ROC AUC
Random Forest	G1 + G2 + G3	84.52	0.893
	G1	85.57	**0.916**
	G2	83.46	0.875
	G3	82.25	0.862
SVM	G1 + G2 + G3	76.69	0.810
	G1	75.71	0.806
	G2	77.97	0.829
	G3	78.20	0.838
Logistic Regression	G1 + G2 + G3	77.21	0.820
	G1	80.33	0.848
	G2	74.69	0.799
	G3	76.18	0.805
Naive Bayesian	G1 + G2 + G3	65.30	0.706
	G1	67.54	0.727
	G2	63.71	0.695
	G3	62.64	0.681

vector classifiers. Here, we reported only the best results from all the algorithms. We trained three classifiers each for *Accuracy, Clarity, and Timeliness*. Tables 5, 6 and 7 shows the accuracy and ROC AUC values for various classification algorithms that we applied on our feature set for *Accuracy, Clarity, and Timeliness*. We also trained the models for the individual type of features and one with all the features combined. All the training models were evaluated using 10-fold cross-validation. To perform all the experiments, we used scikit-learn, a machine learning library for the python programming language.[9] We achieve the best results when all the features are used together to train the model for A-C-T respectively. Random Forest algorithm out-performed other algorithms with the best ROC AUC score of 0.916, 0.875, and 0.851 for A-C-T respectively as shown in Tables 5, 6 and 7.

Also, it is worth stating that even though we used the same feature set to train models for A-C-T, the accuracy peaked for different subsets of the feature set. It means, for A-C-T each feature have different feature importance. For instance, feature which is important for *accuracy* may not be as important to *clarity*. This can be seen from the results. For *accuracy* and *clarity*, page-based features performed the best whereas for *timeliness* the combination of all the features gave the best results. For the same reason, the performance decreases in some cases on the addition of more features.

Comparison with Baseline Model: In the best of our knowledge, there is only one study on the credibility of the Facebook posts [14]. Saikaew and Noyunsan developed two chrome browser extensions. The first extension was used to measure the credibility of each post by asking users to give a score

Table 6. Results for supervised learning experiments for *Clarity* for four classifiers over four different feature sets. Random Forest performed the best with Page-Based features.

Classifier	Feature set	Acc (%)	ROC AUC
Random Forest	G1 + G2 + G3	81.62	0.859
	G1	83.41	**0.875**
	G2	79.38	0.841
	G3	80.19	0.850
SVM	G1 + G2 + G3	75.99	0.812
	G1	74.86	0.801
	G2	78.25	0.836
	G3	74.20	0.794
Logistic Regression	G1 + G2 + G3	65.33	0.706
	G1	67.89	0.728
	G2	75.93	0.810
	G3	69.04	0.758
Naive Bayesian	G1 + G2 + G3	63.13	0.691
	G1	59.81	0.632
	G2	62.42	0.680
	G3	62.56	0.681

---

[9] http://scikit-learn.org.

Table 7. Results for supervised learning experiments for *Timeliness* for four classifiers over four different feature sets. Random Forest performed the best with a set of all features.

Classifier	Feature set	Acc (%)	ROC AUC
Random Forest	G1 + G2 + G3	80.40	**0.851**
	G1	78.82	0.839
	G2	76.91	0.806
	G3	75.94	0.828
SVM	G1 + G2 + G3	73.76	0.784
	G1	71.42	0.763
	G2	67.50	0.727
	G3	69.51	0.759
Logistic Regression	G1 + G2 + G3	67.63	0.729
	G1	64.16	0.692
	G2	65.93	0.708
	G3	65.35	0.706
Naive Bayesian	G1 + G2 + G3	61.23	0.652
	G1	60.71	0.636
	G2	53.72	0.581
	G3	59.16	0.629

from 1 (the lowest value) to 10 (the highest value). These post evaluations were used as a data to train an SVM model. They trained the model on mere 8 features; likes count, comments count, shares count, URL count, images count, hashtag count, video count, is location present. The second extension was used to automatically evaluate the credibility of each post-real-time using the SVM model. On a validation set of 1,348 posts, they report an accuracy of 81.82%. Due to a better feature selection, our models are performing better than this model with the accuracy of 85.57%, 83.41%, and 80.40% for A-C-T respectively. Also, our definition of credibility is more fine-grained which makes our results more promising.

6 Conclusion, Limitations, and Future Work

In this paper, we propose a system which computes the credibility of Facebook posts created by Facebook pages in real-time. Here, we define credibility as a function of *accuracy, clarity, timeliness*. We experiment with 1,056 posts created by 107 *pages* that claim to belong to news-category. We propose a set of 152 features based on post content, source-page and page-history. We use this feature set to train binary classification models each for A-C-T using supervised algorithms. We use the best performing models to develop a RESTful API and a Google Chrome browser extension, named Credibility Investigator (CbI), to rank Facebook posts for its credibility in real-time. To the best of our knowledge, this is the first research work which ranks the Facebook posts for its credibility in terms of A-C-T. The random forest algorithm performed the best and achieved a maximum ROC AUC value of 0.916, 0.875, and 0.851 for A-C-T respectively. There are a few limitations in our proposed system. We do not claim that our dataset represents the entire Facebook news related pages. Facebook does not provide any data about what fraction of information is returned by its API. Due to the Facebook's Graph API restrictions, we can only access public posts. Also, Facebook supports multiple non-English languages too. As of now, our system works only on posts in the English language. In the future, we would like to address this problem. Also, we would like to explore various graph-based techniques to detect the presence of fake pages on the Facebook network.

References

1. Alrubaian, M., Al-Qurishi, M., Hassan, M., Alamri, A.: A credibility analysis system for assessing information on Twitter. IEEE Trans. Dependable Secur. Comput. **15**(4), 661–674 (2016)
2. Castillo, C., Mendoza, M., Poblete, B.: Information credibility on Twitter. In: Proceedings of the 20th International Conference on World Wide Web, pp. 675–684. ACM (2011)
3. Chawla, N.V., Bowyer, K.W., Hall, L.O., Kegelmeyer, W.P.: SMOTE: synthetic minority over-sampling technique. J. Artif. Intell. Res. **16**, 321–357 (2002)
4. Dewan, P., Bagroy, S., Kumaraguru, P.: Hiding in plain sight: characterizing and detecting malicious Facebook pages. In: 2016 IEEE/ACM International Conference on Advances in Social Networks Analysis and Mining (ASONAM), pp. 193–196. IEEE (2016)
5. Dewan, P., Bagroy, S., Kumaraguru, P.: Hiding in plain sight: the anatomy of malicious *Pages* on Facebook. In: Kaya, M., Kawash, J., Khoury, S., Day, M.-Y. (eds.) Social Network Based Big Data Analysis and Applications. LNSN, pp. 21–54. Springer, Cham (2018). https://doi.org/10.1007/978-3-319-78196-9_2
6. Dewan, P., Kumaraguru, P.: Towards automatic real time identification of malicious posts on Facebook. In: 2015 13th Annual Conference on Privacy, Security and Trust (PST), pp. 85–92. IEEE (2015)
7. Gupta, A., Kumaraguru, P.: Credibility ranking of tweets during high impact events. In: Proceedings of the 1st Workshop on Privacy and Security in Online Social Media, p. 2. ACM (2012)
8. Gupta, A., Kumaraguru, P., Castillo, C., Meier, P.: TweetCred: real-time credibility assessment of content on Twitter. In: Aiello, L.M., McFarland, D. (eds.) SocInfo 2014. LNCS, vol. 8851, pp. 228–243. Springer, Cham (2014). https://doi.org/10.1007/978-3-319-13734-6_16
9. Gupta, A., Lamba, H., Kumaraguru, P., Joshi, A.: Faking sandy: characterizing and identifying fake images on Twitter during hurricane sandy. In: Proceedings of the 22nd International Conference on World Wide Web, pp. 729–736. ACM (2013)
10. Haralabopoulos, G., Anagnostopoulos, I., Zeadally, S.: The challenge of improving credibility of user-generated content in online social networks. J. Data Inf. Qual. (JDIQ) **7**(3), 13 (2016)

11. Li, H., Sakamoto, Y.: Computing the veracity of information through crowds: a method for reducing the spread of false messages on social media. In: 2015 48th Hawaii International Conference on System Sciences (HICSS), pp. 2003–2012. IEEE (2015)

12. Mendoza, M., Poblete, B., Castillo, C.: Twitter under crisis: can we trust what we RT? In: Proceedings of the First Workshop on Social Media Analytics, pp. 71–79. ACM (2010)

13. Ratkiewicz, J., et al.: Truthy: mapping the spread of astroturf in microblog streams. In: Proceedings of the 20th International Conference Companion on World Wide Web, pp. 249–252. ACM (2011)

14. Saikaew, K.R., Noyunsan, C.: Features for measuring credibility on Facebook information. Int. Sch. Sci. Res. Innov. **9**(1), 174–177 (2015)

15. Tanaka, Y., Sakamoto, Y., Matsuka, T.: Toward a social-technological system that inactivates false rumors through the critical thinking of crowds. In: 2013 46th Hawaii International Conference on System Sciences (HICSS), pp. 649–658. IEEE (2013)

A Parallel Approach to Detect Communities in Evolving Networks

Keshab Nath[1]([✉]) and Swarup Roy[1,2]

[1] North-Eastern Hill University, Shillong, India
keshabnath@nehu.ac.in
[2] Sikkim University, Gangtok, India
sroy01@cus.ac.in

Abstract. To understand the dynamics, functional and topological aspect of the real-world networks it is necessary to segregate the network into sub-networks, where each member of a sub-network possess analogous characteristics. Numerous number of community finding approaches are proposed in the last few decades to overcome the issues associated with community detection. Although, most of the conventional approaches rely on the premises that networks are static in nature and there won't be any alternation over time. Moreover, all these approaches are single machine approach and hence exhibits poor scalability.

In this work, we propose a new incremental parallel community detection method, PcDEN (Parallel Community Detection approach in Evolving Networks). Our proposed method can detect communities in dynamic distributed networks. We define a new Affinity score based on intra-community strength between nodes and their neighbors. We also derive a new model to perform community merging, based on common high degree nodes present in both the communities. We tested our algorithm on various real-world networks for our experimentation. Results show that, PcDEN produce satisfactory output with respect to various assessment indices.

Keywords: Parallel community detection
Distributed computing · Dynamic graphs · Incremental approach
Evolving networks

1 Introduction

Due to the evolution of the high-speed internet technology more and more peoples are collaborating, tagging, sharing information with each other on a whole different level, forming a massive graph, popularly termed as Social Networks. In terms of graph theory, a social network is a set of members or actors (nodes) and their collaboration or association among them, represented using edges. These elementary components (nodes and links) of a network have their own characteristics and functions. Nodes with similar characteristics or function can be

© Springer Nature Switzerland AG 2018
A. Mondal et al. (Eds.): BDA 2018, LNCS 11297, pp. 188–203, 2018.
https://doi.org/10.1007/978-3-030-04780-1_13

brought together to a single unit (sub-graph), where nodes having high density intra-connection among them, whereas there exist lower inter-connection among the nodes from different unit (sub-graph). These sub-graphs are called communities or modules of a network.

To reveal the internal functionality, formulation and the framework of a network more accurately, it is important to identify the meaningful communities from a network. Community detection assistance in many application, such as search engine rank alteration [1], finding meaningful communities from research publications [2], health care [3], social media analysis [4]. However, due to complex structure and the evolving nature of a network like social networks, transaction networks etc, communities detection become the fundamental problem in network science. Last few decades, numerous methods have been attempted to overcome these problems. However, efficient algorithms to find out an appropriate solution for the same in an evolving network in a reasonable amount of time is inferior. Most of the state-of-art community detection algorithms [5–9] are based on either static or dynamic networks or single machine based approach. Real-world networks such as all biological interaction network, social networks (facebook, twitter etc.) are growing with time. It is infeasible to process the whole network in one machine due to memory and time constraint. An incremental parallel computing approach, which uses the distributed machine with distributed processing power, provides a natural solution to deal with the massive evolving networks. In the last few years, researchers proposed few approaches on parallel community detection to overcome the issues of storage space and execution time. Some works like [10–14] reported distributed approach for both static and dynamic networks. However, most of the distributed approaches are facing scalability problem. This is because, most of the real-world networks are scale-free networks [15], where only a few nodes are having a high degree, but most of the nodes have a low degree [16]. Thus, when a network is divided among different processors, the information loss is comparability more in the scale-free network than small world network [17]. In the small world network (also known as six degree separation), a target node can reach out by using only a few nodes from a source node. On the division of a small world network, performance is marginally affected. In a scale-free network, a parallel approach suffers from a huge communication overload. Communities present in various machine need to share a large amount of information among them to obtain the final outcome. On the other hand, due to the absence of a global structure of the network in each processor, there is a possibility of creation of more small and low-quality communities. Furthermore, a parallel community finding becomes even more challenging when the network is evolving in nature.

Apart from the issues like scalability, the volume of the network, rapid evolvement of the network, it also suffers from the issue like the movement of a node in different communities. Due to rapid alteration in the relationship of community members over time, a node may move from one community [5] to another. Detection of temporal communities in the dynamic network is very crucial, since its presence completely depends on the nodes connectivity strength (CS).

Connectivity strength get affected after every insertion or deletion of new nodes or edges to the existing communities. Moreover, in a parallel approach, where global structure is absent involves information loss due to the division of the network. Finding such temporal communities is a challenging task.

Some works are available on parallel community detection in static networks [14, 18–20], very few attempts have been made for dynamic networks [10, 11]. In this work, we focus on the detection of non-overlapping from evolving networks using a parallel incremental approach. We make the following contributions.

- We design a parallel incremental approach PcDEN, to detect dynamic communities in evolving networks.
- We propose a new affinity measure for each incoming node to the existing network in order to detect a community.
- We adopt a neighbors degree difference (NDD) similarity concept of a node present in various communities to merge any two communities.

We organize our paper as follows. In Sect. 2, we discuss the state-of-art parallel community detection algorithms. Our proposed incremental parallel approach PcDEN is discussed in Sect. 3. The performance evaluation of the proposed model is reported in Sect. 4. Finally, we summarized with concluding remarks.

Next, we will discuss some of the state-of-art parallel community detection algorithms for both static and dynamic networks.

2   Prior Research

Most of the prior community detection algorithms focused on single threaded approach. Real-world networks are large in size and evolving with time. Community finding from such massive networks needs more storage space and computational power. Which is infeasible and expensive to obtain from a single machine. To overcome such issues, in recent years many researchers proposed community detection algorithms in distributed networks. Zhang *et al.* [21] propose a parallel community detection algorithm based on mutual communication between network topology and propinquity. Propinquity is the probability measure for the pair of nodes present in a community. This approach considers the network as a static one. Prat-Perez *et al.* [12] proposed a parallel disjoint community detection algorithm called Scalable Community Detection (SCD). It adopt Weighted Community Clustering (WCC) [22] technique to partition the network. According to the author, SCD produce more accurate and need less computational time compare to other state-of-art algorithms. Galluzzi [11] proposed a real time distributed community detection method in dynamic networks. This algorithm is inspired by the original work [23]. According to the author, the proposed model is lightweight and easily customizable. Pinfomr (Parallel Infomap with MapReduce) [24] is a scalable parallel community detection approach based on single processor algorithm Infomap [25]. Pinfomr uses k-shell decomposition approach for partitioning the network and a Infomap based parallel community detection model is developed in MapReduce framework. This approach is developed only for static networks. Saltz *et al.* [26] propose a node centric distributed approach for community detection using WCC metric. It needs more computational time due to triangle counting for creating new communities for each node. This approach is build for static networks, which is infeasible for real-world networks. Staudt *et al.* [20] propose a distributed algorithm based on Louvain method [27], named Parallel Louvain Method (PLM). Later they extend the proposed approach to Parallel Louvain Method with Refinement (PLMR) by adding a new feature the refinement phase on every level. Further, they also propose a two-phase approach by combining both PLM and PLMR. According to author, propose model is qualitatively strong, fast and suitable for massive data. Palsetia *et al.* [28] propose a distributed-memory based parallel algorithm PMEP, which parallelizes Maximizing Equilibrium and Parity algorithm (MEP) [29]. MEP algorithm run on each sub-network to uncover community by maximizing equilibrium and purity of communities. Disadvantage of PMEP algorithm is that it consider a network as a static one.

Most interestingly, in some research work instead of using a brand new parallel approach, researcher uses the traditional algorithms on each data chunk and finally merge the output. We will discuss some of such traditional algorithms, which are broadly used for community detection in parallel environment. Xie *et al.* [30] proposed LabelRankT, which is a modified version of previous method LabelRank [31]. It needs lower computational cost since, every node requires only local information during label propagation. Intrinsic longitudinal community detection (iLCD) method is presented by [32] for finding communities in dynamic networks. iLCD introduced new nodes to the existing communities according to two adaptive thresholds. A new community is created only when, the new edge added to the network is qualified for the predefine minimal community. Finally, communities which are very much alike to each other are merged together. Markov Cluster Algorithm (MCL) [33] is to find community structures in a graph by using a bootstrapping procedure to compute random walks through out the network. The MCL algorithm uses expansion and inflation operation to simulate the random walk within a network. The expansion represents the power of a stochastic matrix based on matrix squaring. Inflation corresponds with taking the power of Hadamard power of a matrix. SLPAD [34] is a dynamic community detection algorithm, follows the same label propagation model like SLPA [35]. To detect overlapping communities, SLPAD uses the modified version of DSCAN [34] termed as DSCAN overlapping (DSCAN-O) and LabelRankT.

Though there exist a good number of research work on parallel community detection. However, most of the conventional methods are only feasible for detecting disjoint community in static networks. A negligible amount of work has been carried out for detection of communities in dynamic networks. In this work, our motivation is to propose an efficient parallel model to detect dynamic communities in evolving networks. Table 1 represent a comparative analysis of existing parallel community detection algorithms.

Table 1. A comparative analysis of existing parallel community detection algorithms

Algorithms	Incremental	Non-Incremental
Zhang *et al.* [21]		✓
SCD [12]		✓
PSOCD [36]		✓
Galluzzi [11]	✓	
Pinfomr [24]		✓
Clementi *et al.* [10]	✓	
Saltz *et al.* [26]		✓
Staudt *et al.* [20]	✓	
PMEP [28]		✓

3   PcDEN: An Incremental Parallel Community Detection Approach

Real-life networks are generally large in scale and dynamic in nature. Existing static and standalone approaches are limited in handling the situations where networks are growing with time and as a consequence the structure and distribution of the communities also changes. Performance of existing community detection methods largely depends on the volume of the network. As the size increases the quality of the result decreases [37]. Moreover, processing and storing a large network in a single machine is expensive. Hence, a sequential single thread approach is not suitable for handling large real-life networks. As an alternative, a parallel community detection method is a current need that can handle dynamic community structures. The parallel community detection method can be defined as follows.

For a dynamic network $G(V, E)$, where V and E represent the set of nodes and edges respectively. For each incoming edge to the network G, is assign to a worker $w_p \in W$, where $W = \{w_1, w_2, \cdots, w_k\}$ is the set of workers. An incremental approach is run on all the workers simultaneously and produce a set of communities $C_t = \{C_1, C_2, \cdots, C_k\}$ at time t. The results generated by each worker is merged together to obtain the final outcome.

Classical parallel approaches work on static networks. They assume communities are static where no variation of node connectivity over the time is possible. In reality, real-world networks are dynamic in nature and evolve over the passage of time. Hence, any static community detection approach is not suitable for real-world networks. We define next a formal representation of incremental community.

Definition 1 (Incremental Community). *Given an online network $G(V, E)$, for each incoming node v_i, a membership function $f : V \rightarrow \{1, \cdots, k\}$ is defined to assign into any community $C_i(t)$ with respect to time t. Incremental community, $C_i(t)$ contains only those nodes which are assigned to it with* *high membership (for $\forall j = 1, \cdots, k$) at time t. At any time $t + 1$, object $v_q \in C_j(t)$ may shift to any community $C_i(t + 1)$ based on membership function and $C_j(t + 1) = C_j(t) - v_q$. The number of community k is dynamic and may decrease or increase depends on incoming data distributions.*

3.1 A New Similarity Measure for Parallel Community Detection

Performance of an unsupervised community detection algorithm is highly dependent upon the similarity measures. We introduce a new similarity measure for each incoming node to the existing community. We assign a new node to an existing community or create a new community based on `Affinity` score of that node. The `Affinity` score of a node is directly proportional to the proximity of that object towards a community. In order to assign a new node v_i into a community C_k, we calculate a `Affinity` score of v_i with respect to the adjacent node $v_j \in C_k$ of v_i. We call v_i as guest node and v_j as host node. To compute `Affinity` score of v_i towards C_k, we look for the pulling strength of v_j towards C_k based on v_j's degree distribution within the community, termed as intra community strength. Intra community strength of a node represent the cohesiveness of the node with other neighbours of a community and can be defined as follows.

Definition 2 (Intra Community Strength). *Intra community strength, $S(v_j)$, of a host node v_j, is the ratio of number of adjacent nodes of v_j in C_k ($Neigb_{C_k}(v_j)$) and the total number of nodes in C_k.*

$$S(v_j) = \frac{|Neigb_{C_k}(v_j)|}{|\mathcal{V}_k|} \tag{1}$$

where, $Neigb_{C_k}(v_j) = \{v_k | e(v_j, v_k) \in C_k\}$ is the set of adjacent nodes of v_j in C_k.

Definition 3 (Affinity Score). *To define affinity score of a guest node v_i towards C_k due to host node v_j, we use an inverse multiplicative factor Δ with the intra community strength of v_j. Affinity score of v_i with C_k can be defined as follows.*

$$Affinity(v_i, C_k) = S(v_j) \times \frac{1}{\Delta}. \tag{2}$$

$$\Delta = \begin{cases} 1, & \text{if } \deg(v_i) \in \varphi(v_j) \\ & \text{and } \max(\varphi(v_j)) = \deg(v_j) \\ \max(\varphi(v_j)) - \deg(v_i), & \text{if } \max(\varphi(v_j)) < \deg(v_j) \\ & \text{and } \max(\varphi(v_j)) \neq \deg(v_i) \\ \max(\varphi(v_j)) - \deg(v_j), & \text{if } \deg(v_j) \leq \max(\varphi(v_j)), \end{cases} \tag{3}$$

where, $\varphi(v_j)$ is the degree of neighbours of v_j in C_k and can be define as follows.

$$\varphi(v_j) = \{\deg(v_k) | v_k \in Neigb_{C_k}(v_j)\} \tag{4}$$

3.2 A Proposed Parallel Approach

Since almost all real-world network are dynamic in nature, therefore we split each incoming edge on the fly with the help of a central controller. Each incoming node is assigned to its respective workers by a central controller (*see* Fig. 1) based on its host node degree. The central controller is also responsible for maintaining the nodes worker ID (WID) and its degree (δ) value exist per worker (δ/W) and the total degree (T_δ) value. Each worker also maintains local nodes degree list and an adjacency matrix. After each insertion of a node to a worker, it updates its adjacency matrix and the degree list accordingly. Workers continuously synchronized its local degree list with the central controller. To avoid work overload to a particular worker, central controller maintain load balancing. It assigns relevant nodes to a worker up to a certain threshold, if relevant nodes are kept on coming then central controller automatically assign the next incoming node to another appropriate worker. In community merging phase, each worker share the nodes information (NI) including community identifier, high/low degree nodes and linking nodes of all communities individually with the global merger. After accomplishment of each merging (discuss later) phase, global merger concurrently update the nodes information, which includes the total degree of each node (attain from central controller) and degree of each node per community. Total degree and degree of each node per community help in measuring the quality of each community after merging.

3.3 Community Finding in Each Worker

For every insertion of nodes to the workers, there exist few possible scenarios. First, a new edge between two existing nodes may be introduced to the network. Second, a new edge between an existing node and a new node may introduced to the network. Furthermore, a brand new edge may be introduced to the existing network. Each incoming node is assigned to the community C_j with highest Affinity score. A new community C_{new} is created, when there is no shared edge between the new node/nodes with existing node/nodes of a community. A community member v_j may move from one community to another due to continuous growth of connectivity strength in the other community (C_j) rather than its parent community (C_i). If Affinity score of v_j with respect to (w.r.t) community (C_j) exceeds the value obtained from its parent community (C_i), then $v_j \in C_j$ and $V_i - v_j$; $\forall V_i \in C_i$. Given two communities C_i and C_j, we merge the communities if insertion of a new node v_k acts as a linking factor between C_i and C_j. In other words, if v_k become a linking factor, for which v_k having significant number of shared neighbors in both the communities. This can be detected easily by comparing the Affinity score of v_k with respect to both the communities. We merge them if Affinity score difference is less than a threshold value τ i.e. $|\texttt{Affinity}(v_k, C_i) - \texttt{Affinity}(v_k, C_j)| < \tau$. Now, communities formed in all the workers are merged together at global merger (*see* Fig. 1) to obtain the final communities.

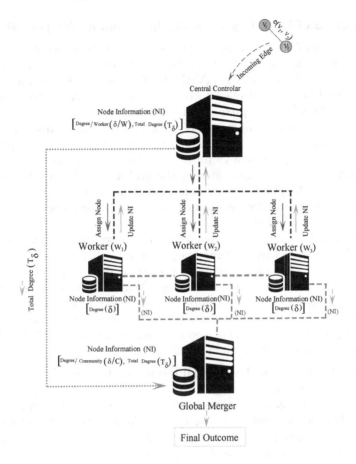

Fig. 1. A proposed parallel approach for dynamic community detection.

3.4   Finding High and Low Degree Nodes

Most of the real-world networks are scale-free networks, where there exist few hub nodes, intimately followed by smaller hub nodes. These smaller hub nodes are followed by even smaller degree nodes and so on and as a whole the network follows a power-law distribution. In our incremental approach, since we are assigning an edge to a worker by the controller based on host node degree (prefer high degree) present on a particular worker. Hence, a high degree node has more possibility to endure as a high degree nodes after removing (splitting) some of its edges. Based on this hypothesis, we perform merging of two communities obtained from different workers. The possibility of two communities getting merged into single one relies on the number of common high degree nodes present on both the communities and number of Linking Nodes (LN). A linking node which has a low degree in one community present in a particular worker and becomes high degree node in another community present in another worker.

3.5 Selection of High Degree Nodes in a Community per Worker

For all community $C_j \in C$, where $j = \{1, \cdots, n\}$ and C is the set of communities formed in a particular worker (say w_1). We select a community C_i and generate a Vector of the High Degree nodes (VHD) for C_i based on the node information table generated by the controller. In this case, we consider the total degree of only those nodes which are present in that community C_i at time t. The total degree is nothing but the summation of the degree of a node present on various workers i.e $\delta = \{\delta_{w_1}, \delta_{w_2}, \cdots, \delta_{w_n}\}$, where δ_{w_1} is the degree of a node present in worker w_1. We also generate a vector of the high degree nodes for C_i based on the node information table generated by the worker (w_1). From this two Vector of High Degree nodes (VHD), we exclude all common nodes and prepare a single high degree nodes vector for the community C_i. A similar process is carried out for other communities of worker w_1.

To detect high degree node within a community, we compute the community degree vector $CD_{C_i} = \{$deg (v_1),deg (v_2), $\cdots$,deg (v_m) $\}$, where deg(v_k) for all the nodes $v_k \in C_i$. For every insertion of a new node in C_i, CD_{C_i} is updated accordingly. First, generate a **V**ector of **H**igh **D**egree (VHD) nodes present within a community (C_i) from CD_{C_i}. In this process, each node select (vote) its all high degree (degree which is greater than a threshold) neighbors and assign to VHD. To obtain high degree neighbors of a node, we define a function called **Neighbor Degree Difference** (NDD) to build a VHD within a community.

Definition 4 (Neighbor Degree Difference). *Neighbor Degree Difference (NDD) is the normalized degree difference of neighbors of the host node and the host node degree itself. It is represented as the ratio of differences in highest degree neighbor of the host node, say v_j, and the average degree of the neighbors $(\overline{\varphi(v_j)})$ including the host node itself and the total number of nodes in its neighborhood (including itself) and represent as follows.*

$$NDD(v_j) = \frac{\max(\varphi(v_j)) - \overline{\varphi(v_j)}}{|\mathtt{Neigb}_{C_k}(v_j)| + 1} \tag{5}$$

In each iteration, after computation of $\overline{\varphi(v_j)}$, the neighbor with the highest degree is discarded (neighbour nodes are processed in descending order of their degree). $|\mathtt{Neigb}_{C_k}(v_j)| + 1$ represents the total number of neighbors of node v_j including itself, which also remains constant for the node v_j at time t. We include all neighbor nodes of v_j into VHD^t at time t, for which NDD$(v_j) \geq \gamma$. Based on this condition, we create VHD^t for the whole network. VHD^t does not contain any nodes which having neighbors with approximately equal in degree.

3.6 Selection of Low Degree Nodes in a Community per Worker

Process of low degree nodes detection is just reverse of Sect. 3.5. Nodes which satisfy NDD$(v_j) \leq \gamma$ are considered as low degree nodes. Two vector of Low Degree Nodes (LDN) are created, one by using controller node information list and second is by using the worker node information list. All common nodes are excluded from these two LDN and a single LDN is created by combining these two vector.

3.7 Merging Communities from Different Workers

For merging communities taken from two different workers, we first find out the HDN of each community. If the number of common nodes present in both HDN is more than a threshold (ζ), then we assign the community ID of one to another. If the number of common nodes is less than ζ, then we generate LDN from both the communities. From these two LDN, we select all common LDN and create a single LDN. A threshold value is measured by taking the mean of final LDN vector. Instead of considering all low degree nodes, we consider only those nodes which have a low degree than the threshold. This will reduce communication cost between two workers while sharing node information (adjacency matrix) during merging. Based on the adjacency matrix, we search for nodes which can be a member of HDN. Moreover, nodes which have a sufficient number of neighbors are already in HDN are also included in HDN. The newly detected HDN vector is the combination of both previously detected HDN (before sharing adjacency matrix) and newly detected HDN (after sharing adjacency matrix). If the number of new HDN is more than a threshold (ζ), then both the communities will have the same ID. If it less than ζ, then the same process will be carried out with the next community. In both the cases, merging using HDN and sharing LDN, we always keep track of Linking Nodes (LN). The more the number of LN, better is the possibility of merging. Most interestingly, due to sharing of the adjacency matrix, there is a possibility of merging (local merging) communities within the worker also. Our approach take care of both type of merging, local (within worker) and global (between worker) simultaneously. Next, we evaluate the performance of PcDEN.

4 Performance Evaluation

To evaluate the effectiveness and accuracy of PcDEN in detecting communities, we consider various validity measures like Dunn [38], Silhouette [39], Connectivity [40], Davies-Bouldin (DB) [41] and Extended Modularity (EQ) [42]. We also consider Normalized Mutual Information (NMI) [43] to assess the performance of our algorithm in detecting communities. NMI is a similarity measures between two different communities, with value 1 indicate total matching and value 0 if they are totally separated.

4.1 Dataset Used

We use six real-world networks, which are available at **UCI**[1] Machine Learning Repository and SNAP[2] for our experimentation. A brief description of the datasets used in our experiment is given in Table 2.

---

[1] http://archive.ics.uci.edu/ml/datasets.html.
[2] https://snap.stanford.edu/data/citnets.

Table 2. Dataset used in the experiments

Dataset	Available	#Nodes	#Edges	#Communities
American College Football		115	616	11 [44]
Dolphin Social Network	UCI[3]	62	159	2 [43]
Books about US Politics		105	441	3 [45]
Arxiv General Relativity (ca-GrQc)		5,242	14,496	--
Arxiv High Energy Physics Theory (ca-HepTh)	SNAP[4]	9,877	25,998	--
Email-Enron		36,692	183,831	--

– Information not available.

Note: All these real-world networks used in the experiment are undirected in nature.

4.2 Experimental Results

The PcDEN algorithm is implemented in R and run in Windows 7 having machine configuration of 4 GB RAM and 2.7 GHz CPU. Table 3 represents the internal quality assessment of PcDEN. We compare our results with some of the well-known community detection algorithms (non-parallel) such as LabelRankT, MCL, SLPAD and iLCD. Though, these algorithms are not parallel, still we used them to analyze our results quality. Because, no matter what approach we adopt, the final result should not be compromised. The best performing scores, in each validity criteria, are highlighted in the table. As a similarity measure between a pre-assume community structure and the extracted structure by the algorithm, we use the Normalized Mutual Information (NMI). We calculate the NMI score for a certain percentage of edge constraints of a network. We keep on increasing the percentage and simultaneously measure the NMI score. Figure 2 shows NMI score with respect to the percentage of incoming edge constraints.

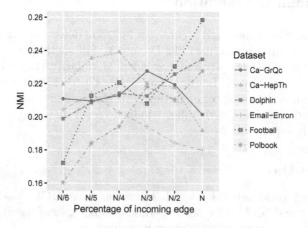

Fig. 2. Performance of PcDEN on Real-World networks with respect to NMI (Y-axis) and Percentage of incoming edge (X-axis)

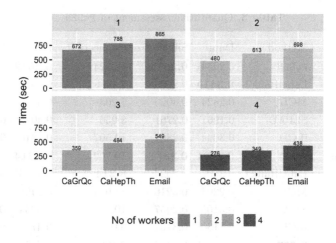

No of workers ■ 1 ░ 2 ▨ 3 ■ 4

Fig. 3. PcDEN execution time w.r.t number of Workers.

4.3 Scalability

We test the scalability of PcDEN with respect to (w.r.t) number of workers and number of threads. First, we experiment scalability w.r.t workers. In this experiment, we consider three large networks namely Ca-GrQc, Ca-HepTH, and Email. Our approach is evaluated against each network by using 1, 2, 3 and 4 workers respectively provided the number of thread is fixed (we use four threads). Figure 3 shows the scalability of PcDEN w.r.t various workers. In second experiment, we measured the total elapsed time for one iteration of PcDEN varying the number of threads from 1 to 15. Figure 4 show that the elapsed time decreased as the number of threads increased. However, in the case of Ca-GrQc network the

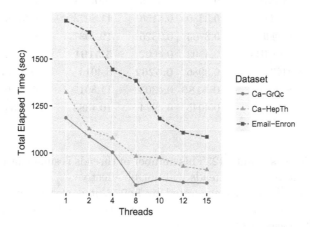

Fig. 4. Scalability of PcDEN w.r.t number of Threads.

Table 3. Qualitative assessment of PcDEN

Dataset	Method	Dunn	Silhouette	Connectivity	DB	EQ
ca-GrQc	PcDEN	0.2054	**0.3312**	12.324	**0.2622**	**0.4187**
	LabelRankT	**0.3414**	0.2205	**7.051**	0.3018	0.3402
	MCL	0.2824	0.3029	7.203	0.3116	0.2143
	SLPAD	0.1961	0.2761	8.056	0.4014	0.2954
	iLCD	0.3094	0.2864	10.507	0.2816	0.3876
ca-HepTh	PcDEN	**0.4201**	**0.3371**	10.702	**0.4214**	0.3521
	LabelRankT	0.3507	0.2524	**8.214**	0.4323	0.4078
	MCL	0.2204	0.3216	11.014	0.5401	0.4264
	SLPAD	0.3105	0.2957	10.605	0.5034	**0.5108**
	iLCD	0.3130	0.3162	9.013	0.4771	0.3708
Email-Enron	PcDEN	**0.4181**	0.2425	**9.761**	0.5975	**0.3682**
	LabelRankT	0.2923	0.3301	10.027	0.4479	0.2723
	MCL	0.3506	**0.4736**	13.372	0.5102	0.3345
	SLPAD	0.2512	0.4036	11.617	**0.3176**	0.2489
	iLCD	0.3987	0.3705	10.781	0.4005	0.3654
Football	PcDEN	0.1051	**0.3420**	14.456	0.3704	**0.4404**
	LabelRankT	0.2512	0.2704	**7.026**	0.5219	0.3325
	MCL	0.3384	0.3175	10.714	0.4129	0.2374
	SLPAD	**0.4218**	0.2507	8.218	0.4808	0.3876
	iLCD	0.2755	0.3207	11.016	**0.3486**	0.4089
Dolphin	PcDEN	0.3801	**0.4836**	12.0716	**0.3208**	0.4761
	LabelRankT	**0.4618**	0.4417	10.1022	0.4729	0.4478
	MCL	0.4416	0.4516	9.1063	0.3558	**0.5098**
	SLPAD	0.3246	0.3829	**9.0073**	0.3211	0.4561
	iLCD	0.3380	0.4156	11.8073	0.5126	0.3875
US Politics	PcDEN	0.3793	0.2325	12.4037	0.4103	0.4391
	LabelRankT	0.4503	0.4507	**5.5191**	0.4213	0.5209
	MCL	0.3986	**0.5262**	8.4014	**0.3971**	**0.5478**
	SLPAD	**0.5186**	0.4305	11.3915	0.5270	0.3576
	iLCD	0.4045	0.4974	10.1385	0.4234	0.4398

elapsed time decreased until 12. This number of threads is sufficient for Ca-GrQc network, and adding more threads will not be effective.

By analyzing the results generated by PcDEN, it is clearly evident that the performance of our approach is quite satisfactory and efficient with respect to various validity measures.

5   Conclusion

We introduced PcDEN, which is capable of detecting dynamic communities from distributed dynamic networks. We used two new concepts, (i) intra-community strength between nodes and their neighbors to calculate the Affinity score and (ii) perform community merging, based on common high degree nodes present in both the communities. We tested our algorithm on various real-world networks. Results show that, PcDEN produce satisfactory output with respect to various assessment indices. Our next extended work will be to propose a method which can handle both overlapping and intrinsic community in growing networks like social and biological networks.

References

1. Schaeffer, S.E.: Graph clustering. Comput. Sci. Rev. **1**(1), 27–64 (2007)
2. Haley, B.M., Dong, A., Tumer, I.Y.: Creating faultable network models of complex engineered systems. In: ASME 2014 International Design Engineering Technical Conferences and Computers and Information in Engineering Conference. American Society of Mechanical Engineers (2014) V02AT03A051-V02AT03A051
3. Jonsson, P.F., Cavanna, T., Zicha, D., Bates, P.A.: Cluster analysis of networks generated through homology: automatic identification of important protein communities involved in cancer metastasis. BMC Bioinform. **7**(1), 2 (2006)
4. Gargi, U., Lu, W., Mirrokni, V.S., Yoon, S.: Large-scale community detection on youtube for topic discovery and exploration. In: ICWSM (2011)
5. Palla, G., Derényi, I., Farkas, I., Vicsek, T.: Uncovering the overlapping community structure of complex networks in nature and society. Nature **435**(7043), 814–818 (2005)
6. Kumpula, J.M., Kivelä, M., Kaski, K., Saramäki, J.: Sequential algorithm for fast clique percolation. Phys. Rev. E **78**(2), 026109 (2008)
7. Zhang, S., Wang, R.S., Zhang, X.S.: Identification of overlapping community structure in complex networks using fuzzy c-means clustering. Phys. A Stat. Mech. Appl. **374**(1), 483–490 (2007)
8. Ren, W., Yan, G., Liao, X., Xiao, L.: Simple probabilistic algorithm for detecting community structure. Phys. Rev. E **79**(3), 036111 (2009)
9. Xie, J., Szymanski, B.K.: Towards linear time overlapping community detection in social networks. In: Tan, P.-N., Chawla, S., Ho, C.K., Bailey, J. (eds.) PAKDD 2012. LNCS, vol. 7302, pp. 25–36. Springer, Heidelberg (2012). https://doi.org/10.1007/978-3-642-30220-6_3
10. Clementi, A., Di Ianni, M., Gambosi, G., Natale, E., Silvestri, R.: Distributed community detection in dynamic graphs. Theor. Comput. Sci. **584**, 19–41 (2015)
11. Galluzzi, V.: Real time distributed community structure detection in dynamic networks. In: 2012 IEEE/ACM International Conference on Advances in Social Networks Analysis and Mining (ASONAM), pp. 1236–1241. IEEE (2012)
12. Prat-Pérez, A., Dominguez-Sal, D., Larriba-Pey, J.L.: High quality, scalable and parallel community detection for large real graphs. In: Proceedings of the 23rd International Conference on World Wide Web, pp. 225–236. ACM (2014)
13. Papadakis, H., Panagiotakis, C., Fragopoulou, P.: Distributed detection of communities in complex networks using synthetic coordinates. J. Stat. Mech. Theory Exp. **2014**(3), P03013 (2014)

14. Riedy, E.J., Meyerhenke, H., Ediger, D., Bader, D.A.: Parallel community detection for massive graphs. In: Wyrzykowski, R., Dongarra, J., Karczewski, K., Waśniewski, J. (eds.) PPAM 2011. LNCS, vol. 7203, pp. 286–296. Springer, Heidelberg (2012). https://doi.org/10.1007/978-3-642-31464-3_29

15. Barabási, A.L., Albert, R.: Emergence of scaling in random networks. Science **286**(5439), 509–512 (1999)

16. Clauset, A., Shalizi, C.R., Newman, M.E.: Power-law distributions in empirical data. SIAM Rev. **51**(4), 661–703 (2009)

17. Watts, D.J., Strogatz, S.H.: Collective dynamics of'small-world'networks. Nature **393**(6684), 440 (1998)

18. Moon, S., Lee, J.G., Kang, M., Choy, M., Lee, J.W.: Parallel community detection on large graphs with MapReduce and GraphChi. Data Knowl. Eng. **104**, 17–31 (2016)

19. Gagnon, P., Caporossi, G., Perron, S.: Parallel community detection methods for sparse complex networks. In: Cherifi, C., Cherifi, H., Karsai, M., Musolesi, M. (eds.) COMPLEX NETWORKS 2017. Studies in Computational Intelligence, vol. 689, pp. 290–301. Springer, Cham (2017). https://doi.org/10.1007/978-3-319-72150-7_24

20. Staudt, C.L., Meyerhenke, H.: Engineering parallel algorithms for community detection in massive networks. IEEE Trans. Parallel Distrib. Syst. **27**(1), 171–184 (2016)

21. Zhang, Y., Wang, J., Wang, Y., Zhou, L.: Parallel community detection on large networks with propinquity dynamics. In: Proceedings of the 15th ACM SIGKDD International Conference on Knowledge Discovery and Data Mining, pp. 997–1006. ACM (2009)

22. Prat-Pérez, A., Dominguez-Sal, D., Brunat, J.M., Larriba-Pey, J.L.: Shaping communities out of triangles. In: Proceedings of the 21st ACM International Conference on Information and Knowledge Management, pp. 1677–1681. ACM (2012)

23. Leung, I.X., Hui, P., Lio, P., Crowcroft, J.: Towards real-time community detection in large networks. Phys. Rev. E **79**(6), 066107 (2009)

24. Jin, S., Yu, P.S., Li, S., Yang, S.: A parallel community structure mining method in big social networks. Math. Probl. Eng. **2015**, 13 pages (2015)

25. Rosvall, M., Bergstrom, C.T.: Maps of random walks on complex networks reveal community structure. Proc. Natl. Acad. Sci. **105**(4), 1118–1123 (2008)

26. Saltz, M., Prat-Pérez, A., Dominguez-Sal, D.: Distributed community detection with the WCC metric. In: Proceedings of the 24th International Conference on World Wide Web, pp. 1095–1100. ACM (2015)

27. Blondel, V.D., Guillaume, J.L., Lambiotte, R., Lefebvre, E.: Fast unfolding of communities in large networks. J. Stat. Mech. Theory Exp. **2008**(10), P10008 (2008)

28. Palsetia, D., Hendrix, W., Lee, S., Agrawal, A., Liao, W., Choudhary, A.: Parallel community detection algorithm using a data partitioning strategy with pairwise subdomain duplication. In: Kunkel, J.M., Balaji, P., Dongarra, J. (eds.) ISC High Performance 2016. LNCS, vol. 9697, pp. 98–115. Springer, Cham (2016). https://doi.org/10.1007/978-3-319-41321-1_6

29. Zardi, H., Romdhane, L.B., et al.: An o (n2) algorithm for detecting communities of unbalanced sizes in large scale social networks. Knowl. Based Syst. **37**, 19–36 (2013)

30. Xie, J., Chen, M., Szymanski, B.K.: LabelRankT: incremental community detection in dynamic networks via label propagation. In: Proceedings of the Workshop on Dynamic Networks Management and Mining, DyNetMM 2013, pp. 25–32. ACM, New York (2013)

31. Xie, J., Szymanski, B.K.: LabelRank: a stabilized label propagation algorithm for community detection in networks. In: 2013 IEEE 2nd Network Science Workshop (NSW), pp. 138–143. IEEE (2013)

32. Cazabet, R., Amblard, F., Hanachi, C.: Detection of overlapping communities in dynamical social networks. In: 2010 IEEE Second International Conference on Social Computing (SocialCom), pp. 309–314. IEEE (2010)

33. Dongen, S.: A cluster algorithm for graphs. Technical report, Amsterdam, The Netherlands (2000)

34. Aston, N., Hertzler, J., Hu, W., et al.: Overlapping community detection in dynamic networks. J. Softw. Eng. Appl. **7**(10), 872 (2014)

35. Xie, J., Szymanski, B.K., Liu, X.: SLPA: uncovering overlapping communities in social networks via a speaker-listener interaction dynamic process. In: 2011 IEEE 11th International Conference on Data Mining Workshops (ICDMW), pp. 344–349. IEEE (2011)

36. Sun, H., et al.: A parallel self-organizing community detection algorithm based on swarm intelligence for large scale complex networks. In: 2017 IEEE 41st Annual Computer Software and Applications Conference (COMPSAC), vol. 1, pp. 806–815. IEEE (2017)

37. Lancichinetti, A., Fortunato, S.: Community detection algorithms: a comparative analysis. Phys. Rev. E **80**(5), 056117 (2009)

38. Dunn, J.C.: Well-separated clusters and optimal fuzzy partitions. J. Cybern. **4**(1), 95–104 (1974)

39. Rousseeuw, P.J.: Silhouettes: a graphical aid to the interpretation and validation of cluster analysis. J. Comput. Appl. Math. **20**, 53–65 (1987)

40. Handl, J., Knowles, J., Kell, D.B.: Computational cluster validation in post-genomic data analysis. Bioinformatics **21**(15), 3201–3212 (2005)

41. Davies, D.L., Bouldin, D.W.: A cluster separation measure. IEEE Trans. Pattern Anal. Mach. Intell. **2**, 224–227 (1979)

42. Shen, H., Cheng, X., Cai, K., Hu, M.B.: Detect overlapping and hierarchical community structure in networks. Phys. A Stat. Mech. Appl. **388**(8), 1706–1712 (2009)

43. Lancichinetti, A., Fortunato, S., Kertész, J.: Detecting the overlapping and hierarchical community structure in complex networks. New J. Phys. **11**(3), 033015 (2009)

44. He, D., Jin, D., Chen, Z., Zhang, W.: Identification of hybrid node and link communities in complex networks. Sci. Rep. **5**, 8638 (2015)

45. Cao, X., Wang, X., Jin, D., Cao, Y., He, D.: Identifying overlapping communities as well as hubs and outliers via nonnegative matrix factorization. Sci. Rep. **3**, 2993 (2013)

Modeling Sparse and Evolving Data

Shivani Batra[1(✉)], Shelly Sachdeva[2], Aayushi Bansal[3],
and Suyash Bansal[4]

[1] Department of Computer Science and Engineering, GD Goenka University,
Gurgaon, India
ms.shivani.batra@gmail.com
[2] Department of Computer Science and Engineering,
National Institute of Technology (NIT), Delhi, India
sachdevashelly1@gmail.com
[3] SDET-1, Quality Engineering, Sumo Logic, Noida, India
aayushi02695@gmail.com
[4] Associate Solution Advisor, Risk and Financial Advisory, Hyderabad, India
suyash.bansal5@gmail.com

Abstract. Existing relational database management system (RDBMS) excels in providing transactional support. However, RDBMS performance declines when sparse and evolving data needs to be stored. Modeling of highly evolving and sparse data is a major issue that needs attention to provide faster and competent technology solutions. This research work is focused on providing a solution to handle sparseness and frequent evolution of data with adherence for the transactional support. Recently, authors propose an extension of binary table approach to overcome the lacking aspects. The proposed approach is termed as Multi Table Entity Attribute Value (MTEAV) model. To make users completely unaware about the underlying modeling approach, MTEAV is augmented with a translation layer. It translates conventional SQL query (as per the relational model) to a new SQL query (as per MTEAV structure) to provide the user friendly environment. In this research, authors extend the functionality of the translation layer to provide support for data definition (creating, reading, updating and deleting schema). Authors have experimented MTEAV for analyzing the effect of sparseness on the performance of MTEAV. Results achieved clearly indicate that the MTEAV performance increases with increase in sparseness.

Keywords: Attribute centric query · Entity centric query · Data models
Frequent evolution · Sparseness · Storage

1 Introduction

Relational model suggested by Codd [1] provides transactional support in terms of ACID (Atomicity, Consistency, Isolation, and Durability) properties. Relational model is adopted very frequently by the user due to its simplicity and availability of query support. Relational model simplicity is a by-product of the fact that schema of the relational table needs to be described beforehand. However, technology shift demands

© Springer Nature Switzerland AG 2018
A. Mondal et al. (Eds.): BDA 2018, LNCS 11297, pp. 204–214, 2018.
https://doi.org/10.1007/978-3-030-04780-1_14

to have a flexible schema to eliminate the need of storing null values and accommodating any future additions to the list of existing attributes. Evolution of schema is an expensive process that demands moving of the existing dataset in the newly defined (encompassing all new attributes evolved) schema. However, schema evolution in the relational model is constrained with an upper limit. For example, upper limit of columns in one table of DB2 is 1012 [2]. In such scenario, extra tables can be constructed and linked to the existing table through references (made by using the primary key and foreign key). Maintaining extra tables require extra effort while retrieving data.

Apart from capturing evolving attributes, relational model exhibits sparseness [3]. Relational model is designed using a fixed schema. Hence, absence of any value is attributed as "NULL" and leads to consumption of storage space. Problem of sparseness become prominent when percentage of null values is much larger than the percentage of non-null values. Healthcare and e-commerce are two domains where sparseness is a challenging issue.

Further, this paper is organized as follows. Existing literature and approaches to store sparse and evolving data are explored in Sect. 2. Section 3 provides an insight of MTEAV. Section 4 provides implementation details of translation layer accommodated over native RDBMS to enhance user friendliness. Section 5 presents the results of various experiments performed in the current research. Section 6 finally concludes the research work.

2   State-of-the-Art

Logical layer of RDBMS is modified to represent data vertically or in the form of binary tables to deal with sparseness and frequent schema evolution. A generic structure is defined that captures only non-null values and can accommodate any future schema evolution without altering the definition (implementation) of the underlying application. Vertical representation of data is commonly known as Entity Attribute Value (EAV) Model [4]. Data representation in EAV is equivalent to space efficient method for storing sparse matrix. It consists of three columns termed as 'Entity', 'Attribute' and 'Value'. 'Entity' and 'Attribute' columns save unique identification. Whereas, 'Value' column reserves the actual data stored corresponding to the desired entity and attribute. In certain situation, unique identifiers stored in 'Entity' and 'Attribute' columns map to other tables (one for the entity and one for attribute) which store details related to the underlying unique identifier [4].

Another well-known logical layer modification of RDBMS suggests segregation of data into multiple binary tables. This technique has been first identified as Decomposed Storage Model (DSM) [5]. DSM divides the n-ary relational table into n-1 tables, each storing combination of the primary key and the underlying attribute. An extension to DSM has been proposed as Dynamic Tables (DT) [6]. DT also (similar to DSM) constructs one unique table corresponding to one attribute (except primary key only). Each table stores pairs of entity identification and corresponding attribute (non-null) value as tuples. In addition, DT constructs a table (not present in DSM) that reserves only the unique identifiers (primary key) of the entity to validate the existence of any entity before scanning all attribute tables.

EAV and DT structures are assumed to be generic in terms of capturing any future schema amendments without changing underlying definitions. However, generic nature of EAV and DT is achieved at a certain cost. EAV compromises on search efficiency and DT lacks support for faster entity centric queries. Apart from other limiting aspects, a major objection in adaptability of EAV and DT is due to unavailability of dedicated database management system (DBMS) to create, manage and enquire underlying database.

3 Modeling Sparse and Evolving Data

Multi Table Entity Attribute Value (MTEAV) [7] is proposed as an extension of DT to provide an improvement over limiting aspects (entity centric query, and query support) of DT. Among existing structures (EAV and DT), DT is chosen for extension because its performance is observed better in comparison to EAV [8]. MTEAV is a logical level modification of existing relational model. A sample dataset designed according to MTEAV is exhibited in Fig. 1.

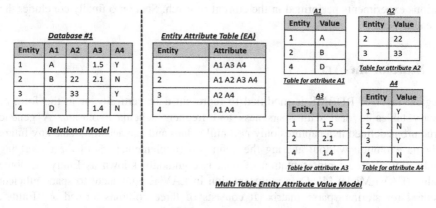

Fig. 1. Sample MTEAV database (proposed in [7])

MTEAV segregates the relational table into multiple tables, each corresponding to one attribute of the relational table. For instance, left side of Fig. 1 presents a sample relational table constituting five columns (Entity, A1, A2, A3 and A4). Entity column is considered as the primary key. Equivalent MTEAV representation is shown on the right side of Fig. 1. Count of tables in MTEAV representation is equivalent to the number of columns in the relational table (i.e., five in the case depicted in Fig. 1). Segregation of relational table in MTEAV resolves sparseness issue. Each attribute table in MTEAV stores only non-null entries corresponding to the underlying entity. Presence of 'Entity' column aids in identifying the entity for which data is stored.

In addition to all attribute tables, a special table EA is defined. EA plays a vital role in optimization of queries that investigate data related to one specific entity. EA table encompasses two columns termed as 'Entity' and 'Attribute'. 'Entity' column stores the unique entity identifier, i.e. primary key and 'Attribute' column defines the list of attribute names (separated by a single whitespace) that correspond to non-null entries for the underlying entity. Utility of EA table is of three-fold. Firstly, it enhances the speed of extracting entity specific data by specifying the attribute tables containing (non-null) data related to the underlying entity. Secondly, only scanning EA table can verify the presence or absence of an entity data. Also, EA table aids in performing lossless and reduced cost JOIN operation. MTEAV is considered as 'generic' and can accommodate any future schema amendments without making any changes at the application level. Addition of any new knowledge (attribute) can be incorporated by simply inserting a new table at the database level. In contrast, addition of new attributes in relational model demands creation of a new schema and moving old data in newly defined schema. No such data movement is required in MTEAV. As MTEAV is a logical modification of the relational model. The structure of a standard query (such as, SQL) changes for MTEAV. This change in structure makes user uncomfortable, since users are habitual of adopting standard query structure. To amplify user friendliness, MTEAV is complemented with a translation layer. It enables the user to pose a data manipulation query using the standard structure of SQL. The query posed by the user (standard SQL structure) is converted into an equivalent MTEAV query. In our earlier efforts, translation layer provides functionality of data manipulation [7]. Current research aims to make a translation layer fully functional along with the capabilities of data definition. Further subsections explain how MTEAV can (1) handle sparseness, (2) build translation layer, and (3) deal with complex queries.

3.1   Incapacitating Sparseness

In MTEAV one attribute corresponds to one table. Each attribute table of MTEAV stores only non-null attributes and eliminates the need for storing null values. Information regarding non-null and null values can be attained using EA table. EA table acts as a communication medium between various attribute tables. To make user unaware about the approach (MTEAV) sparseness is handled, query translation should be made invisible. Thus, MTEAV is incorporated with a translation layer (for data manipulation operations) that helps in abstracting the communication between attribute tables (of MTEAV) from the user.

3.2   Building Translation Layer

MTEAV is built over existing RDBMS and utilizes SQL for query formulation. Modifications done at the logical level (to build MTEAV structure) demands restructuring standard SQL query structure. For instance, a query formulated for relational table is as:

SELECT Entity, A1, A2, A3 FROM RELATIONAL_TABLE

should be restructured for MTEAV as

SELECT EA.Entity, A1.Value, A2.Value, A3.Value FROM EA
  LEFT OUTER JOIN A1 ON EA.Entity = A1.Entity
  LEFT OUTER JOIN A2 ON EA.Entity = A2.Entity
  LEFT OUTER JOIN A3 ON EA.Entity = A3.Entity

3.3   Handling Complex Queries

Restructuring of queries (for MTEAV) involve JOIN operation. Each attribute table participating in MTEAV stores only non-null entries. However, when JOIN is performed on these attribute tables; it may cause loss of information. Availability of EA table aids in retrieving lossless JOIN. EA table stores list of all entities and the corresponding list of attributes containing non-null value for the underlying entity. Performing left outer join of the attribute table with EA table (constraining equality of the entity) will provide details about null entries also. This results in to a lossless JOIN operation. Involvement of JOIN makes data access costlier. However, examining various query scenarios (as listed below), JOIN is acceptable in MTEAV.

Extracting One Attribute. When an attribute needs to be extracted, specific table of MTEAV can be retrieved instantly. In contrast, in the relational model we need to perform a complete table scan to extract specific column information when relational table follows row-oriented approach at the physical level. However, if relational table follows a column oriented approach at physical level, data of a specific attribute is retrieved much faster in comparison to the relational table stored as per row-oriented approach. MTEAV performance is also improved in case of column-oriented RDBMS. Section 6 presents experimental comparison of MTEAV over row-oriented and column-oriented approach.

Extracting One Row. EA table is utilized to identify list of attributes containing non-null entries for the underlying entity. Scanning of data for underlying entity is performed only in the attribute table specified by the list of attributes (retrieved from EA table). Since, only one row is involved, JOIN cost is negligible. Moreover, if UNION is performed on various resultant data entries rather than JOIN, result is visually more suitable [4].

Extracting Multiple Attributes. In the real scenario, multiple attributes are extracted for performing some analytical task. Results retrieved after analytics can be used to build an efficient decision support. However, analytics is not required to be performed in real time. Thus, delay due to JOIN is acceptable [4].

In addition, query is also optimized in MTEAV at two layers as stated below:

- Structure of MTEAV is designed by considering efficient data retrieval through segregation of relational table in multiple tables and maintaining EA table.

- RDBMS employs its query optimization techniques to improve overall query performance. Since, we are using standard SQL (although restructured through the translation layer), RDBMS will use inbuilt query optimization techniques to execute the query in the best suitable manner.

4 MTEAV: Extended Functionality

Primary motive of MTEAV is to eliminate sparseness and provide support for schema evolution. Data Manipulation Language (DML) specifies the operations on data stored, while Data Definition Language (DDL) defines the schema related operations. In translation layer of MTEAV, DML commands such as Projection, Selection, and Insertion are facilitated to handle sparse data. To achieve the motive of schema evolution, MTEAV is extended with DDL functionality. Extending translation layer with DDL support enables flexibility of evolving schema without having a priori knowledge about the non-null distribution. In essence, MTEAV will handle sparseness and complications of query and updates.

4.1 Need of Extension

Primary aim of MTEAV is to abstract logical modifications done on RDBMS and facilitate the user with a traditional query environment based on standard SQL structure. DML commands enable the user with intended abstraction. However, administrator of database must be aware about the underlying logical modifications (i.e., adoption of MTEAV structure) and learn the restructuring required in standard SQL structure. Current research extends translation layer of MTEAV with the functionality of DDL operations.

4.2 Abstracting Modeling Details

The translation layer translates the standard SQL query (hereafter termed as relational query in this research) to a restructured MTEAV query (hereafter termed as MTEAV query in this research). Based on the relational query provided by end user, it makes a distinction between different types of operations (by extracting first keyword of relational query). After identification of the type of operation (such as, Create, Alter, Drop, and Truncate), appropriate conversion module is executed. Various procedures used in algorithms are:

Reg_Match: It considers the generalized regular expression of underlying calling function. It scans the relational query for any matches. Matched objects are returned as an array in output. Many DBMS provide an inbuilt function for Reg_Match functionality.

RemoveDup: Remove duplicate values from an array.

Create: It generates multiple create statements (for MTEAV model) corresponding to a single create statement of the horizontal model. It calls the Reg_Match procedure which parses the original statement and returns the values in an array A[], to get the table name in A[1] and the attribute list in A[2]. List of attributes in EA table entry is extracted in an array B. Attribute list is identified as sub-strings separated by a delimiter (i.e., 'comma') in A[2]. Similarly, presence of attribute and its corresponding data type is identified through the presence of delimiter, (i.e., 'space'). This information is stored in a two dimensional array termed as C[][]. For each indexed value of array C[][], a sub query is constructed for creating an attribute table (of MTEAV). In addition, sub-query for constructing an EA table is also formulated. This will give a total of 'n' create statements, where 'n' (including primary key) is the attributes in the relational create query. All the sub-queries are concatenated using semicolon (;) and returned as output.

Alter_Add: For the new column to be added, it generates a create statement for the corresponding attribute table. Using Reg_Match, table name is collected in A[1], column name is collected in A[2], and data type and constraint are collected in A[3]. A new query is thus formulated to construct a new attribute table.

Function *Create* (rel_c)
{A[ ] <= Call *Reg_Match* (reg_ex, rel_c)
T <= A[1]
B = φ
foreach occurrence of delimiter in A[2]
    store the string in B
C[ ][ ] = φ
foreach element x of B
    C <= break on the first occurrence of space
con_c<= string for EA creation
remove the first index of C
foreach row y in C
    temp<= string for creation of attribute table
    y[0] using y[1] as datatype and constraints
con_c<= Concatenate con_q& temp
return con_c }

Function *Alter_Add* (rel_c)
{A[ ] <= Call Reg_Match (reg_ex, rel_c)
T <= A[1]
N <= A[2]
C <= A[3]
con_c<= Create table statement for table N with constraints C
return con_c }

Function *Trun_Table* (rel_c)
{A[ ] <= *Reg_Match* (reg_ex, rel_c)
T <= A[1]
A <= Get all attributes in the EA table
B <= *RemoveDup* (EA table)
C[ ] = φ
foreach element x of B
 C <= Truncate table statement for x
con_c<= Implode the array to sting
return con_c }

Function *Alter_Modify* (rel_c)
{A[ ] <= *Reg_Match*(reg_ex, rel_c)
T <= A[1]
N <= A[2]
C <= A[3]
con_c<= Alter table statement for table N with constraints C
return con_c }

Function *Drop_Table* (rel_c)
{A[ ] <= *Reg_Match* (reg_ex, rel_c)
T <= A[1]
A <= Get all attributes in the EA table
B[ ] <= *RemoveDup* (EA table)
C [ ]=φ
foreach element x of B
    C <= Drop table statement for x
con_c<= Implode the array to sting
return con_c }

Function *Insert* (rel_c)
{A[ ] <= *Reg_Match* (reg_ex, rel_c)
Attrs<= A[1]
Values <= A[2]
FAttr<= First element of Attrs array
FValue<= First element of Values array
B[ ] = Prepare EA insert statement using FValue and Values
C[ ] = φ
foreach element x of Attrs and y of Values
    C <= Prepare insert statement for every table x
        having value y
con_c<= Implode the array to sting and add B
return con_c }

Alter_Modify: For the column to be modified, it generates an alter statement for the corresponding attribute table. It works similar to the Alter_Add function. An Alter Modify statement (of standard SQL) is constructed to make the desired modification in 'Value' column of attribute table (specified as the attribute to be modified by relational query).

Drop_Table: It generates multiple drop statements for 'EA' table and all the attribute tables. Distinct attribute information is collected through RemoveDup function from 'Attribute' column of EA table. Relational DROP statement is then converted into multiple DROP statements, each corresponding to one of the attribute table participating in MTEAV structure.

Trun_Table: It generates multiple truncate statements for 'EA' table and all the attribute tables. It works similar to the Drop_Table function. Distinct attribute information is collected through RemoveDup function from 'Attribute' column of EA table. Relational TRUNCATE statement is then converted into multiple TRUNCATE statements, each corresponding to one of the attribute table participating in MTEAV structure.

Insert: It generates multiple insert statements for each attribute table. Corresponding to each data value in Insert relation query, an MTEAV insert sub-query is constructed. Each sub-query corresponds to the insertion of a pair comprising of primary key, and value into the desire attribute table.

4.3   Flexibility to Evolve Schema

Incorporation of DDL in MTEAV facilitates future evolution of schema without any intervention of DBA. MTEAV segregates underlying database and do not pose any burden on the schema designer. It frees the user completely from schema designing and evolution overhead. Each attribute table in MTEAV has the same schema (Entity and Value) and require no normalization process or knowledge of the relative sparseness. User can write an ALTER TABLE ADD query (standard SQL) without knowing the structure of the database (MTEAV or Relational). Translation layer will convert the query into CREATE TABLE QUERY. So, user is unaware how data evolution is handled at backend. Also, MTEAV is unaffected by the upper limit on the number of columns a table can have in RDBMS. Since each table in MTEAV corresponds to same two column structure (i.e., Entity and Value). EA table handles relationships. Hence, any changes in the schema will not lead to a change in application definition (coding).

5   Experiments and Results

The processor of the machine on which experiments are performed is Intel® Core™ i5-3230 M (2.6 GHz). RAM available in the machine is 8 GB DDR3 (1600 MHz) and hard disk is of Toshiba 500 GB SATA HDD (5400 rpm). Authors have chosen Apache 2.4.25, MariaDB 10.1.21, and PHP 7.0.15 for maintaining our database and executing queries on it. The operating system used is Windows 10 (Version 1607).

5.1   MTEAV Versus Existing Models

MTEAV performance is tested under various database operations such as, projection, selection, insertion, modification, deletion, statistical, and grouping operations. It has been observed that MTEAV excels under all database operations [7]. Table 1 presents a comparative analysis of MTEAV, EAV and DT.

Table 1. Comparative analysis of various logical modifications of RDBMS

Challenge	EAV	DT	Solution in MTEAV
Storing NULL values	No	No	MTEAV stores only non-null values
Accommodation of evolving features	Yes	Yes	MTEAV can accommodate any evolution in schema without altering underlying application definition (implementation code)
Able to store heterogeneous data	No	Yes	MTEAV can store any type of data supported by underlying RDBMS
Faster extraction of attribute data	No	Yes	Accessing attribute is the easiest and fastest task in MTEAV. Performance is equivalent to DT
Faster extraction of tuple data	No	No	MTEAV access tuple data faster than all other
Coding/Decoding of data	No	No	No coding/decoding mechanism is incorporated in MTEAV
Availability of query support	No	Partial	Translation layer is defined to support data manipulation operations

5.2   Effect of Sparseness

MTEAV is designed to handle sparseness. Performance of MTEAV depends upon the amount of sparseness present in the underlying dataset. To record the behaviour of MTEAV with respect to amount of sparseness, same set of sample queries is executed on the same dataset (having 14000 records) with varying sparseness. To increase the sparseness, desired amount of non-null entries are updated to null.

Table 2. MTEAV behavior with respect to Sparseness

Operation	Attributes▼	Conversion Time (milliseconds)			
	Sparseness➤	40%	65%	75%	85%
Selection	2	60.7	12.8	2.9	2.5
	3	11.1	4	3.2	2.5
	4	18.7	4.3	3.3	3.1
Projection	2	7000	3714	3387	2401
	3	18030	13419	10039	5683
	4	38000	28760	20631	11566
Entity Selection	All	50	26	14	3.2

Complied result of the experiment performed is presented in Table 2 for various operations (such as selection, projection and entity selection). It is observed that MTEAV performance increases with increase in amount of sparseness.

6 Conclusions

Storage formats of current relational systems are inefficient for handling sparse datasets. Moreover, accommodating any future evolution in schema requires redefining of the underlying application. Restriction of pre-defining the schema in the relational model is relaxed to eliminate the need for (1) storing null values attributing towards extra memory consumption, and (2) modifying the application definition to accommodate schema evolution. Alternative logical modifications of relational model such as EAV and DT have been introduced by well-known researchers to overcome the negative side of the relational model. However, none of the proposed approaches (EAV and DT) is observed to be the best under all query scenarios. Absence of query support hinders adoption of EAV and DT. MTEAV proposed (as an extension of DT) by authors overcomes the limiting aspects of existing approaches. Experimental observations portray the effectiveness of MTEAV in comparison to EAV and DT under various database operations (such as projection, projection with the condition, selection, insertion, modification, deletion, statistical and grouping operations). To enable query support, standard SQL is restructured. To make user unaware about the restructuring of standard SQL, a translation layer is implemented with the functionality of data definition and data manipulation operations. Using translation layer, user can pose a query assuming relational model (i.e., standard query structure), that will be automatically translated to an equivalent MTEAV query (i.e., restructured SQL query). MTEAV and proposed query translation approach are worth exploration. However, its success heavily relies on the amount of sparseness present. Current research experimentally analyzed the behavior of MTEAV and relational model with respect to sparseness. It has been observed that with increase in sparseness, MTEAV performance is improving.

References

1. Codd, E.F.: Relational database: a practical foundation for productivity. Commun. ACM **25** (2), 109–117 (1982)
2. Agrawal, R., Somani, A., Xu, Y.: Storage and querying of e-commerce data. In: VLDB, vol. 1, pp. 149–158, September 2001
3. Abadi, D.J., Madden, S.R., Hachem, N.: Column-stores vs. row-stores: how different are they really? In: Proceedings of the 2008 ACM SIGMOD International Conference on Management of Data, pp. 967–980. ACM, June 2008
4. Dinu, V., Nadkarni, P.: Guidelines for the effective use of entity–attribute–value modeling for biomedical databases. Int. J. Med. Informatics **76**(11), 769–779 (2007)
5. Copeland, G.P., Khoshafian, S.N.: A decomposition storage model. In: ACM SIGMOD Record, vol. 14, No. 4, pp. 268–279. ACM, May 1985

214   S. Batra et al.

6. Corwin, J., Silberschatz, A., Miller, P.L., Marenco, L.: Dynamic tables: an architecture for managing evolving, heterogeneous biomedical data in relational database management systems. J. Am. Med. Inform. Assoc. **14**(1), 86–93 (2007)
7. Batra, S., Sachdeva, S.: Anomaly free search using multi table entity attribute value data model. Int. J. Comput. Sci. Eng. **16**(4), 363–377 (2017)
8. Batra, S., Sachdeva, S.: Suitability of data models for electronic health records database. In: BDA, pp. 14–32, December 2014

Big Data Systems and Frameworks

Polystore Data Management Systems for Managing Scientific Data-sets in Big Data Archives

Rashmi Girirajkumar Patidar, Shashank Shrestha,
and Subhash Bhalla$^{(\boxtimes)}$

Department of Information Systems, University of Aizu, Aizu-Wakamatsu,
Fukushima, Japan
{d8202102, d8201104, bhalla}@u-aizu.ac.jp

Abstract. Large scale scientific data sets are often analyzed for the purpose of supporting workflow and querying. User need to query over different data sources. These systems manage intermediate results. Most prototypes are complex and have an ad hoc design. These require extensive modifications in case of growth of data and change of scale, in terms of data or number of users. New data sources may arise to further complicate the ad hoc design. The polystore data management approach provides 'data independence' for changes in data profile, including addition of cloud data resources. The users are often provided a quasi-relational query language. In many cases, the polystore systems support distinct tasks that are user defined workflow activity, in addition to providing a common view of data resources.

Keywords: Big data analytics · Cloud-based databases · Heterogeneous data
Distributed data · Polystore data management · Query language support
Scalability

1 Introduction

Modern day scientific data archives require the use of multiple data stores and data models within the same application. The database approach is gradually becoming impractical while handling huge amount of data. The time for modeling the data in relations and using only relational database is no longer a possible solution. Recently, many scientific domains are using different data storage techniques to manage large amount of data. With the intensification of data increase, the database community has proposed "Polystores", as a solution. Polystore is a new view of federated databases which manages multiple data models within multiple data stores. It provides a unified single query language over various data models. Since past few years, there have been some research works going on this novel approach. Many educational institutions and big organizations have been working on Polystores. MIT's BigDAWG architecture provides a query facility over various data models for medical dataset MIMIC II [1], [2]. In the same way, UCB's SparkSQL provides SQL like query language with a nested data model where SQL can be performed atop Apache Spark [3, 4]. PolyBase

© Springer Nature Switzerland AG 2018
A. Mondal et al. (Eds.): BDA 2018, LNCS 11297, pp. 217–227, 2018.
https://doi.org/10.1007/978-3-030-04780-1_15

[5] system by Microsoft is a new technology that integrates Microsoft's MPP product, SQL Server Parallel Data Warehouse (PDW). It has the capability of querying over PDW connected with Hadoop from RDBMS using Hadoop Distributed File System (HDFS) module.

Polystores is a new topic. It traces its origin back to the multi-database systems or federated database systems. It is also called multistore systems [22]. Polystores provide integrated access to multiple, heterogeneous data stores as well as well as cloud data stores. The CloudMdsQL Polystore [6] presents all query building blocks as functions with a functional SQL-like query language to query multiple data sources (relational, NoSQL and HDFS) in a cloud. Other applications of Polystores can be represented through mediator/wrapper architecture. Mediator provides a Global View to Applications and users (GVA). Wrappers can encapsulate the details of component DBMS to manage communication with Mediator. Native APIs or any other general APIs can be deployed as a wrapper to migrate data within applications using multiple stores.

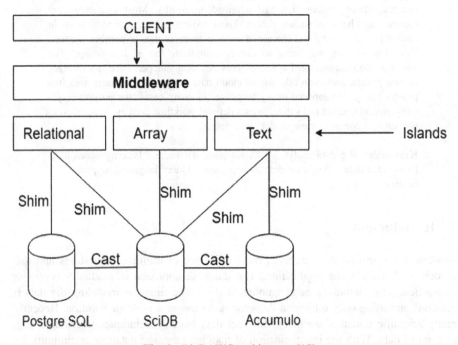

Fig. 1. BigDAWG architecture [15]

In this report, we present a mediator/wrapper architecture to provide a web query system for managing large astronomical data over cloud. The initial query support system presents an Information Requirement Elicitation (IRE) interface [20]. It presents users with a simple query language. The users can select the attributes and then relate the objects selected for the purpose of querying on the system. The prototype system aims to support astronomers to query over cloud data sources of Palomar Transient

Factory (PTF) [8, 11]. PTF is a Astronomical science data repository that contains big data of astronomical observations, covering the northern sky, over a period of 2009 to 2016.

2 Palomar Transient Factory (PTF) Data Repository

The astronomical data is provided by Palomar Transient Factory (PTF). PTF (Palomar Transient Factory) has a collection of telescopes monitoring the Northern Sky for any changes in the astronomical bodies [7]. The study of changes in astronomical bodies with respect to time is termed as "Time-domain astronomy". Any changes in astronomical bodies is recorded and indexed in their database in real-time. PTF deals with two kinds of data processing (real-time and archival). Like many other astronomical data repositories, PTF contains high-resolution images, key-values, relations and unstructured text files.

The archival data for PTF is available publicly through IRSA (NASA/IPAC Infrared Science Archive) cloud database [8]. The IRSA/IPAC has developed an image archive, a high-quality photometry pipeline and a searchable database (relational) of observed astronomical sources. There have been three data releases so far between the years 2009-2016. All those data releases have highly calibrated epochal images and photometric catalogs which have the information of the imagery data.

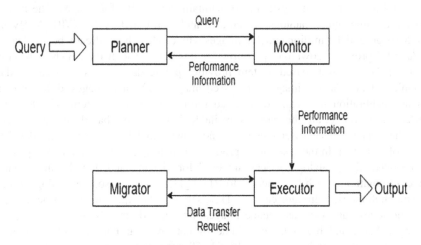

Fig. 2. BigDAWG middleware [15].

2.1 PTF Data Processing Requirements

Captured data stream in real-time, is used for detection of transient astronomical sources. It is also used to identify Near Earth Objects for example asteroids or comets. Infrared Processing and Analysis Center (IPAC) provides image processing and curates a data archive for the PTF data. IPAC also provides science operation, data management, data archives, community support for astronomy and planetary science mission, including numerous missions like IRSA, ISO, Spitzer, WISE, 2MASS and so on. The 48-inch Oschin Schmidt telescope at Palomar Observatory takes high-resolution images of the night sky that represent the raw data. The raw data goes through a detection pipeline [9], which compares the old image with the new one. If there is any change in the images, the images are indexed and stored in the archives or else it is removed. The images that are indexed after the images subtraction are stored in the archives. NASA/IPAC Infrared Science Archive (IRSA) curates the archive data. IRSA has stored all the raw data, processed data and advance data archives with data exploration tools with public release.

Image data products have been released in three different times; each released data have different features. The first data was release (DR1) in 2014 with specific fields of galaxy taken by the PTF collaboration during 2009-2012. DR1 was released in Flexible Image Transport System (FITS) format [10]. DR1 includes epochal (single exposure) image and Photometric catalogs. The PTF second data was released (DR2) on 2015. DR2 covers whole of the northern sky including all the g and R band data. The PTF announced third Data Release (DR3) on 2016 which have a combination of DR1 and DR2 including g-band and R-band data where low-quality data in the galactic plane were omitted. A lightcurve and source database was also included in DR3.

2.2   IRSA Cloud Service and Archive

PTF alongside its real-time system also maintains an archive for images. The images with the accompanying catalogs are curated and distributed by IPAC/IRSA. By the year 2016, around 4.1 million epochal images with catalogs have already been released. IRSA/IPAC provides download platform for this data through their own web based systems [11]. The web based system provides public data access to different data available in PTF. Camera images, processed images, reference images, different catalogs and calibration files can be accessed through their web system. The Program Interface data product are the same as available from the GUI based images services. From the image services user can view and download PTF images from all public release of PTF data. In these services, products include g and R band single exposures and coadded tiles. Catalog services are used for viewing and downloading catalog image information, which have access to PTF lightcurve table, source catalogs, object list, and photometric calibrator catalog. Time series tools require manual processing to plot time series data, view associated images, find period, and phase fold.

The IRSA cloud has a large volume of data with around 80 billion detected astronomical sources with around 4 million FITS image files. There are varieties of data ranging from binaries, relations, texts, graphs and images. The rate of increase in the size of the data is also very high with 4×10^4 alerts per night. Therefore, the archive updates frequently.

2.3   Querying Over Cloud-Data Resources of IRSA

All image metadata, structured and text data have been downloaded into a local database. Due to the large size of image data, the local database only includes the header files which include HTML tags required to map the URL links provided by the remote IRSA cloud for the images. It also includes, the key-values (catalogs/links) and FITS header information available in the IRSA archives. The catalogs include attributes for approximately 4 million FITS images. The PTF catalog data is downloaded from the IRSA archives where the data is stored in Hierarchical File System (HFS). Data is processed by using the Astropy API [12] which is a Python script for formatting astronomical data. The data is reorganized by a process which transforms the catalogue enteries (key-values) into entity and relations. These are to be utilized for querying the cloud data resources.

3 Polystore Data Management System

Polystore data management systems can work on different databases irrespective of the underlying DBMS as they are developed over multiple, heterogeneous and integrated storage engines. Querying over multiple data models uniformly can be done using polystores. Polystores are needed to manage information over various data models easily and efficiently. Thus, polystores are used for large different datasets or data models to provide data management solutions [15].

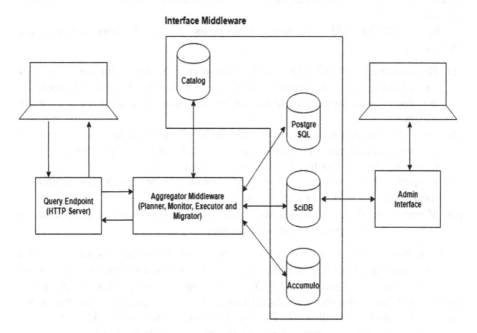

Fig. 3. A user's-eye view of BigDAWG software components.

3.1   Common Architecture of Polystore Database System

A typical polystore system architecture consists of four distinct layers (see Fig. 1) [14].

  (i)   Database and Storage Engines
 (ii)   Islands
(iii)   Middleware and API
(iv)   Applications

BigDAWG initial release supported 3 open source databases alongside for help for relational, array and text islands. The three databases are: PostgreSQL (SQL), Apache Accumulo (NoSQL), and SciDB (NewSQL) [15]. Abstraction of data models and query language are provided by Islands in conjunction with a set of candidate database engines. A shim is used to connect databases with islands. The shim does the work of translating queries - island activities to native language of the respective database. The work of receiving a query and passing it to the appropriate island is done by the middleware. To communicate with other systems on the island - a connector has to be written. BigDAWG interface is the middleware that underpins a typical application programming interface to various databases by the help of islands. Processing should take place in databases which are suitable to the features of data - this is the primary function of polystore database. Queries will produce results best suited for a particular database. Thus, BigDAWG needs capability to move data between databases and these are done by Casts.

The BigDAWG middleware has 4 components: Planner, Monitor, Migrator and Executor [16].

(1)   Planner - the query planning module - this module parses the incoming query of collection of objects and creates possible query plan tree sets which highlight the possible engines for each collection of objects. Then these trees are sent to the Monitor.
(2)   Monitor - the performance monitoring module - this module uses the existing performance information to determine a tree with the best engine for every collection of objects. The tree is then sent to the executor.
(3)   Executor - the query execution module - this module decides the best method of combine the collection of objects and then executes the query.
(4)   Migrator - the data migration module - if required, this module is used by executor to move objects between islands and engines.

These components are shown in Fig. 2. Query Endpoints interacts with users at a very basic level, accepts queries, routes them to aggregation middleware and responds back with results. This shown in Fig. 3. The catalog is a PostgreSQL engine which contains metadata about other engines, islands, datasets and connectors - all of these are managed by interface middleware. The initial release depends on the Docker to simplify the installation and startup experience, the interface middleware can also run on a server and connect to the existing database engines [15].

3.2   Need for Polystore Databases

Polystore is needed to provide data management solutions to different datasets which have different underlying data and programming models. Polystores support multiple query languages and different DBMSs [16]. Polystores are designed to unify querying over multiple data models. So with the polystore systems we can access multiple storage engines through a single interface. It is an approach to integrate data in complex systems [16, 17].

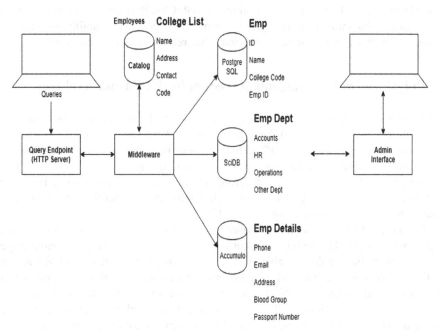

Fig. 4. Example of query in Example 1.

3.3   Query Using Polystore Data Management Systems

Existing polystore data management systems can be classified into 3 groups as- (i) Loosely coupled, (ii) Tightly coupled, and (iii) Hybrids systems [18, 19]. CloudMdsQL is a query language which is functional SQL-like language. It can be used to query and integrate data from various heterogeneous data stores within a single query. It satisfies all legacy requirements needed for a common query language such as schema independence, support of nested queries over data stores, optimization, etc. CloudMdsQL query can utilize the full power of local data stores by allowing local data store queries - this is the major innovation of CloudMdsQL. It uses bind join which is an efficient method to perform semi-joins over various heterogeneous data stores which use subquery rewriting to push join conditions.

Example 1. (refer ro Fig. 4 above) [19],

```
empdept(
filter(
dept(
emp_details{ select min (salary) from (select distinct salary
from emp order by salary desc) where rownum <=2;)))
```

CloudMdsQL query engine has a distributed architecture so that communication can take place between the nodes by exchanging query plans and data. This distributed architecture is different from the traditional one where mediator/wrapper are centralized. Hence, this architecture can be optimized well. CloudMdsQL has a relational data model due to its intuitive data representation, wide acceptance, ability to integrate various datasets by using joins, unions and other relational algebra operations.

4 Polystore Data Management System for PTF Data Archives

The proposed system enables users to access various PTF data (relation data, imagery data, light curves, and text data) through a Graphical User Interface (GUI) with image visualizer named JS9 [13]. The PTF data is stored in a local PostgreSQL database. Managing the entire cloud data resources in a single database system, may have low performance and low efficiency. Therefore, the current study proposes a Polystore System to manage complex problem of such heterogeneous data for getting efficient result across multiple storage engines (see Fig. 5).

The Proposed system is a web based information system with workflow based query management. The webpage layout supports dynamic querying and multi-stage querying from the catalogue table. Using the Information Requirement Elicitation (IRE) process, a user interacts with the system to generate the queries. Users can select the objects and assign the object value to execute the query via Graphical User Interface (GUI). Each time a value is assigned and searched, the system generates a catalogue query in SQL. The SQL is saved in the local database and returns the query result to the system. The result is displayed in a tabular format. To refine the result (multi-object search) users have the option to add another object and execute the query. After appending another object, the system generates a subsequent query. At this stage, the system performs join operation using the objects of interest, and returns the query result. At a second stage, to view the image of the selected object from the result table, the image queries (SQLs) are formed by joining the results (of the formulated queries) with cloud data directories (i.e., ProcImages table and the raw Images table) which have header information about the images (stored at the local database). Generated Image query (SQLs) is connected to IRSA web server through the URL links to retrieve the image of selected object. The resulting query answers are displayed and the image data is visualized through a visualizer software JS9 using its APIs.

4.1   Data Integration Processes and Data Independence

Many sources of data need to be integrated to work together. Users preferably wish to have a single view of the data. The goal is to allow users to express his/her queries or workflow needs without a regard for the heterogeneity of underlying data resources. In a polystore system the data is kept at the source. Systems containing data may be different. These may include, relational databases, Object-oriented databases, XML databases, HTML page repositories (e.g., MedlinePlus Encyclopedia), web services APIs and other systems supporting data downloads. The systems pose many semantic heterogeneity challenges. The heterogeneity may occur in the form of differences in schema, data types, language, completeness, aggregation, and taxonomy. Thus, conflicts that arise may have differences at syntactical, structural or semantic levels. The earlier approaches considered creating a federated architecture with wrappers. Alternatively, the Data warehousing approach was considered with copy and store models to bring a uniform interface for access by users.

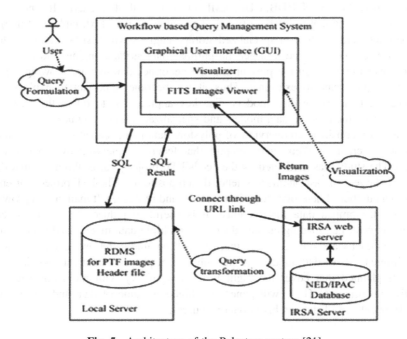

Fig. 5. Architecture of the Polystore system [21].

More recently, a mediation approach has been considered by prototypes for multi-database systems [22, 23]. A generic framework for integration has been evolved that considers various schemas at data sources. These are subjected to schema transformation based on common judgements and rules. In the subsequent processing the correspondence investigation is carried out for creating an integrated schema. Recent research activity on polystore systems is focused on creating a direct link with the underlying data sources through a 3-level data dictionary. It considers supporting the users by proving a query language interface or a partial workflow support. For its purpose a catalogue of resources is prepared that aims to connect with the polystore data resources at the physical level. It provides a conceptual view of the data for each user's workflow or query needs. At the top level it presents the user with a functional interface or a quasi-relational query language [21].

This is in contrast to the former proposals. The database, semantic web, and linked data communities have proposed solutions that federate queries over multiple data sources [24]. These solutions use a single model.

5 Summary and Conclusions

The data retrieval requirements originating from domains, such as- astronomy, healthcare and life sciences are changing the trends. The idea of federating query over a single data model has been popular, primarily due to the simultaneous use of different data models (CSV, JSON, RDB, RDF, XML, etc.) in a real-life scenario. It is no longer practical to assume that the variety (graph, key-value, stream, text, table, and tree) of high volume data residing in specialized storage engines will first be converted to a common data model, stored in a general-purpose data storage engine, and finally be queried over the Web. In the present era, the genomics datasets are growing. These grow from peta-scale to exa-scale. Thus, it has become important to explore data resources in their native data models. The key approach is to query the vast data resources from their native data models and specialized storage engines.

This study considers an overview of polystore data management systems. Its main goals are to create a Web query support that federates querying over native data models. The study was motivated by the needs PTF project that is aiming to link data across the large-scale astronomical science data repositories. The PTF project currently refers to a number of heterogeneous data systems and resources. It maintains a growing pool of astronomical data for reference. It is useful to reduce the expensive data conversion cost, i.e., loading curated data from multiple data models and sources to a centralized data warehouse for querying; and still be able to retrieve complete results from different native data models. In the subsequent phases, larger astronomical projects are planned that cover the night time view of sky with more powerful and sensitive instruments. These will generate a higher volume of data and will require more frequent use of data archived data resources.

References

1. Duggan, J., et al.: The bigdawg polystore system. ACM Sigmod Rec. **44**(2), 11–16 (2015)
2. Saeed, M., et al.: Multiparameter Intelligent Monitoring in Intensive Care II (MIMIC-II): a public-access intensive care unit database. Crit. Care Med. **39**(5), 952 (2011)
3. Armbrust, M., et al.: Spark sql: relational data processing in spark. In: Proceedings of the 2015 ACM SIGMOD International Conference on Management of Data. ACM (2015)
4. Apache Spark. https://spark.apache.org/documentation.html

5. What is PolyBase. https://docs.microsoft.com/en-us/sql/relational-databases/polybase/poly base-guide?view=sql-server-2017
6. Kolev, B., Bondiombouy, C., Valduriez, P., Jiménez-Peris, R., Pau, R., Pereira, J.: The cloudmdsql multistore system. In: Proceedings of the 2016 International Conference on Management of Data. ACM (2016)
7. Law, N.M., et al.: The Palomar Transient Factory: system overview, performance, and first results. Publ. Astron. Soc. Pac. **121**(886), 1395 (2009)
8. Information on IRSA. http://irsa.ipac.caltech.edu/about.html
9. Laher, R.R., et al.: IPAC image processing and data archiving for the Palomar Transient Factory. Publ. Astron. Soc. Pac. **126**(941), 674 (2014)
10. Pence, W.D., et al.: Definition of the flexible image transport system (fits), version 3.0. Astronomy & Astrophysics **524**, A42 (2010)
11. IRSA web based system. http://irsa.ipac.caltech.edu/applications/ptf/
12. Robitaille, T.P., et al.: Astropy: a community Python package for astronomy. Astron. Astrophys. **558**, A33 (2013)
13. Information on JS9 FITS image viewer. https://js9.si.edu/
14. http://istc-bigdata.org/index.php/istc-releases-open-source-code-for-bigdawg-polystore-system/
15. Gadepally, V., et al.: The BigDAWG polystore system and architecture. In: 2016 IEEE High Performance Extreme Computing Conference (HPEC), Waltham, MA, pp. 1–6 (2016)
16. Elmore, A., et al.: A demonstration of the BigDAWG polystore system. Proc. VLDB Endow. **8**. 1908–1911 (2015). https://doi.org/10.14778/2824032.2824098
17. Kolev, B., Valduriez, P., Bondiombouy, C., Jiménez-Peris, R., Pau, R., Pereira, J.: CloudMdsQL: querying heterogeneous cloud data stores with a common language. Distrib. Parallel Databases **34**(4), 463–503 (2016)
18. Kolev, B., et al.: Design and Implementation of the CloudMdsQL Multistore System, 4 July 2016. https://hal-lirmm.ccsd.cnrs.fr/lirmm-01341172/document
19. O'Brien, K.: Polystore Systems for Complex Data Management. HPEC 2017. https://bigdawg.mit.edu/sites/default/files/documents/20170910r3-BigDAWG_Details.pdf
20. Sun, Jun: Information requirement elicitation in mobile commerce. Commun. CM (CACM) **46**(12), 45–47 (2003)
21. Shashank, S., et al.: PDSPTF: polystore database system for scalability and access to PTF time-domain astronomy data archives. In: International Workshop on Polystores and Other Systems for Heterogeneous Data (Poly'2018) co-located with VLDB 2018 (2018)
22. Tamer Özsu, M., Valduriez, P.: Principles of Distributed Database Systems. Springer, 2018-19
23. Valduriez, P., Danforth, S.: Functional SQL, an SQL Upward Compatible. Database Programming Language. Information Sciences (1992)
24. Khan, Y., Zimmermann, A., Jha, A., Rebholz-Schuhmann, D., Sahay, R.: Querying Web Polystores. In: 2017 IEEE International Conference on Big Data (Big Data), December 2017

MPP SQL Query Optimization
with RTCG

K. T. Sridhar[1,2]([envelope]), M. A. Sakkeer[1], Shiju Andrews[1], and Jimson Johnson[1]

[1] XtremeData Technologies, Bangalore, India
sridhar@xtremedata.com
[2] XtremeData, Inc., Schaumburg, USA

Abstract. Analytics database dbX is a cloud agnostic, MPP SQL product with both DSM and NSM stores. One of the techniques for better micro optimization of SQL query processing is runtime code generation and JIT compilation. We propose a RTCG model that is both query aware and hardware conscious extending analytics SQL query processing to a high degree of intra-query parallelism. Our approach to RTCG, at system level targets to maximize benefits from modern hardware, and at use level focuses on typical, industry type SQL, somewhat different from standard benchmarks. We describe the model, highlighting its novel aspects, techniques implemented and product engineering decisions in dbX. To evaluate the efficacy of the RTCG model, we perform experiments on desktop and cloud clusters, with standard and synthetic benchmarks, on data that is more commensurate in size with industry applications.

Keywords: SQL · Query processing · Micro optimization
RTCG · JIT

1 Introduction

Motivated primarily by performance gains, *Runtime Code Generation (RTCG)* has been in use for several decades, across a variety of software products and is referred to by multiple synonyms: *Just-In-Time (JIT)* compilation, *dynamic compilation, specialization*, etc. The technique essentially defers to runtime the translation, of either source code or intermediate representation (IR) such as bytecode, to machine instructions for generating optimal code using auxiliary data, such as program dependent runtime data, programmer annotations and heuristics. Balancing runtime compilation cost to performance gains from optimized code is a trade-off. Interest in RTCG and techniques have evolved with advances in hardware: faster CPUs, more cores, cache, faster/larger memory.

John McCarthy was the first [4] to use RTCG in LISP and was followed by Ken Thompson's publication of an application in regular expression processing. Research prototype System R [3] from IBM was the pioneer in databases to use compilation for query processing. Researchers from several disciplines of

© Springer Nature Switzerland AG 2018
A. Mondal et al. (Eds.): BDA 2018, LNCS 11297, pp. 228–249, 2018.
https://doi.org/10.1007/978-3-030-04780-1_16

computing focused on JIT: programming languages [8,10,18], compilers [13,14], operating systems [24], graphics [23], databases [11], etc. Poor performance of early Java bytecode interpreters gave RTCG research a further impetus. Despite Aycock's observation *"JIT compilation systems are completely unnecessary"* [4], there is continuing interest in study of JIT, and with advent of Big Data more so from the database community [9,15,19–22,26,27,34] and other areas too: FPGA hardware [5] and IoT [25]. The importance of JIT compilation in databases has been emphasized: *"JIT compilation for SQL is a fresh and potentially phase-shifting approach to a core database research topic: query processing"* [34].

In this paper we outline use of RTCG in an industry product, XtremeData's dbX, an MPP analytics database that supports both columnar and row stores, and is cloud agnostic. We propose an RTCG model that is both query aware and hardware conscious, tailoring it for requirements from two ends of query processing: system level dealing with modern hardware, and use level focusing on industry practices for SQL. In implementing the model within dbX, emphasis is on maximizing benefits from modern hardware for better micro optimization of SQL query processing by extending JIT code to a threaded environment, adequately balancing resource contentions. At the same time, we also pay considerable attention to industry practices to avoid ill-effects due to RTCG.

The rest of this paper is organized as follows. Next two sections examine RTCG related work: Sect. 2 in computing domains other than databases, and Sect. 3 on JIT in databases. Section 4 presents a brief overview of MPP product *dbX*. Section 5 proposes our runtime code generation model, and Sect. 6 discusses factors influencing implementation of RTCG in an MPP database product. Section 7 presents experimental evaluation results using both standard and synthetic benchmarks. Section 8 compares and places in context work reported here to related work in databases and also summarizes our contributions.

2   Background

Origins of RTCG may be traced back to programming languages and compilers research that views it as a form of *partial evaluation*: source-to-source transformation of a program with respect to parts of its input leading to a *specialized* version that executes more optimally [7]. An elegant explanation of a partial evaluator, historically called *mix*, from [17]: *"takes the code of a function f and an argument x, and returns optimized code for the result of applying f to x"*.

$$mix(f, x) = f_x \quad \text{where } f : \tau_1 \to (\tau_2 \to \tau_3) \text{ and } x : \tau_1 \tag{1}$$

$$f_x : (\tau_2 \to \tau_3) \quad \text{and } f_x = f(x) \tag{2}$$

RTCG is like *mix*, translating a piece of code into another form using runtime data to generate an equivalent but more optimal code. The specialization performed by RTCG takes advantage of runtime data dependent optimization that has been termed *value-specific data dependent optimization (VSO)* [14]. VSO based RTCG uses static compiler optimization techniques, e.g., constant propagation, dead code elimination, loop unrolling, inlining, etc.

The basic issue of identifying parts of a program to be transformed was addressed either imperatively, where programmer explicitly specifies runtime code, or declaratively with *annotations* using program invariants, identity variables, etc. Example of an imperative style system is $'C$ (Tick-C) [10] that extended ANSI C with programmer directives in special syntax: any C statement preceded by a single quote $'$ may be dynamically generated. Annotations were specified for C programs, with their denotational semantics, proved correct by structural induction [8], and implemented in Tempo using GNU C for runtime transformations. Annotations were translated to *templates* and dynamic runtime code generated from templates filling its *holes*. Functional language ML was extended with annotations in Fabius [18], and DyC [13] targeted C language programs with a more expressive annotation language. Aycock's survey [4] evaluates state-of-the-art of JIT from the point of view of programming languages and proposes a classification of JIT systems:

- **invocation**: termed *implicit* if JIT compilation is done without user knowledge, else *explicit*
- **executability**: every JIT system has two languages, source and target; termed *monoexecutable* if it supports only one language, else *polyexecutable*
- **concurrency**: not considered *concurrent* if rest of program pauses for JIT compilation to complete

Applications of RTCG in other domains of computing include kernel level *read* call optimization of HP Unix [24], digital compositing of images [23], circuit simulation and verification [5] on a softcore FPGA processor, efficiency gains in IoT devices [25] of a Wireless Sensor Network, databases (Sect. 3), etc.

3   Related Work in Databases

Traditionally, DBMS query processing used a tuple-level interpreter on a query plan tree, traversing downwards from root, pulling rows from lower to higher levels for result generation using an *iterator* model with a simple abstract interface: *open*, *next* and *close* calls [12] for query operators. Such an iterator is termed [34] *macro optimization* and is contrasted with *micro optimization* that enhances query performance through RTCG and JIT compilation.

Earliest use of compilation for query processing was in System R, viewed as a *"compiler for data manipulation statements"* [3], albeit in a context different from current day RTCG: for SQL in host languages like COBOL and PL/I, the *embedded* SQL style of DBMS access that is uncommon today. SQL was translated to machine code library calls during host language compilation and, later extended to handle ad-hoc queries from a terminal. Despite Codd terming it *"ingenious"* in his Turing Award lecture [6], RTCG went no further than System R; DB2, a later IBM product, used the iterator model without RTCG [26].

Advances in memory technology and Java VM rekindled interest in JIT to build in-memory database JAMDB [26], SQL virtual machine in Java, with both a conventional iterative SQL interpreter and dynamically generated JIT code in Java for a subset of SQL. JIT version outperformed interpreted version.

Recent database research on SQL RTCG explores micro optimization alternatives, more suited for modern hardware. Dealing with a subset of SQL, HIQUE uses JIT compilation of C code generated from templates for query processing on NSM storage: *holistic query evaluation* [15] with compiler optimization across entire query. Data centric query processing [21], with a push model to stream data across query operators, is coupled with LLVM framework for JIT compilation in HyPer, an in-memory DBMS that adopts a mixed execution model: pre-compiled C++ for complex part of query processing and LLVM assembler code that is JIT compiled for low compile time and better runtime. For efficient query processing in a uniform language framework, [20] explores LINQ for in-memory DBs with different RTCG strategies in managed runtime of .NET.

Vectorization and JIT compilation have been studied; combining JIT with vectorization is advocated [27] for better performance with case studies on projection, selection filter and hash join probe; performance impact of query selectivity and use of loops in RTCG are investigated in [22]; *Data Blocks* [16] of HyPer use a judicious mix of both JIT and vectorization for column compression.

Microsoft's Hekaton [9], memory optimized OLTP engine of SQL Server, translates stored procedures with imperative constructs in T-SQL into C code, with VC++ runtime compilation. Generated code uses iterator model with the constraint that T-SQL must access only memory resident Hekaton tables. NoSQL product Impala of Cloudera running on Hadoop clusters applies RTCG to a limited set of its Hive like SQL operators and functions [35] using LLVM through front-end Clang. Other products that use RTCG for query processing include Amazon Redshift [2], embedded SQL engine SQLite [28].

The reawakened interest and attention on JIT and RTCG for query processing from DB community, prompted IEEE Data Engineering Bulletin to publish a special issue *When Compilers Meet Database Systems* [33] comprising versions of some of the papers discussed above [9, 20, 21, 35] and others.

4   dbX Architecture

The architecture of analytics database dbX is shared nothing, MPP running on a cluster comprising a head node connected to multiple data nodes through a high speed network. Persistent data goes into a hybrid store comprising (a) modern, compressed [31] column store of DSM pages storing each column separately without any surrogate key overheads and (b) a row store of uncompressed NSM pages. Intra-query parallelism across data nodes is by data distribution through placement options: (a) round robin for near equal distribution (b) hashing by one or more columns and (c) single node placement, typically for small tables. Data in both stores may be range partitioned based on column values.

Query execution does not use the standard tuple-at-a-time, iterator model. Planner generates a sequence of *macroQ* ops; each macroQ op is a set of one or more *microQ* ops scheduled in parallel with block-level dataflow between ops.

As explained in Sect. 5, dbX query execution model is micro optimized with thread-safe RTCG in C and JIT compilation for all or part of microQ ops. Large memory of modern systems may be configured to support an in-memory cache for user table buffers and materialized intermediate data.

dbX implements agile techniques [32] for loading and extract of bulk data with rates in TBs/hr. Transactions are ACID compliant and are implemented with MVCC. Product engineering [29,30] is cloud agnostic with deployments on Amazon AWS, Microsoft Azure, private clouds virtualized through VMware and bare metal systems. On both AWS and Azure, storage may be node attached, or network storage that decouples compute and storage for cost benefits.

5 Code Generation: Model

The non-iterator query execution model of dbX takes advantage of modern CPUs by multi-threading, and intra-query parallelism is realized at two levels: (1) across the cluster with parallel data nodes running query ops on distributed data (2) within a data node, microQ ops run as threads pipelining block-level data. Threaded execution is also extended to communication (socket threads) and IO, implemented over Linux asynchronous (aio) layer with parallel read/write.

dbX query processing gains in performance with VSO based RTCG for several query operators as described in later sections. Runtime thread-safe, C code is generated for all or parts of microQ ops. For brevity of space, we illustrate our RTCG model through the simple SELECT query of Table 1.

Table 1. Query: Simple SELECT

SELECT a2+a3 FROM t1 WHERE a1 > 100 and a3 < 500;

The example query generates three microQ ops: scan ($\Uparrow$), filter (σ) and projection (π); ops σ and π may be combined as a single op. All three ops run in parallel on data nodes, with scan op using Linux async-IO threads for disk reads. Code for evaluating filter condition, project expression and type based data conversion during scan are all candidates for RTCG. Dataflow from scan to filter microQ ops is at block-level, and filter JIT code is a tight loop running on an aligned array of data values with pointer arithmetic to move across rows of block.

As microQ ops are run in parallel, JIT code generated must be thread-safe and use query dependent VSO based RTCG. Thread safety is essential, even if there is a single microQ op for a query, as multiple threads of the same op running on different data may be scheduled concurrently.

In this example, filtering and projection are through JIT code, while JIT code for scan microQ op is only piecemeal: aio scan threads overlapped with tuple data conversion. Runtime code is not monolithic in the sense that it contains the entire implementation of SQL in C. Such monolithic code does not suit our threaded, pipelined query execution model. Instead, candidates for RTCG are carefully chosen from repetitive tasks that can exploit VSO and can concurrently execute in a threaded world.

We follow the same approach outlined for query of Table 1 in more complex cases that use joins, grouping, aggregation, windowing, CTEs, bulk load/extract, etc. JIT compilation at query runtime is by GNU GCC C/C++ compiler.

5.1 SQL Opportunities

Table 2 lists different stages of SQL query processing that we take advantage of for runtime code generation in dbX query execution model. First column labeled *QryOp* uses standard relational algebra symbols for several rows; second column labeled *SQL* relates it to SQL keywords or query execution phases, and the last column labeled *Notes* briefly explains the context of use.

Table 2. Query processing: RTCG opportunities in dbX

QryOp	SQL	Notes
σ	SELECT: WHERE	WHERE filter clause for all SQL expression ops
π	SELECT: project	Projection clause for all SQL expression ops
Δ	Partitioning in hash join, group, set ops, subquery, DISTINCT	Hashing in MPP data partitioning
$\bowtie$	Hash join probe: inner and outer joins	Probing and join filter using both tables
$\cap, \cup, -$	UNION, INTERSECT, EXCEPT	Hash table management for set operators
γ	GROUP BY and aggregate functions	Hash table management and evaluating group aggregation functions
γ_w	WINDOW and SQL analytic functions	Window partitioning and evaluating windowing analytic functions
$\in, \notin$	IN, NOT IN	Existence subquery probe; optional row expression
$\subset, \subseteq$	ANY/SOME, ALL with comparison op	Subset subquery probe; optional row expression
$\Uparrow_c, \Uparrow_r$	Table scan	Column type based tuple conversion on data read by aio from DSM and NSM tables
$\mapsto$	Result generation	Type based binary to formatted ASCII conversion
$\Rightarrow$	COPY TO	Bulk data extract to file
$\Leftarrow_{RP}$	COPY FROM file to range partitioned table	Distribute bulk data to child partitions of RP table
λ_{SP}	SQL stored procedures	Imperative loops with SQL; multiple invocations; cache JIT code library to avoid repeated compilations

The runtime code generation opportunities listed in Table 2 may in some cases implement a complete microQ op and in other cases only some steps, or piecemeal parts, of it. A SQL query is composed of multiple microQ ops that cover several rows of Table 2. For example, an aggregate GROUP BY query would use JIT code for opportunities that span across several rows of the table.

- for scanning the column table ($\Uparrow_c$)
- filtering data scanned (σ)
- hash partitioning by GROUP clause (Δ)
- hash table management for grouping (γ)
- aggregation (γ) with any project expression (π) arguments
- finally, type based result generation in text form ($\mapsto$)

5.2   Modern Hardware

The inability of row DBMS products, unlike HPC systems, to take advantage of advances in hardware technology was studied [1] to identify CPU or cache stall issues in SQL query processing. Subsequent work [15, 21, 27, 34] highlighted the unsuitability of tuple-at-a-time, iterator model of query processing to take advantage of modern hardware, with pre-fetching of data and instructions into a hierarchy of CPU caches, and proposed alternatives. Salient aspects of modern hardware that SQL query processing could utilize are summarized below.

- cache hierarchy of CPUs to exploit data locality and instruction pipelining; prefetching that is not affected by mispredictions due to branches and jumps
- multi-core CPUs that can support intra-query parallelism
- faster and cheaper memory that makes in-memory databases viable

We examine our JIT driven execution model in the context of modern hardware capabilities, and show that the model is hardware conscious.

- Better data cache coherency: microQ ops run on block-level, not tuple, data.
- Generated JIT code as well as pre-compiled parts of executor do not jump across contexts: widely varying processing steps as in the iterator model. Code is small, context based and even piecemeal for a microQ op; also, thread-safe to take advantage of multi-core CPUs.
- RTCG code for loop oriented tasks in small tightly optimized fragments that appear to work on aligned arrays with minimal branching aids instruction prefetching. C pointer arithmetic to move across rows of block to reduce address dereferencing overheads.
- Large memory available on modern systems is used as in-memory cache to materialize temp data.

5.3   Optimization Techniques

Standard compiler optimization techniques used in generated C code of dbX are summarized in Table 3. Despite compilers being able to automatically support some of these optimizations, templates used to generate thread-safe JIT code are written taking these optimizations into account.

Table 3. RTCG code optimization

Technique	Notes
Constant propagation	RTCG function arguments are mostly constants; most pointer arguments cannot be aliased, and rarely updated
Constant folding	Used with scalar subquery expressions
Dead code elimination	JIT code has no unused functions or vars; no null check code if all columns are non-nullable; only VSO based code
Macros	Frequently used in RTCG for repeatedly used pieces of code; e.g.: 3-state AND/OR logic code, error checks, etc.
Inlining/registers	Small functions marked for inlining; within blocks local vars may use C register declarations
Minimize branches	Avoid SWITCH and nested IFs; early exit with GOTO for bool expressions; due to VSO, no data type based branching and contextual annotations for block reordering
Loop unrolling	Constants as loop limits; in some cases, unroll loops

6 Code Generation: Meta-Questions

We address meta-questions on RTCG to understand it better for its implementation in a product, and to facilitate better product engineering.

- are there SQL queries that are not good for RTCG?
- when and where to generate code, particularly for MPP?
- what if SQL is not written by humans, but by tools?
- how can RTCG benefit from compiler technology?
- what alternatives to code generation method adopted?
- what if compilation is interrupted by user or other faults?

6.1 To RTCG or Not?

The cost of runtime compilation is a deterrent for RTCG, and break-even points have been addressed by several researchers in compilers and earliest DBMS user of runtime compilation: System R [3]. With advances in both hardware and compilers, compilation cost has significantly reduced and further improved with micro optimization techniques of Sect. 5. Yet, for cases listed below RTCG is not beneficial and could be detrimental; dbX avoids RTCG for such cases.

Code Size: If size of generated code is large, system attracts multiple cost penalties: compile time, generated library size, its runtime loading cost and execution time. Generating monolithic code for a query, tantamount to VSO optimized interpreter, incurs such overheads. As dbX generates code for individual microQ op or even parts of it, code size is less of an issue in our model than monolithic JIT code; Sect. 7.1(a) experimentally evaluates impact of code size.

Point Queries: JIT code for a filter or project expression may be avoided for point queries that use index scan on row store tables. In other cases, similar to Java JVM heuristics to decide on method compilation, cost and row counts estimated by planner may be used to decide about resorting to RTCG.

Compressed Data: Use of RTCG for decompressing column table data compressed by lightweight methods does not show significant performance gains over pre-compiled code for decompression [31]. Further, cost overheads of JIT compilation with multiple compression methods as in dbX worsens performance for short queries. Consequently, we do not use RTCG for decompression.

Scalar Subquery Expression: A scalar subquery returns a single value and SQL allows such scalar subqueries as part of WHERE/HAVING clauses and project expressions along with other dyadic operators. The first query in Table 4 is an example of a scalar subquery using TPC-H tables. It picks only those groups of a join that have an average value for *s_acctbal* less than its average in table *supplier*. A naive approach of inlining the scalar subquery incurs overhead of multiple subquery executions. Even the iterative query interpreter may avoid multiple evaluations with a state value that is set after first evaluation.

Table 4. Query: scalar subquery expression

```
SELECT n_name, sum(s_acctbal) FROM supplier, nation
   WHERE s_nationkey = n_nationkey GROUP BY n_name
   HAVING avg(s_acctbal) < (SELECT avg(s_acctbal) FROM supplier);
⟹
SELECT n_name, sum(s_acctbal) FROM supplier, nation
   WHERE s_nationkey = n_nationkey GROUP BY n_name
   HAVING avg(s_acctbal) < 4518.25;
```

The query may be transformed as shown in the second query of Table 4, provided result 4518.25 of scalar subquery expression is available. This cannot be done at plan time. At runtime, the scalar subquery is to be first evaluated and its result value used for constant folding in JIT code for HAVING clause. The scalar subquery execution itself may in turn use JIT code for its query: aggregation (γ) in this example. Such a scalar subquery expression is not suited for direct JIT code; it must first be evaluated by its SQL execution, and then, in its use context, resort to RTCG with constant folding.

Big IN Queries: In industry context, queries tend to use IN expressions in WHERE clauses enumerating a large number of constants, sometimes 100s and even 1000s. TPC-DS has query with template number 08 listing about 400 zip codes, possibly, influenced by industrial use of such clauses; this is unlike TPC-H benchmark with nice and simple clauses. An example SQL query, with 50 constants in its IN expression, checking for zip code strings with TPC-DS table *customer_address*, is shown in Table 5 as the first query.

If standard OR expression code is generated for a nullable column, which needs 3-state OR logic, the generated IN expression code is likely to be huge. And execution cost is also very high, particularly for a value that fails the IN condition, and average cost is 50% of number of OR operators in expression. Such IN expressions are not suited for JIT code.

Table 5. Query: Big IN translation

```
SELECT ca_location_type, count(*) FROM customer_address
  WHERE substr(ca_zip,1,5) IN ( '24512', '24613', '22233', '64826', '20611', '84770',
    '39901', '51052', '11122', '13437', '49481', '14673', '71933', '33010', '39549',
    '17725', '66755', '36784', '14356', '13513', '24382', '44248', '39872', '90794',
    '48152', '19480', '16137', '75628', '36853', '38245', '92431', '87145', '87833',
    '37905', '50116', '84831', '58188', '52713', '16098', '28481', '75780', '71536',
    '65511', '32173', '89944', '56235', '81497', '37717', '22191', '30373')
  GROUP BY ca_location_type;
⟹
SELECT ca_location_type, count(*) FROM customer_address, t1_000052EF00010000_50
  WHERE substr(ca_zip,1,5) = c1_000052EF00010000 GROUP BY ca_location_type;
```

Our RTCG model does not generate JIT code for such IN expressions. Instead, planner transforms IN expression into an appropriate inner join query. The enumerated constants on right side of IN are stored in an automatically managed, in-memory temp table. Second query of Table 5 is equivalent SQL for planner generated inner join, with system generated table ($t\_000052EF00010000\_50$) and column ($c1\_000052EF00010000$) names that avoid clashes within a user transaction. Section 7.1(d) experimentally evaluates RTCG for big IN expressions.

When an IN expression contains a small number of items (say, less than 16), the join query overheads may be more than filter evaluation through JIT code. dbX supports a user controllable configuration parameter for the number of enumerated constants to trigger IN transformation to join. The transformation for performance gains is not applicable to NOT IN expressions; such negated disjunctive expressions are translated to conjuncts using $\neq$.

6.2   When and Where to RTCG?

The engineering question of at what point of query processing and where, more so in an MPP framework, should RTCG be performed is important. Clearly, RTCG can be resorted to only after query plan generation, and subsequently there are multiple options that we discuss below.

Incremental: Runtime code may be generated from plan query ops incrementally during execution on a need basis. Such an approach involves multiple compiler invocations, but each pass is on a small piece of code. In a multi-threaded, concurrent system too many compiler invocations, consequent to OS resources taken, may impact both intra-query and inter-query parallelism; it may also affect aio scans as compilers too access persistent storage. The alternative is to generate code upfront after planning, for all query ops that require RTCG. This incurs a small pre-processing overhead of examining all ops of a query to generate runtime code for some of them.

MPP: Head or Node?: In an MPP system, the question of generating code on head or data node of the cluster is important. As plan generated on head node is available to data nodes, RTCG may be done either on head or data node.

Table 6. Query: JIT code differs

```
SELECT nodeid(), r_regionkey FROM region;
```

If code is generated and compiled only on head node, either runtime library must be transferred over the network to all data nodes, or a shared nothing system must resort to sharing through NFS for data nodes to access generated library. Node based JIT code generation involves repetitive work across the data nodes: though JIT code is generated on all nodes in parallel, it must be done once on each data node.

In an MPP system, we can find SQL examples, as shown in Table 6, where JIT code differs across nodes: function *nodeid* is node specific and is evaluated at plan time. For most queries, such node based differences are unlikely; node based code generation handles such special cases trivially; head based is harder. By conducting experiments for all options discussed above, we choose the best option of generating code upfront on all data nodes in parallel. Experimental results in Sect. 7.1(c).

6.3   What if SQL Is Machine Generated?

Industry makes use of GUI tools that generate SQL, and are available from several vendors, for querying analytic databases. Often, machine generated SQL is not optimized for even routine and mundane issues such as code repetition, big INs that come from a GUI list box selection, dead code, etc.

As almost all SQL code used in production systems is machine generated, a newer issue of poor optimization due to repetitive SQL code in a query has to be addressed by backend databases that resort to JIT compilation. Even if the analytics DBMS supports CTEs (common table expressions), the tool may not generate WITH code that avoids code repetition. In an operational context, where database must co-exist with tools, an MPP analytics database promising performance must find alternatives to eliminate repetitive code, if not at a subquery level suitable for CTEs, at least at expression level.

During runtime code generation, we implement *SQL code pattern mining* at query op level to look for *similar* patterns in the generated query op sequence and reuse RTCG code that may already have been generated. A query op is viewed as a function that transforms its input to an output based on data types and expression operators. So long as two, say, add expressions are operating on same two input data types to produce same output data type, we consider them to be similar expressions irrespective of source tables or column names. Our definition of similarity is not purely syntactic, but takes advantage of polymorphic overloaded nature of several SQL expression ops to reduce JIT code size.

Obviously, real world expressions are not as simple as addition expressions and need a more elaborate comparison technique for matching expressions. dbX implements context driven code pattern mining techniques to reduce size of JIT code; the techniques are applicable to almost all RTCG opportunities listed in Table 2. Gains are measured experimentally in Sect. 7.1(f).

6.4   How to Compile?

Optimization levels supported by GNU C/C++ compiler GCC are explored for runtime invocation of compiler with experimental results in Sect. 7.1(b). By default we use GCC's *-O2* as optimization level. Higher levels do not provide any significant gain; poorer performance may be observed with no optimization level. In addition to optimization level, GCC also supports compiler invocation switches; some of these switches are used in JIT invocation of GCC. Only those code generation optimization switches used by JIT invocation of GCC, and not product pre-compilation, are listed below.

- *loops*: loops are unrolled if the number of iterations is known (*-funroll-loops*); also, move out branches with loop invariants out of loop (*-funswitch-loops*)
- *alignment*: bypass GCC's default inlining only for 4-byte aligned destinations (*-minline-all-stringops*); also, bypass GCC's default of aligning inlined string options (*-mno-align-stringops*)

JIT code generated is made to conform to ISO C standard (*-std = c99*). For a binary distributed product, we forgo optimization that may be applicable to target specific hardware.

- *target CPU*: no special tuning for CPU of target machine; code tuning is for generic CPUs (*-mtune = generic*)
- *runtime loadable library*: even if there is additional cost, code is generated for shared *so* libraries (*-fpic*), and loaded at runtime with Linux system calls

Yet another optimization opportunity with GNU compiler is to explore its compiler macro directives, defined using __builtin_expect, that improve machine generated code. This is like specialization through annotations of Tempo [8] or DyC [13]. A few important directives routinely used in generated C code:

unlikely/likely: These are hints to the compiler for optimizing conditional branches through block reordering. An *unlikely* directive tells the compiler that the condition wrapped by the macro is most of the time not true, e.g., error checking conditionals; *likely* means the opposite: almost always true.

inline: Used with small and simple functions to provide a hint to GCC that function body may be used *inline* at place of call avoiding its call cost. Compilers may further optimize inline code in its call context, more so, if arguments to it are constants as is often the case in RTCG code.

__restrict: Directive is used with pointer arguments to a function to indicate that argument points to a distinct memory location, even when there are multiple pointer arguments; during its lifetime within the function block, it cannot be aliased; at most it can be used directly to access its pointed location or with pointer arithmetic. Code generated is more optimized and likely to be smaller in size when there are multiple pointer arguments to the function; all pointer addresses passed in RTCG code conform to this behavior.

6.5   Why Not libJIT or LLVM?

Generating native C code directly is perhaps harder than using tools like libJIT or LLVM. But we have opted to generate native C code compiled directly by GCC, instead of tools. Tools are of great value to anyone starting on RTCG today, but not so for one with investment in the hard part before tools became popular. Also, industry tends to be conservative in basing its product on nascent tools; dbX development is contemporaneous to rise in popularity of tools.

Better performance, weighed with integration effort, could be a motivation to switch path to tools. To address this question, we evaluate in Sect. 7.1(e) both libJIT and LLVM tools against dbX native code generated; experiments show that our approach to JIT by C code generation outperforms the tools.

6.6   Who Takes the Kill?

An important implementation issue to consider is runtime compiler invocation from DBMS code, more so in a multi-threaded, parallel, MPP system; simplest way is to use Linux *system* call. If on a long drawn query, user issues a Ĉ (SIGINT) DBMS must trap it, perform exception handling and gracefully end the query and become ready for next query [30]; such a signal may arrive during runtime compilation. GNU compiler traps Ĉ, handles it and returns back to *system*, which may not deliver it to DBMS, and database misses a signal meant for it. Even if compilation failure is detected, a wrong message may be reported, not user interrupt; worse, it may continue further and report other errors.

User interrupt through Ĉ is only an illustration of the problem: MPP database may miss other signals, e.g., network partitioning in cluster, fatal SIGSEGV on another node, simpler error from peer node like missing compiler installation, etc. The issue lies in exception handling during compilation. As standard tools like GNU compiler and Linux are beyond our scope, dbX must implement robust exception handling. dbX never invokes the compiler directly using Linux *system* call; instead, it spawns a child process that in turn uses the *system* call and reports signals to its parent, dbX. All errors including compiler faults can now be handled as exceptions by the database.

7   Experimental Evaluation

Results of experiments to evaluate efficacy of proposed RTCG model, and validity of our product engineering decisions towards achieving optimal query performance are given. Two sets of queries are used in the experiments: (1) *microbenchmark* with synthetic queries, some modeled on production environments, or borrowed pieces from TPC benchmarks and (2) *macrobenchmark* with TPC-H and TPC-DS queries at specified scale factors. All experiments are with dbX software; platforms used for micro and macro benchmarks are:

- desktop system: Intel Xeon E3 1225, 3.3 GHz, 4 cores, 16 GB RAM, 1 TB disk; Linux Cent OS 7
- cloud: MPP clusters on AWS and Azure of multiple nodes running Linux

In order to run experiments outlined below comparing RTCG and non-RTCG versions of dbX, some modifications to dbX code are necessary, as standard product does not support non-RTCG query processing; necessary modifications were made to dbX for SQL query processing without use of RTCG. Benchmark queries were selected to ensure that the only difference between non-RTCG version and standard product was in use/non-use of RTCG. All runtime values were averaged over three runs and obtained with DBMS instrumentation using Linux system calls; runs were on "cold" data without a user table cache, in a single client environment: no inter-query parallelism.

7.1 Microbenchmarks

Queries and data used for microbenchmarking vary across experiments with details in relevant subsections; platform is desktop system, unless stated otherwise.

(a) Compile Time Cost & Size: TPC benchmarks target query execution complexity with small and compact queries; hence compile time and library size are not significant (average TPC-H: 294 mSec, 37.1 KB; TPC-DS: 417 mSec, 61.8 KB). To illustrate RTCG cost issue, which affects query processing time, we use a synthetic benchmark evolved by us, modeled on industry practice, with star schema: long, repetitive SQL with union, join, grouping, big INs, etc.

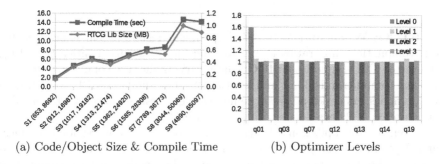

(a) Code/Object Size & Compile Time   (b) Optimizer Levels

Fig. 1. RTCG: Compilation

Figure 1(a) shows JIT code compile time and runtime library size for synthetic queries. X axis labels give number of lines in SQL query (no blank lines or dead code; Linux *wc* value) and generated JIT C code (logical statement count from USC's *UCC*). Both time in sec and code size in MB are significant for synthetic SQL queries in Fig. 1(a).

Note that neither time taken for compilation (left Y axis) nor size of runtime library (right Y axis) are dependent on just length of SQL or generated C code; S3 to S4: with increase in size both values decrease; S6 to S7: only size decreases; S8 to S9: time dips slightly, but size decrease is steeper. The non-linear behavior with code size is due to code dependent compiler optimizations. Examining JIT C code for queries S8 and S9, we observe that S8 has a number of small functions suitable for inlining and GCC increases object size.

(b) Compiler Options: Effect on query performance, for selected TPC-H queries on SF = 10 (size 10 GB), varying compiler optimization levels for JIT compilation is shown as ratios in Fig. 1(b). No optimization (level 0, -O0) performs worst; chosen default level -O2 has best average performance, though variations are insignificant across levels 1 to 3 of GCC, even for level 0 excluding q01.

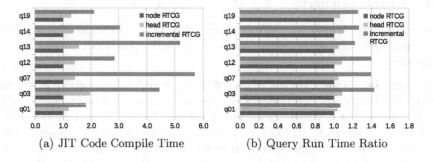

(a) JIT Code Compile Time   (b) Query Run Time Ratio

Fig. 2. RTCG: When and where?

(c) RTCG: When and Where?: Experiment results as bar charts for JIT code compilation time ratios (Fig. 2(a)) and query execution ratios (Fig. 2(b)) for selected TPC-H queries on 10 GB data, varying the stage of query processing at which compilation is done, and location of RTCG node in MPP system: head or data nodes. As desktop is a single node system, test is run on Azure for MPP configuration of 2 data nodes with premium IO.

Incremental compilation, always on data nodes, is the worst for both measures. Upfront JIT compilation is better on data nodes than head. Bigger difference could be due to part-work being done by data nodes for head compilation (Sect. 6.2); efforts required for full head RTCG are expensive. Runtime of head compiled and node executed RTCG is slightly poorer due to shared nothing MPP becoming shared system through NFS for library access. dbX uses node RTCG.

(d) non-RTCG SQL: Figure 3(a) shows query performance ratios, with expression RTCG and equivalent join rewrite without expression RTCG, for SQL queries using big IN with many enumerated constants. TPC-DS query 08 (*dsq*08) with 400 constants and four synthetic grouping queries are used. Synthetic queries vary constant data type (string or number) and constant count (50 or 1000), and are appropriately named in graph.

All queries use 10 GB TPC-DS data; synthetic queries on *customer_address*. The benefit from not generating JIT code for such queries is clearly highlighted by the graph for both synthetic and TPC-DS queries. RTCG code for 1000 string constants is 43x worse off compared to join transformation.

(e) Using Tools for JIT: We run tests, outside dbX, to validate use of GCC with direct C code generation instead of using tools like libJIT or LLVM. A stand-alone program, restricted to simplest case of RTCG in SQL: WHERE clause filter expression was developed. Table 7 lists WHERE clause of three synthetic queries, *F*1, *F*2 and *F*3, on TPC-H *lineitem* used for the experiment.

Table 7. Tools for JIT: SQL WHERE clauses

$\Longrightarrow$ F1: int32 (date), double
WHERE l_shipdate $\geq$ date '1-jan-1996' AND l_quantity < 25
AND l_shipdate < (date '1-jan-1996' + interval '1 year')
AND l_discount BETWEEN .07 - 0.01 AND .07 + 0.01;
$\Longrightarrow$ F2: int64, char, char *(varchar)
WHERE l_suppkey > 2000 AND l_linestatus = '0' AND l_shipinstruct = 'NONE';
$\Longrightarrow$ F3: int64, char, int32:strlen(char *)
WHERE l_suppkey > 2000 AND l_linestatus = '0' AND length(l_shipinstruct) = 4;

With a *select count(*) from lineitem* prefix on all three WHERE clauses, JIT code for three queries was generated from dbX along with binary dump of data scanned for queries from a single node (29,984,970 rows). Use of JIT code from dbX maps SQL data types to C data types: 4 or 8 bytes int, double or strings.

The stand-alone program to read dumped binary data and measure runtime values, was extended with JIT C code from dbX for filters. API calls of libJIT and LLVM were added to it using program logic of JIT C code from dbX. All three methods, dbX JIT code, libJIT and LLVM worked within same framework, used same data and compiler options to ensure that there was no bias. libJIT provides two options to call JIT-ed function: APIs using returned pointer of JIT-ed function or a generic call; function pointer method works faster and was used; LLVM also uses a function pointer. Clang of LLVM was not used.

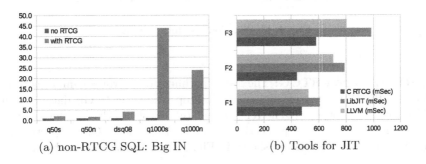

(a) non-RTCG SQL: Big IN   (b) Tools for JIT

Fig. 3. Choosing RTCG & Tools

Results of filter runtime values excluding IO cost, averaged across five runs, are given in Fig. 3(b) in milliseconds along X axis; dbX generated JIT C code performs the best, followed by LLVM and worst of all three is libJIT.

(f) SQL Code Pattern Mining: Experiment shows gains from code pattern mining of Sect. 6.3 for nine synthetic queries of Sect. 7.1(a). For JIT code, we measure compile time (Fig. 4(a) in sec), library size (Fig. 4(b) in MB) and total number of functions (Fig. 4(c)), with and without code pattern mining. For all three measurements, RTCG with pattern code mining is way better.

With additional instrumentation in dbX code, we measure reuse frequency of code pattern mined JIT functions for synthetic query S7 (2,789, 36,733), and depict frequencies as a pie chart (Fig. 4(d)). There is one JIT function used 194 times, another at 99 times, with frequency dropping quickly to 1; just once used are 374; *"used"* is from static analysis, not runtime invocations. Clearly, code pattern mining helps enormously in RTCG and query execution for repetitive, machine generated SQL.

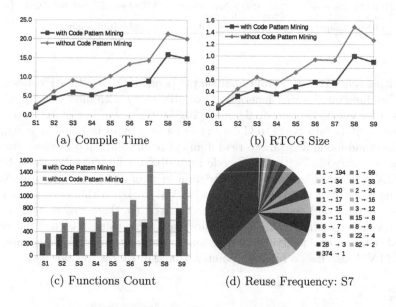

(a) Compile Time

(b) RTCG Size

(c) Functions Count

(d) Reuse Frequency: S7

Fig. 4. RTCG: Code pattern mining

(g) RTCG vs non-RTCG: The last experiment of microbenchmark is to run selected TPC-H queries on single node desktop system on 10 GB data: (a) with RTCG for all microQ ops (b) no RTCG at all (c) disable RTCG only for filters (σ) (d) disable RTCG only for projection (π) and (e) disable RTCG for partitioning (Δ) and join ($\bowtie$). Results are shown in Fig. 5 as performance ratio bars.

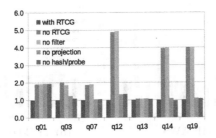

Fig. 5. RTCG: Query processing opportunities

As seen in Fig. 5, it is advantageous to utilize all RTCG opportunities identified in Table 2. Among the opportunities, most beneficial case for JIT code is expression processing (both σ and π) as most queries use them. Three queries with no RTCG, or disabling only σ, show gains from 4x to 5x; all three do inner join on biggest table *lineitem*; RTCG gains improve with more complex expression operators used in SQL query: CASE ($q12$, $q14$), OR with ANDs ($q19$).

The impact of generating JIT code for join ($\bowtie$) and partitioning (Δ) in an MPP system depends on the query: greater impact in $q12$ (1.4x: inner $\bowtie$, γ). Disabling JIT for partitioning (Δ) also affects grouping (γ) queries, e.g., $q01$ (2x, γ with aggregation functions), which has grouping without join.

7.2   Macrobenchmarks: TPC-H & TPC-DS

The macrobenchmark uses standard TPC benchmark queries on cloud platforms with larger sizes. Figure 6 shows results of running queries on TPC scale factor of 100 (100 GB), on a 4 data nodes AWS cluster of *i3.8xlarge* using EBS storage. Selected TPC-H (Fig. 6(a)) and TPC-DS (Fig. 6(b)) queries are run on AWS cluster varying RTCG generation as done in Sect. 7.1(g). Unlike TPC-H, two TPC-DS queries, $q43$ and $q99$, are more sensitive to disabling projection RTCG.

Results of 100 GB runs on Azure 4 nodes cluster of *Standard_E32s_v3* nodes with premium IO are shown for TPC-H in Fig. 6(c) and TPC-DS in Fig. 6(d). Bar charts are similar to AWS, but gains are lower in Azure; probably, due to differences in cloud environments.

Results of scaling up TPC benchmark queries to SF = 1000, data size 1 TB, on same AWS cluster are shown in Fig. 7(a) for TPC-H and Fig. 7(b) for TPC-DS. As data size is now ten-fold, we may expect higher gains from RTCG optimization; gains from RTCG are more in TPC-DS compared to TPC-H indicating that nature of the query and data also have a bearing in data scale-up.

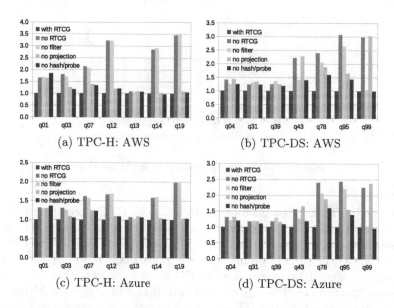

(a) TPC-H: AWS   (b) TPC-DS: AWS

(c) TPC-H: Azure   (d) TPC-DS: Azure

Fig. 6. RTCG: Macrobenchmarks (SF = 100)

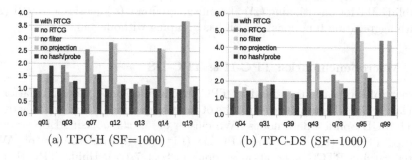

(a) TPC-H (SF=1000)   (b) TPC-DS (SF=1000)

Fig. 7. RTCG: Macrobenchmark 1 TB on AWS

8   Discussion and Conclusion

As per ontology of RTCG systems by Aycock [4], our system may be classified as: (a) *implicit invocation* as SQL user does not initiate JIT (b) *polyexecutable executability* as dbX RTCG deals with C and SQL (c) *concurrent*, even if compilation must complete DB server is not stalled, exceptions are handled, etc.

All database use of RTCG discussed in Sect. 3 employ VSO techniques but are primarily *manual code generation* in compiler JIT terminology: either C, C++, C#, or assembly templates with or without a tool. dbX code generation technique uses C templates that are hardcoded in dbX source code; again manual, but we also use annotation techniques for GCC directives (Sect. 6.4). None of the DBMSs use special purpose compiler extended for annotations; even libJIT and LLVM tools are used through API calls.

With reference to related work in databases, our approach to micro optimization for RTCG is not *holistic* in the sense of [15]. We do not generate monolithic code and it would not suit our execution model. Even though all our query ops work on block-level data, we do not use specialized vectorization techniques or resort to optimizing for SIMD as done in [27].

Our model for SQL query processing and RTCG are closest to work reported in [21]. Independently, we too have adopted an approach similar to the elegant metaphoric analogy of *cogwheels* and *chain* for JIT code given in [21]. However, granularity of code tradeoff between pre-compiled and RTCG is likely to differ. Unlike HyPer, despite using memory caches, dbX does not target OLTP applications; also, dbX is an MPP product with threaded JIT code for parallelism.

We generate JIT C code and others have reported [20] that C code performs best. Experiments of Sect. 7.1(e) show that dbX generated C code outperforms API based libJIT or LLVM. We did not use LLVM assembler, which may do better. Unlike academic projects, industry is unlikely to use assembly due to maintenance and portability issues: LLVM Clang, not assembler, is used in [35].

Microsoft's Hekaton [9] is an add-on to SQL Server and its JIT system targets T-SQL code only for in-memory applications, not general purpose SQL. Unlike dbX RTCG, Hekaton code generation appears to use standard tuple-level iterator model. dbX too supports imperative SQL stored procedures: extended version of *PL/pgSQL*. Though PL/pgSQL is interpreted, SQL in stored procedure is JIT compiled, binary code is cached and loops reuse JIT code without recompilation.

Web forum (stackoverflow.com) discussions on first run cost and product documentation [2] indicate that Redshift uses JIT for iterative, monolithic code, different from our model. Caching JIT code across sessions benefits re-runs; dbX does not cache JIT code across sessions, but caches JIT code of PL/pgSQL procedures. RTCG work reported in [35] on Impala is for SQL-like constructs on NoSQL, and may not be same as standard SQL.

We have outlined a query aware and hardware conscious model for RTCG in SQL analytics with an industry product. Our approach is tailored for requirements from two ends of query processing: modern hardware and industry SQL. To the best of our knowledge, we have not found any industry product that utilizes RTCG for micro optimization on modern hardware addressing industry SQL practice as done in dbX. Results from both micro and macro benchmarks of Sect. 7 demonstrate performance gains from our model.

Our contributions may be summarized as (1) developing query aware and hardware conscious RTCG for micro optimization in a cloud agnostic, industry product (2) identifying optimal opportunities for RTCG in SQL query processing, splitting tasks into simple, threaded JIT code for high degree of intra-query parallelism on multi-core CPUs balancing resource contention issues (3) addressing RTCG specific issues in an MPP system (4) tackling issues in industry-type SQL, not seen in standard benchmarks, including optimized RTCG by code pattern mining on machine generated SQL (5) experiments on data sizes commensurate to production environments.

Acknowledgment. We thank several people; at Bangalore: Pramod Sahu for testing JIT modules and SQL code; Dipanjan Deb and Prajeesh for operational cloud support; at Schaumburg: Jim Benbow for dbX cloud deployment scripts.

References

1. Ailamaki, A.A., Dewitt, D.J., Hill, M.D., Wood, D.A.: DBMS on a modern processor: where does time go? In: Proceedings of 25th VLDB, pp. 266–277 (1999)
2. Amazon: Redshift (2017). http://docs.aws.amazon.com/redshift/latest/dg/c-query-performance.html
3. Astrahan, M.M., et al.: System R: a relational data base management system. Computer **12**, 42–48 (1979)
4. Aycock, S.: A brief history of Just-In-Time. Comput. Surv. **35**, 97–113 (2003)
5. Becker, A., Sirowy, S., Vahid, F.: Just-In-Time compilation for FPGA processor cores. In: ESLsyn Conference, pp. 1–6 (2011)
6. Codd, E.F.: Relational databases: a practical foundation for productivity, Turing award lecture. Commun. ACM **25**, 109–117 (1982)
7. Consel, C., Danvy, O.: Tutorial notes on partial evaluation. In: 20th POPL, pp. 493–501. ACM (1993)
8. Consel, C., Noel, F.: A general approach for Run-Time Specialization and its application to C. In: 23rd POPL, pp. 145–156. ACM (1996)
9. Diaconu, C., et al.: Hekaton: SQL Server's memory optimized OLTP engine. In: SIGMOD 2013, pp. 1243–1254. ACM (2013)
10. Engler, D.R., Hsieh, W.C., Kaashoek, M.F.: 'C: a language for high-level, efficient and machine-independent dynamic code generation. In: 23rd POPL, pp. 131–144. ACM (1996)
11. Freytag, J.C., Goodman, N.: Translating aggregate queries into iterative programs. In: Proceedings of 12th VLDB, pp. 25–28 (1986)
12. Graeffe, G.: Query evaluation techniques for large databases. Comput. Surv. **25**, 73–170 (1993)
13. Grant, B., et al.: DyC: an expression annotation-directed dynamic compiler for C. Theor. Comput. Sci. **248**(1–2), 147–199 (2000)
14. Keppel, D., Eggers, S.J., Henry, R.R.: Evaluating runtime-compiled value specific optimizations. Technical report 93-11-02 (1993)
15. Krikellas, K., Viglas, S.D., Cintra, M.: Generating code for holistic query evaluation. In: Proceedings of 26th ICDE, pp. 613–624. IEEE (2010)
16. Lang, H., et al.: Data blocks: hybrid OLTP and OLAP on compressed storage using both vectorization and compilation. In: SIGMOD, pp. 311–326. ACM (2016)
17. Leone, M., Lee, P.: A declarative approach to run-time code generation. In: Proceedings of WCSSS, vol. 73, p. 10 (1996)
18. Leone, M., Lee, P.: Optimizing ML with run-time code generation. SIGPLAN Not. **31**, 137–148 (1996)
19. Murray, D.G., Isard, M., Yu, Y.: Steno: automatic optimization of declarative queries. SIGPLAN Not. **46**(6), 121–131 (2011)
20. Nagel, F., Bierman, G., Viglas, S.D.: Code generation for efficient query processing in managed runtimes. In: Proceedings of 40th VLDB, vol. 7, pp. 1095–1106 (2014)
21. Neumann, T.: Efficiently compiling efficient query plans for modern hardware. In: Proceedings of 37th VLDB, vol. 4, pp. 539–550 (2011)
22. Pantela, S., Idreos, S.: One loop does not fit all. In: Proceedings of SIGMOD 2015, pp. 2073–2074. ACM (2015)

23. Pike, R., Locanthi, B., Reiser, J.: Hardware/Software trade-offs for bitmap graphics on the BLIT. Softw. Pract. Exp. **15**, 131–151 (1985)
24. Pu, C., et al.: Optimistic incremental specialization: streamlining a commercial Operating System. In: Proceedings of SIGOPS, vol. 29, pp. 314–321. ACM (1995)
25. Queva, C., Courousse, D., Charles, H.: Self-optimisation using runtime-code generation for wireless sensor networks. In: Proceedings of ICDN, p. 6 (2016)
26. Rao, J., Pirahesh, J., Mohan, C., Lohman, G.: Compiled query execution engine using JVM. In: Proceedings of 22nd ICDE, pp. 23–23. IEEE (2006)
27. Sompolski, T., Zukowski, M., Boncz, P.: Vectorization vs. compilation in query execution. In: Proceedings of 7th DaMon, pp. 33–40 (2011)
28. SQLite: The SQLite Bytecode Engine (2017). https://www.sqlite.org/opcode.html
29. Sridhar, K.T.: Modern column stores for big data processing. In: Reddy, P.K., Sureka, A., Chakravarthy, S., Bhalla, S. (eds.) BDA 2017. LNCS, vol. 10721, pp. 113–125. Springer, Cham (2017). https://doi.org/10.1007/978-3-319-72413-3_8
30. Sridhar, K.T.: Reliability techniques for MPP SQL database product engineering. In: 2nd International Conference on System Reliability (ICSRS), pp. 180–185. IEEE (2017)
31. Sridhar, K.T., Johnson, J.: Entropy aware adaptive compression for SQL column stores. In: Kozielski, S., Mrozek, D., Kasprowski, P., Małysiak-Mrozek, B., Kostrzewa, D. (eds.) BDAS 2018. CCIS, vol. 928, pp. 90–104. Springer, Cham (2018). https://doi.org/10.1007/978-3-319-99987-6_7
32. Sridhar, K.T., Sakkeer, M.A.: Optimizing database load and extract for big data era. In: Bhowmick, S.S., Dyreson, C.E., Jensen, C.S., Lee, M.L., Muliantara, A., Thalheim, B. (eds.) DASFAA 2014. LNCS, vol. 8422, pp. 503–512. Springer, Cham (2014). https://doi.org/10.1007/978-3-319-05813-9_34
33. Sudarshan, S. (ed.): Special Issue on When Compilers Meet Database Systems, IEEE Data Engineering Bulletin, vol. 37. IEEE (2014). http://sites.computer.org/debull/A14mar/issue1.htm
34. Viglas, S.D.: Just-in-time compilation for SQL query processing. In: Proceedings of 39th VLDB, vol. 6, p. 2 (2013)
35. Wanderman-Milne, S., Li, N.: Runtime code generation in Cloudera Impala. IEEE Data Eng. Bull. **37**(1), 31–37 (2014)

Big Data Analytics Framework for Spatial Data

Purnima Shah[✉] and Sanjay Chaudhary

School of Engineering and Applied Science, Ahmedabad University,
Ahmedabad, India
purnima.shah@iet.ahduni.edu.in,
sanjay.chaudhary@ahduni.edu.in

Abstract. In the world of mobile and Internet, large volume of data is generated with spatial components. Modern users demand fast, scalable and cost-
effective solutions to perform relevant analytics on massively distributed data
including spatial data. Traditional spatial data management systems are
becoming less efficient to meet the current users demand due to poor scalability,
limited computational power and storage. The potential approach is to develop
data intensive spatial applications on parallel distributed architectures deployed
on commodity clusters. The paper presents an open-source big data analytics
framework to load, store, process and perform ad-hoc query processing on
spatial and non-spatial data at scale. The system is built on top of Spark
framework with a new input data source NoSQL database i.e. Cassandra. It is
implemented by performing analytics operations like filtration, aggregation,
exact match, proximity and K nearest neighbor search. It also provides an
application architecture to accelerate ad-hoc query processing by diverting user
queries to the suitable framework either Cassandra or Spark via a common web
based REST interface. The framework is evaluated by analyzing the performance of the system in terms of latency against variable size of data.

Keywords: Big data analytics · NoSQL database · Geospatial data
Spatial analytics

1 Introduction

Large amount of data is generated and collected from multiple data sources such as
scientific applications, GPS enabled devices, observations from satellites, real-world
applications, government agencies, etc. by using advanced data acquisition and collection tools and technologies. These massively distributed data are equipped with
spatial attributes and characteristics. Exploiting such spatial characteristics in data
analytics can provide more interesting and readable solutions to the end user. Spatial
data are complex and multi-dimensional. Compared to non-spatial data, spatial data are
characterized by special data types, index structures, operations and functions. It
requires specializing systems and architectures to manage spatial data.

The hundreds of millions of users demand fast and efficient analytics on massively
distributed data. The design and implementation of low-latency query processing on

© Springer Nature Switzerland AG 2018
A. Mondal et al. (Eds.): BDA 2018, LNCS 11297, pp. 250–265, 2018.
https://doi.org/10.1007/978-3-030-04780-1_17

exponential growth of geospatial data has made geospatial analytics more challenging and complex for traditional systems. The state-of-the-art big data storage and processing frameworks such as, NoSQL databases, Hadoop[1] and Spark [1] are highly scalable, high available, fault tolerant and efficient to handle massive scale geospatial data. However, these frameworks provide very limited geo-functionality and OGC (Open Geospatial Consortium) [2] standard methods compared to traditional systems. Hence, it is highly enviable to exploit such frameworks to store and process geospatial data at scale.

1.1 Novelty and Contributions

To the best of our knowledge, there is no open source big data analytics framework available that provides end-to-end solutions for spatial data i.e. from data loading to data retrieval through Spark and Cassandra integration. The innovative idea is, we are able to store and process spatial data without changing the underlying architecture of either Spark or Cassandra. The framework is not only able to perform the described analytics operations but also offers an infrastructure to perform more sophisticated and complex analytics on spatial data. It can be extended to enhance the analytical capability.

The main contributions of the paper are:

1. To present an open-source big data analytics framework to load, store, process and perform ad-hoc query processing on spatial and non-spatial data at scale.
2. To realize the proposed system by developing big data analytics pipeline on top of open-source big data architectures; Spark and Cassandra.
3. To build and implement Cassandra based spatial data storage framework.
4. To implement analytics operations like selection, filtration, aggregation, exact match, proximity and KNN search.
5. To implement an application architecture to accelerate ad-hoc query processing by diverting user queries to the suitable framework either Cassandra or Spark via a common web based REST interface.
6. To evaluate the performance of the system in terms of latency against variable size of data.
7. To compare the performance of the system with the baseline technology i.e. Cassandra for low-latency queries.

The paper is organized as follows. Section 2 presents the systems available for spatial data management, Sect. 3 discusses about the seamless integration of big data frameworks; Spark and Cassandra, Sect. 4 explains the proposed framework, and Sect. 5 presents the experimental results.

---

[1] http://hadoop.apache.org/.

2 Related Work

2.1 Traditional Databases for Spatial Data

The traditional spatial database management systems are the extension to the relational database systems [14]. They manage spatial data with new data types, operators and index structures. They perform spatial operations based on OGC standards and methods on geometry instances. Though such systems perform well while dealing with fairly large spatial datasets, they are less efficient to store and process massive scale geospatial data because of poor scalability. Spatial databases implemented on single node suffer from limited computational power, storage and single point failure. Although Parallel DBMS solutions may increase the storage and computation capability, they are very expensive due to advance hardware configurations and proprietary license.

2.2 NoSQL Databases for Spatial Data

Big data storage frameworks like NoSQL databases are schema-free, horizontally scalable, massively distributed, fault tolerant, fast and highly available. They are capable to execute large scale operations on big datasets with multiple insertions/retrievals. Some NoSQL databases include support for geospatial data either natively or with an extension. The contemporary NoSQL databases like, MongoDB [3] and Cassandra [4] are able to manage geospatial data with limited functionalities. In contrast to relational databases, NoSQL databases store and model data based on data duplication and de-normalization. They offer limited support for advance analytics operations like joins, aggregation and group_by. The comparison analysis of state-of-the-art spatial databases is shown in Table 1.

Table 1. State-of-the-art databases for spatial data

Database	Supported geometry objects	Supported geometry functions	Spatial index	Horizontal scalable
PostGIS	Point, LineString, Polygon, MultiPoint, MultiPolygon, MultiLineString, geometry collection	OGC standard methods on geometry instances	B-tree, R-tree, GiST	No
MySQL	Point, LineString, Polygon, MultiPoint, MultiPolygon, MultiLineString, geometry collection	OGC standard methods on geometry instances	2d plane index, B-trees	No
MongoDB	Point, LineString, Polygon, MultiPoint, MultiPolygon, MultiLineString, geometry collection	Inclusion, intersection, distance/proximity	2dsphere index, 2d index	Yes
Cassandra	Point, Polygon, LineString	Distance/proximity intersects, iswithin, isdisjointto	Solr/lucene	Yes

2.2.1 Cassandra for Spatial Data

Cassandra is an open-source and column oriented NoSQL database. It is a fast, write durable and high available distributed database. It has master-less architecture with no single point failure. It is the best suitable database to store structured datasets (also, vector based spatial data). It outperforms HDFS (Hadoop Distributed File System) in many aspects. HDFS suffers from single point failure because of its hierarchical master-salve architecture. Cassandra is considered as more reliable and fault tolerant compared to HDFS because of its peer-to-peer architecture. It allows varying consistency level at each read and write operations. It has specializing indexing techniques like secondary index to do fast retrieval from massive scale data. It provides convenient SQL like interface called CQL. It has achieved excellent read and write latency compared to HDFS. Though both Cassandra and HDFS are designed for big data storage, they need to integrate with big data computational frameworks (Spark/Hadoop) to perform advance analytics operations like join, aggregation and group_by.

Cassandra is considered as the best choice for developers to store and process massively distributed data. It has no native support for spatial data. Cassandra-solr-spark provides limited spatial support (no joins) with no performance evaluation. Ben Brahim et al. [5] has demonstrated the spatial data extension for Cassandra with extended CQL support. The framework has used numeric Geohash index. The index attribute has been modeled as clustering key because of CQL restriction on partition key for range queries. This may degrade the performance of the database during data retrieval. There is no documentation available to use this framework in GIS applications. The literature found that Cassandra has limited support for spatial data compared to other NoSQL databases.

2.3 Big Data Computational Frameworks for Spatial Data

Hadoop and Spark are very commonly used frameworks for big data processing. Though these frameworks don't offer native support for spatial data, the extension systems built on top of Hadoop and Spark have been emerged to manage storage and processing of spatial data at scale. As an exception, SpatialHadoop [15] is a MapReduce based framework and provides a native support for spatial data. The disk-based (SpatialHadoop and HadoopGIS [7]) and in-memory based (GeoSpark [8], Magellan [9] and SpatialSpark [10]) spatial frameworks have been developed as an extension on top of Hadoop and Spark respectively. A list of computational frameworks that support spatial data is depicted in Table 2.

2.3.1 Spark for Spatial Data

Spark is an in-memory computational framework. Spark is fast compared to Hadoop framework. It provides libraries for interactive analysis, machine learning and graph processing. Resilient Distributed Dataset (RDD) is the main abstraction provided by Spark. Data analytics applications are developed by performing transformation and action operations on generic RDD. Spark does not have native support for indexing on generic RDD. The extension system is required to integrate indices on generic RDD. Spark is the best suitable architecture to achieve fast and efficient big data processing compared to Hadoop. It has not its own storage system. It can perform data processing

Table 2. Big Data Computational Frameworks for spatial data

Framework	Index support	Input format support	Spatial geometry operators supported	Spatial geometry objects supported	Language support/interface
Proposed framework	Geohash	Cassandra	Circle range query, KNN query, point query, attribute query	Point	R [11]/REST
Disk based systems					
SpatialHadoop	Grid/R-tree/R + tree	HDFS compatible input formats	Spatial analysis and aggregation functions, joins, filter, box range query, KNN, distance join (via spatial join)	Point, LineString, Polygon	Pigeon [12]
Hadoop-GIS	Uniform grid index	HDFS compatible input formats	Range query, self-join, join, containment, aggregation	Point, LineString, Polygon	HIVEQL
Memory based systems					
Geospark	R-trees, Quad-trees	CSV, GeoJSON, shape files and WKT	Box range query, circle range query, kNN, distance join	Point, Polygon, Rectangle	Scala, Java
Magellan	z-order curve (default precision 30)	ESRI, GeoJSON, OSM-XML and WKT	Intersects, contains, within	Point, LineString, Polygon, MultiPoint, MultiPolygon	Scala
SparkSpatial	Grid/Kd-tree	Form of WKT in Hadoop File System (HDFS)	Box range query, circle range query, KNN, distance join (point-to-polygon dist), point-in-polygon	Point, Polygon	Impala

on datasets accessible from external data sources like HDFS, Cassandra, MongoDB, S3, etc. It is very important to select appropriate data source for spatial analytics. Spark has no native support for spatial data. The existing Spark based extension frameworks are limited to perform spatial operations on datasets available in text based file formats stored in HDFS or local disk.

2.4 Shortcomings of the Existing Systems for Big Spatial Data

Traditional spatial database management systems are less efficient to store and process massive scale geospatial data. The set of geospatial operations supported by the state-of-the-art big data storage and processing frameworks (NoSQL/Hadoop/Spark based) are very limited compared to the standard GIS products, such as ArcGIS. The existing extensions to the NoSQL databases for spatial data management lack the support for spatial aggregation and join operations. Spark has no native support for spatial data. The existing Spark/Hadoop based frameworks are only able to execute spatial operations on datasets that are available in text based file formats (CSV/GeoJSON/shape files and WKT) and stored in HDFS or local disk. In comparison with the existing frameworks, the proposed framework is able to provide efficient and scalable solutions for spatial data via Spark and Cassandra integration.

3 Integration of Big Data Frameworks – Spark and Cassandra

In modern application development, only one big data tool would not be able to store and process massive scale data efficiently and effectively. Hence, it is highly desirable to propose frameworks and architectures built on top of more than one technology to develop robust and powerful applications. Each big data framework has different architectural design, data distribution and partitioning techniques, indexing techniques, performance tuning parameters, etc. Hence, the seamless integration of big data architectures and tools poses lots of challenges for the developers. It requires efficient integration of technologies and expertise to avoid the unnecessary complexities in the system.

The integration of NoSQL datastore (Cassandra) with big data computations framework (Spark) enables efficient storage and fast analytics on large volume of spatial data. It also enables fast and efficient ad-hoc query processing on spatial and non-spatial datasets.

3.1 Spark-Cassandra Connector

Spark does not offer built in support for Cassandra. Beyond the advantages of Cassandra, it provides seamless integration with Spark using Spark-Cassandra connector [13]. The Spark-Cassandra connector is a client program developed by DataStax and implemented in Scala language. It allows Spark context to access keyspaces exist in Cassandra database. It imports the Cassandra table into Spark memory in the form of Spark object; RDD or dataframe. It also offers convenient APIs to store Spark objects into Cassandra partitions.

3.1.1 Architecture Implementation
Cassandra and Spark partitioning data based on their own partitioning methods. Spark reads data from many different data sources and each data source follows different

approach to partition the data. The number of Spark partitions and their size are defined based on external data source from which data is being accessed.

Spark-Cassandra connector performs data distribution differently than HDFS. Cassandra partitions are different than Spark partitions. Normally, Cassandra partitions are more than Spark partitions. There is a one-to-many mapping between Spark and Cassandra partitions. It is very important to create sufficient number of Spark partitions aligned to Cassandra partitions to increase parallelism and improve the performance of the Spark. Fortunately, Spark-Cassandra connector loads Cassandra data into Spark memory and automatically defines Spark partitions aligned to Cassandra partitions. The number of Spark partitions and their size are estimated by reading the metadata provided by Cassandra cluster. Figure 1 shows how connector distributes data and aligned Cassandra partitions with Spark partitions.

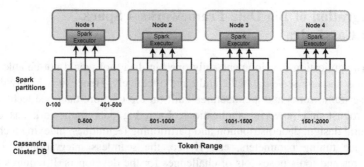

Fig. 1. Data distribution by Spark-Cassandra connector

The connector is widely used to perform three operations, (i) Read Cassandra partitions into Spark memory, (ii) Perform analytics and (iii) Store analytical results back into Cassandra. The write operation from Spark to Cassandra database is challenging. It is very easy to store results back into Cassandra after performing one-to-one transformation (i.e. map, filter, etc.) with the same partition key. Whereas, it is very challenging if Spark performs wide transformation on data (i.e. shuffle) and then store the results back into Cassandra with a different partition key.

4 Proposed Framework

A Spark based big data analytics framework is proposed to load, store, process and query spatial and non-spatial data by exploiting the features provided by standard storage and processing frameworks like Cassandra and Spark. The proposed framework is shown in Fig. 2. It is divided into four layers; (i) Spatial data storage layer, (ii) Spark core layer, (iii) Spatial data processing layer and (iv) Application layer.

4.1   Spatial Data Storage Layer

Massive scale data including spatial data are available at heterogeneous distributed data sources. These data are in different formats with lots of inconsistencies. To prepare such data for analytics is a very challenging and tedious task for developers. The data preparation services can be implemented to fetch consistent and clean data from the disparate sources. However, implementation of data preparation services is not the scope of this paper.

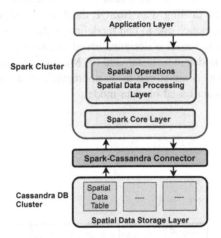

Fig. 2. Big data analytics framework for spatial data

The implementation of a Cassandra based spatial data storage framework is divided into five phases. The implementation architecture is depicted in Fig. 3.

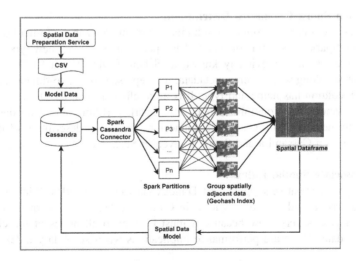

Fig. 3. Implementation architecture of spatial data storage framework

4.1.1   Data Loading

Raw data are fetched from data preparation services in CSV format. The fastest way to load data into Cassandra is bulk loading. Raw datasets including spatial components are loaded into Cassandra using bulk loading. Bulk loading is performed in two steps, (i) Generate SSTables from raw dataset, and (ii) Load SSTables into Cassandra using sstableloader utility.

4.1.2   Data Storage

The data are stored based on Cassandra data model. Data modeling is considered as a very import aspect while dealing with NoSQL database. Most of the applications fail and degrade their performance due to bad modeling of data. Cassandra suggests query based data model. Its data model is designed based on two rules, (i) Distribute the data evenly amongst all the nodes in a cluster, and (ii) Minimum partitions to be read during data retrieval. Raw datasets are stored using appropriate data model. Figure 4 depicts how Cassandra stores data based on data model.

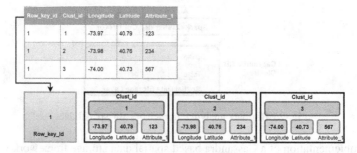

Fig. 4. Cassandra data storage

4.1.3   Align Spark-Cassandra Distribution

Raw datasets are imported from Cassandra into Spark memory in the form of dataframe object using Spark-Cassandra connector. The Spark dataframe is a new type of Spark data structure which is previously known as SchemaRDD. It is a RDD based distributed dataset along with schema metadata. It is represented as two-dimensional table where each column has name and type. It is more efficient than RDD.

The Cassandra partitions are mapped into Spark partitions based on data distribution performed by Spark-Cassandra connector. The detail description of data distribution between Cassandra and Spark is given in Sect. 3.1.

4.1.4   Associate Spatial Index

Cassandra provides native support for 1-D index via secondary index. Primary index is global whereas secondary index is local in Cassandra database. The query based on secondary index is very slow because it must be sent to all nodes of the cluster to retrieve the data. There are performance implications when secondary index built on high cardinality column values. Hence, instead of building secondary index on spatial column values, data duplication and de-normalization of data will be a potential approach to store spatial data in Cassandra.

We have used Geohash[2] data structure for spatial index. It is quite simple and efficient to implement compared to complex data structures like R-tree. Geohash is a hierarchical data structure used for indexing spatial data. Once the number of partitions in Spark are identified, the Geohash character string is associated with each record available in Spark partitions. It performs one-to-one transformation on data and maintains data locality. The resultant Spark dataframe is called as spatial dataframe and is to be stored back into Cassandra.

4.1.5   Store Spatial Dataframe into Cassandra

Each record of a spatial dataframe is associated with Geohash code. Then, each rectangular area represented by Geohash code is mapped to a Cassandra row. Geohash is used as a partition key in the spatial data model. Each row represents a rectangular area and columns store spatial objects fall within that rectangular area. The spatial dataframe is stored using the spatial data model given in Fig. 5. The spatial attributes are stored in WKT format.

Row Key	TimeStamp	Column Family: Spatial Object				
		WKT	attribute_1	attribute_2	……….	
Index_ID1	…………..	POINT(….,….)		……………..		………….
Index_ID2	…………..	POINT(….,….)		……………..		
		…………….				

Spatial Data Table

Fig. 5.   Cassandra data model for spatial data

The write operation from Spark to Cassandra is very challenging. It needs to collect all Spark partitions into memory and store back into Cassandra. The write performance from Spark to Cassandra is optimized by creating micro batches of the data available in Spark partitions. The number of batches and batch size is determined from the metadata available on each node of a Cassandra cluster. The connector converts the data into number of batches and then writes each batch into Cassandra based on partition key.

4.2   Spark Core Layer

Spark core layer provides user friendly APIs for machine learning, graph processing and structured data processing using Spark SQL. It performs fast data processing on large datasets by exploiting data parallelism and partitioning techniques. However Spark doesn't support native partitioning for spatial data; for example proximity based

---

[2] Geohash WG. Geohash. https://www.en.wikipedia.org/wiki/Geohash.

partitioning for spatial data. The developers would require to write wrappers to perform spatial operations. The dataframe APIs are built and implemented to perform spatial and non-spatial operations using standard spark core APIs.

4.3 Spatial Data Processing Layer

The spatial data processing layer creates an interface to query spatial and non-spatial data available in Spark dataframe object. Spatial operations are complex and compute intensive. The aggregated and complex queries are implemented on top of Spark dataframe. The distributed APIs are implemented for spatial operations.

4.3.1 Proximity Search

Proximity search queries are widely used in many analytics applications. Circle range queries are used to find the number of spatial objects within a specified circular range from a given query point. A Spark dataframe API *circle_range_query(spark_ object, query_point, range)* is implemented to perform proximity search within circular boundary. Circle range query is implemented by identifying the number of Geohash bounded within circular area based on query input. The ProximityHash algorithm is used to find the bounded Geohash. Finally, filter the partitions based on bounded Geohash values and collect the spatial objects belongs to those partitions. This will give approximate number of spatial objects within the range.

4.3.2 KNN Search

K nearest neighbor search queries are used to find K nearest spatial objects from a given query point. A Spark dataframe API KNN_*query(spark_ object, query_point, K)* is implemented to find K nearest neighbors. It is implemented by using selection and merge method. First, select the large enough circular search space defined by r. Find the Geohash covered by circular search space. Sort subspaces based on the distance between query point and Geohash. The distance is calculated between a query point and the center point of a rectangular area represented by Geohash. Now, traverse each subspace and find k' nearest neighbor points from the query point until k' >= K or end of the list of Geohash. Merge points from all subspaces and find k' nearest neighbor points. if k' < K, then expand search space based on Geohash precision. If new search space is out of the scope then return an error else repeat the procedure till success.

4.3.3 Point Query

Point query is designed to find a particular point within dataset along with attribute data. The Spark dataframe APIs point_*query(spark_ object, query_point) and* point_*query_with_index(spark_ object, query_point)* are implemented to find the details at a given location. It is implemented by searching each spatial object within dataset i.e. full table scan or using Geohash index. To find a point without using Geohash index required to scan each partitions of dataframe whereas using Geohash index it searches within a single partition defined by Geohash index.

4.4   Application Layer

The framework provides a convenient web based REST interface to the end user. Cassandra performs fast retrieval of data based on partition key and clustering key compared to Spark. The framework has implemented an application architecture such that the end user can execute ad-hoc queries on suitable framework either Spark or Cassandra via a common user interface. The low-latency or simple queries will be executed on Cassandra and aggregated and spatial queries will be executed on Spark framework. The application architecture is shown in Fig. 6.

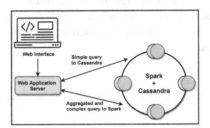

Fig. 6. Application architecture

5   Experimental Results and Discussion

The framework is evaluated by analyzing the performance of the system in terms of latency against variable size of data. The performance of the system is compared with the baseline technology i.e. Cassandra for low-latency queries.

5.1   Experimental Setup

All experiments are conducted on a cluster consisting of four nodes. Each node runs Ubuntu 14.0.4 with Spark 2.1.0 and Cassandra 3.10. Each node is equipped with an Intel(R) Core(TM) i7-4770 CPU @ 3.40 GHz, 1 CPU, 4 physical cores per CPU, total 8 logical CPU units with 16 GB RAM. The Spark cluster is deployed in the standalone mode. The proposed framework is implemented and deployed using Sparklyr[3] framework. The web based RESTful ad-hoc APIs are built and implemented to explore analytical results on top of the shiny[4] framework.

5.2   Description of Dataset

NYC[5] taxi dataset which is an open database containing 167 million records in 30 GB. Each record describes a taxi trip made by a particular driver at the particular date and

---

[3] http://spark.rstudio.com/.

[4] http://www.shiny.rstudio.com.

[5] http://www.andresmh.com/nyctaxitrips/.

time. Each record has 16 attributes: hack_license, two 2D coordinates which represent pickup location and drop-off location, pickup date_time, drop-off date_time, and other nine attributes represent other related information. All experiments are done on a sample dataset containing about 30 million records in 5 GB storage.

5.3 Results

We have generated datasets of variable size (5 million to 30 million records; 800 MB to 5 GB) to evaluate the performance of the system. All experiments are performed on 2-dimensional vector dataset. Pickup location is considered as 2D spatial object. We have designed different test cases to evaluate the performance of various queries like attribute query, point query, circle range query, KNN search query and simple query. The performance of different queries is evaluated in terms of average latency against the variable size of datasets.

5.3.1 Load Data into Cassandra

The performance of loading phase is evaluated using loading time vs. data size and shown in Fig. 7.

5.3.2 Establish Analytics Pipeline

The big data analytics pipeline consists of three phase, (i) Read data from Cassandra, (ii) Associate Geohash index to each record of data and (iii) Write indexed data back to Cassandra. The total elapse time is calculated to establish the analytics pipeline against variable size of datasets. The result is shown in Fig. 8.

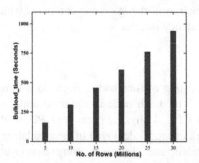

Fig. 7. Bulk load performance (effect of data size)

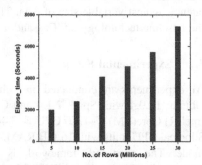

Fig. 8. Analytics pipeline (effect of data size)

5.3.3 Attribute Query Analysis

In spatial analytics non-spatial data are equally important. They provide the characteristics of spatial data. We have executed 25 sample attribute queries containing selection, filtration, aggregation and group_by operations and find the average latency. The attribute query performance in terms of latency against different size of datasets is shown in Fig. 9.

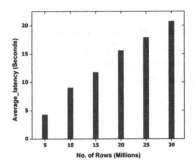

Fig. 9. Attribute query performance (effect of data size)

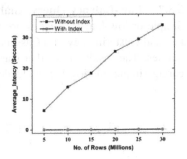

Fig. 10. Point query performance (effect of data size)

5.3.4 Point Query Analysis

Point queries are used to find the characteristics at a particular location. Twenty five tests are performed against pickup locations with and without Geohash index each. Spatial index plays an important role in fast retrieval of data. It is observed that point queries are very fast when they referred with Geohash index because Geohash is used as partition key in spatial data table stored in Cassandra. To find a particular location without index requires full table scan hence, it is very slow. The performance of point query is shown in Fig. 10.

5.3.5 Circle Range Search Analysis

We have performed ten test cases with varying range from 500 m to 3000 m from a query point. The query point is selected using sampling method. The results show that proximity search using Geohash index offers outstanding performance. It will find the approximate number of pickup locations within a specified range from a given query point. The performance of circular range query is shown in Fig. 11.

5.3.6 KNN Search Analysis

Ten test queries are performed to find K (K = 5 to 25) nearest pickup locations from a given query point. Different query points are selected from high density areas using sampling method. The performance of KNN search query is shown in Fig. 12.

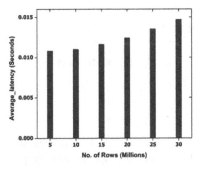

Fig. 11. Circle range query performance (effect of data size)

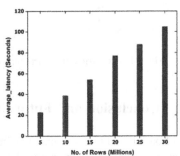

Fig. 12. KNN query performance (effect of data size)

5.3.7   Low-Latency Query Analysis

Simple or low-latency queries are based on partition key or combination of partition key and clustering key of a Cassandra table. We have executed 25 test cases of simple queries. The result shows that Cassandra is very fast compared to Spark when executing simple queries. Figure 13 depicts the query performance.

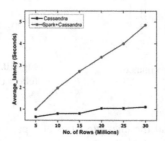

Fig. 13. Simple query performance (effect of data size)

5.3.8   Implementation of Application Layer

The framework provides a convenient web based REST interface to perform analytics on top of suitable architecture; Spark or Cassandra via a common interface. The results of analytical operations are shown in Figs. 14 and 15 by means of Restful ad-hoc APIs. Figure 14 lists the taxi trips made by a particular driver. This query is executed on Cassandra DB by performing filtration on partition key. Figure 15 shows K nearest pickup locations from a given location. The operation is performed using KNN query executed on Spark framework.

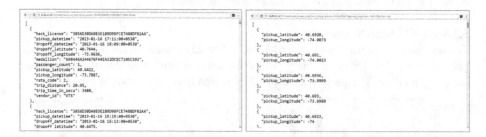

Fig. 14. Taxi trips of a particular driver   **Fig. 15.** KNN search

6   Conclusion and Future Work

A Spark based big data analytics framework is implemented to load, store, process and query spatial and non-spatial data at scale. The framework is implemented by performing analytics operations like selection, filtration, aggregation, exact match, proximity and KNN search. The results are explored through Restful ad-hoc APIs. The proposed framework is able to accelerate ad-hoc query processing by diverting the user queries to the suitable framework like low latency query can be diverted to Cassandra and aggregation and complex queries can be executed on Spark framework. The experimental results show that, framework has achieved the efficient storage management and high computational processing through Spark and Cassandra integration. It would be interesting to extend the work by considering spatial operations like spatial join and kNN join; and applications like spatial aggregation and spatial auto-correlation.

Acknowledgement. This work is a part of a research project on 'Developing Data Analytics Architecture, Applications in Agriculture', funded by NRDMS and NSDI, Department of Science and Technology, Govt. of India, year 2017–2019.

References

1. Zaharia, M., Chowdhury, M., Franklin, M.J., Shenker, S., Stoica, I.: Spark: cluster computing with working sets. HotCloud **10**(10–10), 95 (2010)
2. Open Geospatial Consortium. http://www.opengeospatial.org/
3. Website of MongoDB. http://www.mongodb.org
4. Lakshman, A., Malik, P.: Cassandra: a decentralized structured storage system. ACM SIGOPS Oper. Syst. Rev. **44**(2), 35–40 (2010)
5. Ben Brahim, M., Drira, W., Filali, F., Noureddine, H.: Spatial data extension for Cassandra NoSQL database. J. Big Data **3**(1), 11 (2016)
6. Eldawy, A., Mokbel, M.F.: Spatialhadoop: a MapReduce framework for spatial data. In: 2015 IEEE 31st International Conference on Data Engineering (ICDE), pp. 1352–1363. IEEE (2015)
7. Aji, A., et al.: Hadoop gis: a high performance spatial data warehousing system over MapReduce. Proc. VLDB Endowment **6**(11), 1009–1020 (2013)
8. Yu, J., Wu, J., Sarwat, M.: Geospark: a cluster computing framework for processing large-scale spatial data. In: Proceedings of the 23rd SIGSPATIAL International Conference on Advances in Geographic Information Systems, p. 70. ACM (2015)
9. Website of Magellan. https://github.com/harsha2010/magellan. Magellan - https://hortonworks.com/blog/magellan-geospatial-analytics-in-spark/; https://github.com/harsha2010/magellan
10. Website of Spatialspark. http://simin.me/projects/spatialspark/
11. R Core Team: R: a language and environment for statistical computing. In: R Foundation for Statistical Computing, Vienna, Austria 2013 (2014)
12. Eldawy, A., Mokbel, M.F.: Pigeon: a spatial MapReduce language. In: 2014 IEEE 30th International Conference on Data Engineering (ICDE), pp. 1242–1245. IEEE (2014)
13. Website of spark-cassandra-connector. https://github.com/datastax/spark-cassandra-connector
14. Güting, R.H.: An introduction to spatial database systems. VLDB J. Int. J. Very Large Data Bases **3**(4), 357–399 (1994)
15. Eldawy, A., Li, Y., Mokbel, M.F., Janardan, R.: CG_Hadoop: computational geometry in MapReduce. In: Proceedings of the 21st ACM SIGSPATIAL International Conference on Advances in Geographic Information Systems, pp. 294–303. ACM (2013)

An Ingestion Based Analytics Framework for Complex Event Processing Engine in Internet of Things

Sanket Mishra[✉], Mohit Jain, B. Siva Naga Sasank, and Chittaranjan Hota

BITS Pilani Hyderabad Campus, Hyderabad, Telangana, India
{p20150408,f20130675,f20140050,hota}@hyderabad.bits-pilani.ac.in

Abstract. Internet of Things (IoT) is the new paradigm that connects the physical world with the virtual world. The interconnection is generated by the optimal deployment of sensors which continuously generate data and streams it to a data store. The concept drift and data drift are integral characteristics of IoT data. Due to this nature, there is a need to process data from various sources and decipher patterns in them. This process of detecting complex patterns in data is called Complex Event Processing which provides near real-time analytics for various IoT applications. Current CEP deployments have a inherent capability to react to events instantaneously. This leaves room to develop CEPs which are proactive in nature which can take the help of various machine learning (ML) models to work together with CEP. In this paper, the usage of Complex Event Processing (CEP) engine is exhibited that allows the inference of new scenarios out of incoming traffic data. This conversion of historical data into actionable knowledge is undertaken by a Long Short Term Memory (LSTM) model so as to detect the occurrence of an event well before time. The experimental results suggest the rich abilities of Deep Learning to predict events proactively with minimal error. This allows to deal with uncertainties and steps for significant improvement can be made in advance.

Keywords: Complex Event Processing · Internet of Things · LSTM

1 Introduction

Internet of Things is a blossoming domain offering incessant opportunities and caters to the need of multiple domains ranging from smart homes [10], smart energy [17], smart traffic [12], precision agriculture [19], etc. With the advancement in technology, availability of low cost hardware and Internet penetration pervasive computing has reached new avenues. Estimates claim that the number of IoT devices will rise to nearly 20 billion by 2020. IoT will play a pivotal role

The authors would like to thank TCS Foundation for supporting the first author through PhD fellowship.

© Springer Nature Switzerland AG 2018
A. Mondal et al. (Eds.): BDA 2018, LNCS 11297, pp. 266–281, 2018.
https://doi.org/10.1007/978-3-030-04780-1_18

in analyzing the data from heterogeneous devices which are placed at various geographical locations and create actionable knowledge out of it.

In today's world, IoT footprint has been present in almost every domain thus, catalyzing humans to live a comfortable life. In IoT, we follow a three tier generic architecture consisting of terminal nodes, gateway and server layer. The terminal/leaf nodes capture the data after interacting with the environment and send it to the gateway where it is smoothened and forwarded to the server for further analytics. The sensors when capture the data from interacting with physical environments are called simple events. The simple events are fired by individual sensors but in IoT domain the usage of multiple sensors is exhibited for the fulfillment of a particular use-case. Correlation of simple events from various sensors [4] and detecting the underlying patterns in them is called Complex Event Processing (CEP) [6]. The domain of CEP takes care of the analysis and identification of composite patterns from events arriving from heterogeneous sources. These complex patterns are complex events which have to be processed in real-time and on-the-fly analytics has to be executed on the streaming data generated from IoT sensors. The ability to process multi modal data allows CEP to be used in most IoT usecases to generate new complex scenarios.

Complex Event Processing (CEP) is one of the most dominating and widely researched fields in IoT data analytics. It is a principal component in executing analytics, on-the-fly processing and deriving meaningful insights from streaming data in real-time. It plays a major role in identifying relationships, patterns, etc. amongst unrelated, heterogeneous data and thus, triggering an outcome.

But in today's world, there exists a lot of open source CEP engines like Cayuga [2], T-REX [3], etc. which can do the above mentioned operations without fail. But the outcomes of these CEP engines are reactive in nature. The reactive nature, though intuitive, still lacks functionality in critical scenarios like traffic management, fire outbreaks, fraud detection in online transactions, battlefield, etc. The idea of proactive event processing though conceptualized in the works [5,7] lack the implementation and a proper functional framework for stream processing. Event forecasting allows us to detect the events well before time and take proactive steps for the same. This helps in smooth functioning of the system and disallows any faults that may occur because of late outcomes. The CEP follows a Event Processing Language and has a SQL like syntax. Rather than querying data with the (Event Processing Language) EPL queries, data is passed over to the queries which detect patterns and identify simple as well as complex events. CEP needs to be coupled with predictive analytics and historical data to forecast events to add the proactive nature to the reactive CEP which increases throughput, decreases I/O and puts less load on network bandwidth. Several ML based methods known to provide unique and innovative results have been used mostly in the areas of traffic management, smart energy, etc. But they are unsuitable for a wide array of problems and lack scalability.

Our contributions in this paper are as follows:

– Design of a generic IoT architecture using open-source components that is fault tolerant and feasible for deployment.

- Implementation of Deep Learning techniques, such as, Artificial Neural Networks (ANN), Long Short Term Memory Networks (LSTM), etc. to predict/forecast future events with greater accuracy in near real-time.
- Heuristic based functionality to tune the model and make it immune to data drifts.

Rest of the paper is organized as follows: Sect. 2 gives an overview of the existing works in the areas of CEP and Predictive Analytics. Section 3 deals with the proposed CEP engine and detailed overview of its modules. Section 4 demonstrates the LSTM model that has been used in the proposed work to give pro-activity to the application (in this work the CEP engine). Section 5 elaborates on the results procured from various experiments carried out in this work. Section 6 summarizes the work done and discusses future directions.

2   Related Work

2.1   Complex Event Processing

Sase [8] is an event matching language which can add a structural perspective to queries which follows a SQL structure. Though it was helpful in creating complex events out of simple events but it failed to create more complex events out of the set of complex events. Internet of Things comprises of sensors which send data continuously over a networking medium, such as, Zwave, ZigBee, LAN, WiFi, LoRAWan, etc. This data needs utmost care and processing to derive meaningful insights from it. Data stream processing deals thus, with data coming from various sources in order to give birth to new data streams as output which might reflect anomalies or detect disasters [13]. T-Rex [3] proposed their ideas on a complex event processing engine that contain APIs to publish new events and subscribe to complex events. The engine also helps to identify complex events by virtue of its well defined specification rules. Aleri Streaming platform [21] exhibits a graphical language for defining the rules meant for processing. It does not follow temporal data mining because the basis for ordering of sequences is not particularly based on timestamps. The CEP engine proposed in this work has the ability to function in multiple cores or multiple machines. Oracle [15] constructed an event driven skeleton on top of which it integrated BEA's WebLogic Server in 2008 finally constituting what is known today by the name of Oracle CEP. CQL is the standard of rule processing language used in the engine. The advantage it furnishes is the ability to manipulate production and consumption policies. Esper [20] is a prominent open source CEP provider which makes use of Event Processing Language. EPL uses operators of SQL with ad-hoc constructs for defining windows and generation of outputs. Esper can be deployed in centralized and distributed scenarios. Cayuga et al. [2] refers to an event monitoring platform. It's based on Cayuga Event Language (CEL) which has SELECT, FROM and PUBLISH clauses with the help of conventional SQL constructs like selection, projection, renaming, union and aggregates. It makes the use of zero consumption policy which facilitates recurrent detection of complex events. It is built to deal with large scale data. It takes the help of automaton to detect complex events. Cayuga has one shortcoming as it doesn't allow distributed processing. Next CEP [16] happens to be a CEP engine in a distributed fashion. Detection of complex events takes place by conversion of rules into non deterministic automata which is in some way similar to that of Cayuga. Mostly, Next CEP focuses on rule optimization and uses query rewriting which helps in minimal usage of CPU resources and optimal delays in processing. Our proposed CEP engine is based on plug-in modules in which each module can be constructed independently and have its own functionality. These modules collaboratively constitute the CEP engine and give it the much needed functionality so as to analyze data and translate them into actionable knowledge.

2.2 Predictive Analytics

The area of Complex Event Processing is not complete without the inclusion of predictive analytics. The base of predictive analytics helps CEP achieve various complicated functionalities, such as, regression, classification, clustering, etc. These operations simplify the way in which CEP detects and manages events. The necessity of predictive analytics can be attributed to the fact that it allows us to forecast events in advance and steps can be taken in that direction. Jin et al. [11] suggested the usage of Timed Petri Net to capture uncertainty in events from RFID streams which might serve an important cause in predictions. Akbar et al. [1] proposed a generic Adaptive Moving Window Regression model integrated with Esper CEP. Though the model was able to be tuned with varying error rates still the preprocessing discarded data which might be otherwise useful in predictions. Hofleitner et al. [9] proposed hydrodynamic traffic theory coupled with a Dynamic Bayesian network in order to predict the intensity of vehicles on the road. Li et al. [14] developed a CEP framework based on a tree structure and optimized the proposed algorithm by grouping of events. Tommasini et al. [18] proposed an ontology based CEP that can discover temporal relationships by executing a deductive reasoning algorithm. But newer scenarios and advance inferences are some limitations of this work.

3 Proposed Architecture

Figure 1 shows the proposed architecture and the components in it. The individual components are explained in greater details in the following sections.

3.1 Data Ingestion

Telegraf is a server agent for collecting and reporting several performance metrics. Telegraf has a plugin which helps it to act as a source for a variety of metrics directly from the system it is running on. The plugins also support to pull metrics from third party APIs and also listen for metrics via Kafka consumer services. To read data from a Kafka topic, the Kafka input plug-in could

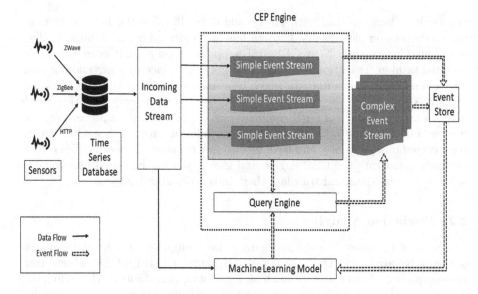

Fig. 1. CEP engine

be used. So, Telegraf would keep polling messages from the specified topic and store then in any output data store like InfluxDB. The traffic data can be stored in InfluxDB by streaming it over a Kafka broker and finally storing it in the InfluxDB with the help of Telegraf.

3.2 Apache Kafka

Kafka is a streaming module that works on Producer Consumer paradigm. Being an open source, it is easily available and consists of three entities Kafka producer, that fetches data from external sources and streams it into the second component that is Kafka Broker. Kafka Broker handles the data collected, and processes it. Once processed, Kafka Consumer collects the data from Broker and sends it to the desired sources. The most interesting part about Kafka is, its scalability and versatility. It is compatible with almost all kinds of streaming engines, and can be a producer or consumer to heterogeneous data streams. Moreover, it is capable of deploying Kafka Broker over multiple machines (collectively known as Kafka Cluster) rather than just centralized ones, hence processing can be done in a distributed manner. Also, one Kafka module can have multiple consumers or producers and it uses Kafka-Topics, associated with each data stream, in order to correctly send data to the desired consumer. Kafka has been used for streaming the data from database into Siddhi CEP. Data is published to topics using Kafka producer.

3.3   Complex Event Processing

WSO2 Siddhi performs Complex Event Processing by listening to events from data streams and then detects complex conditions which can be described via a Streaming SQL Language provided by Siddhi. Siddhi is a lightweight, fast complex event processor. WSO2 Stream processor is an open source analytics solution that allows a developer to build and deploy applications that capture data from different streaming sources, analyze it and detect patterns in real time. It works closely with its specific query language called Siddhi QL in which user can write rules in the predefined format, and take actions based on the response of queries. It uses Siddhi in its core for analyzing the event streams. The Kafka component of Siddhi CEP is used to consume the events from the appropriate topics. InfluxDB is used for storing the results from Siddhi CEP. InfluxDB acts as the sink for the simple and complex events. The machine learning module consumes historical data via another Kafka consumer and predicts the next data point (event) of intensity or velocity. This prediction is published under a topic back to the CEP in the form of a tuple via REST services. The simple events or predicted events are received by the Kafka consumer of the CEP through a REST prediction. Once events are created and collected, with the help of pattern matching, it will be able to correlate simple events and create complex events.

3.4   Dataset Used

For the purpose of evaluation of our approach the dataset concerning the traffic in the city of Madrid[1] is taken into account. Data is collected from sensors located at 3902 locations in Madrid. Many of them faced glitches with respect to large number of missing values while few locations were associated with very less missing data points. This data is in XML format, which is a standard in IoT data. The traffic data has two attributes, average traffic intensity and average traffic velocity, which are logged and can be later used in processing. The traffic intensity gives us the measure of density of vehicles porting at the concerned location while the traffic velocity gives us the measure of the speed of vehicles at a particular time. Figure 2 shows a JSON object that is communicated by Apache Kafka carrying information about a location.

Both the features are used for detection of congestion well before time. This adds pro-activity to the CEP and helps in taking preventive measures at the earliest, such as, choice of alternative routes, limiting the number of vehicles on the locations, etc. In order to work over the data, firstly dataset is segregated location wise and missing values as per the time period are changed using standard cubic interpolation. Than it is segregated into 2 parts, the 2/3 part for training purposes, while 1/3 part for testing. Starting from the very simple linear regression, we extended our work to test applicability of Deep Neural Network and its hybrid to achieve maximum accuracy.

---

[1] http://informo.munimadrid.es/informo/tmadrid/pm.xml.

{
'LocationID':'PM10341',
'Date':'2014-03-01',
'Time':'03:45:00',
'Intensity': 360,
'Speed':94
}
{
'LocationID': PM40753,
'Date':'2014-03-03',
'Time':' 19:45:00',
'Intensity': 120,
'Speed': 76
}

Fig. 2. Sample JSON object from the dataset

4   Proposed Model

LSTM or long short term memory are another form of recurrent neural networks, where current output of the machine learning module will affect the next value. Unlike RNNs, LSTMs are resilient to the problem where current data is not affected by near past data but rather far behind in time. It does so by saving the states of the past in its memory cells. Most of the time series data, especially IoT ones are often having this autoregressive relationships between present and far past and LSTM is one of the most promising solution to such a problem. IoT datasets often possess similar behavior, and hence LSTM can be applied to such datasets, in order to check for the possibilities of improving the results.

Multiple forms of LSTMs are available in the literature, while simplest ones take input for 1 time period others take value from multiple data sources and for multiple time period. A multi-layered or multi-vector LSTM, is used in this work which takes inputs for the last three time periods and forecasts the upcoming value of a required variable. Previous sections shed light on the architectural description of the model presented in this work while this section explains about the methodology used for forecasting the value of parameters. Numerous statistical and machine learning models were experimented over Madrid data in order to find the most suited model and finally concluded with using a window based LSTM where previous three time period values were considered as input and next time period is predicted.

Despite being highly praised, RNNs fail to solve the problems like exploding gradient or vanishing gradient. Exploding gradient occurs when weights of a neural network unit becomes so large and so does their updates that final result appear to be garbage. On the other hand, vanishing gradient might occur when gradient tends to zero and no update happens in the weights. Adding to this, RNN also fails when far past is contributing to the result much more than the near past. As simple RNN works, from every node, there is an output to itself for getting updated for the next time period, and an output towards the next hidden layer (in case of deep network). A more enhanced version of RNN are Long Short

Term Memory. LSTMs are a variant of RNNs, and much more accurate over time series data than a simple ANN or RNN. It not only is resilient to problems like exploding or vanishing gradient but also able to work well with data where data possess far past dependencies. The key to this success is the structure of a normal node, which not only possess an input and output gate but also a forget gate. Each node or cell in an LSTM network has a unique structure as shown below in Fig. 3:

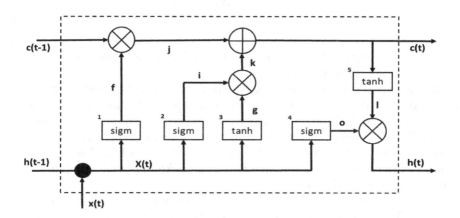

Fig. 3. LSTM Cell

In Fig. 3 for every node, x(t) stands for the input coming at time t, while h(t) is the output of the node. Unlike simple networks, where only these two fields are present with a standard updating function present in the middle of the node, a LSTM node also consists of factor c(t), which is the forget factor. It is the value of c(t) that decides the degree of importance given to the past value and while going out, it is also updated in a way to decide how much importance will be given to the current output.

For every node, the x(t) represents the tuple of inputs, which is concatenated with the output of the previous time period to form a new tuple. This tuple is passed through multiple nodes. The calculations for each node is as shown below:

$$x(t) = <x_1, x_2, x_3, ...> \ (input \ at \ time \ t) \tag{1}$$

$$h(t) = <h_1> \ (output \ at \ time \ t) \tag{2}$$

$$c(t) = c_1 \ (output \ as \ forget \ node \ at \ time \ t) \tag{3}$$

Now, for a time t,

$$X(t) = <X_1, X_{2,3}, ..., h1_{t-1}> \tag{4}$$

Now, X(t) is passed through several nodes in order to get the value of output. In the above figure,

$$f = sig(W_1.X(t) + b_1) \tag{5}$$

$$j = c(t-1) * f \tag{6}$$

$$i = sig(W_2.X(t) + b_2) \tag{7}$$

$$g = tanh(W_3.X(t) + b_3) \tag{8}$$

$$k = g * i \tag{9}$$

$$c(t) = k + j \tag{10}$$

$$\begin{aligned} o &= sig(W_4.X(t) + b_4) \\ l &= tanh(W_5.c(t) + b_5) \\ h(t) &= o * i \end{aligned} \tag{11}$$

This is how, c(t) and h(t) are calculated, at time t and passed onto the next time period and the next layer.

In the experiments performed using the traffic data, the best results were obtained when the input vector is taken as a tuple of 3 values which includes traffic velocity in the last three time periods and targeted to predict the upcoming traffic. The same experiments were ran to calculate the upcoming traffic intensity as well. The structure of the neural network is a LSTM network with 4 nodes in the only layer present and output is backpropagated to minimize the error matrix and results were obtained.

Adding to the above knowledge, all neural networks are slow but powerful learners. Once learned, there is no change in the output despite the error being generated. Knowing about the variation in the IoT dataset, also known as Data Drift, the network needs to learn continuously but retuning the network again and again, for every single value would not be a wise choice and rather, would be very expensive. The below defined heuristics would be much more efficient option than the naïve continuous retuning option. The heuristics are defined in Algorithm 1.

Algorithm 1. Tuning the LSTM Model

Step 1 Count = 0;
Step 2 while *error* **do**
   if *error* < *10* **then**
     | goto Step 1;
   else
     count = count + 1;
     if *count* > *11* **then**
       | goto Step 3;
     else
       | goto Step 2;
     end
   end
end
Step 3 Retune the LSTM network for last 360 periods(3days);
Step 4 goto Step 1;

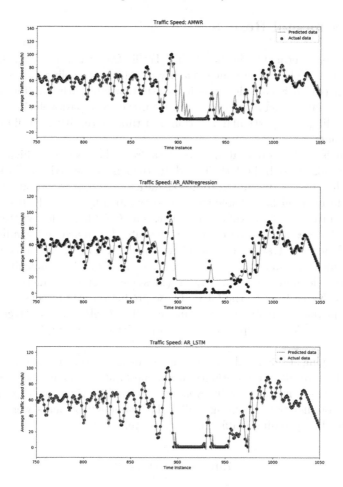

Fig. 4. Comparision between AMWR, ANN Regression and LSTM

Figure 4 exhibits the events captured and predicted by corresponding models. It is observed that AMWR fares the poorest in event capture while LSTM performs better. AMWR has been used in [1] and performs well if we preprocess the data and make it free from outliers which are important in IoT data. The algorithm has to consider the outliers that add to the theory of concept drift. For example, for capturing a congestion event, data nature captured at 3pm might be different from data seen at 3am. ANN though can work without any issues on noisy data but their inability to memorize recent events and inability to tune its weights disallows them to capture all events. LSTM has the intrinsic characteristic to find optimal results by analyzing long term dependencies making it the most suitable model which can not only handle noisy, non-linear data but also contributes to greater accuracy in predictions.

5   Experimental Results

We conducted experiments on the Madrid Traffic Dataset. Data from the month of March, 2014 is chosen, in which velocity of vehicles traveling across the location in next period is considered as fore-casted variable, while different combinations of variables including time of day, current and past velocities, number of cars traveling at a particular location and time (intensity of traffic) etc. are taken as inputs.

Telegraf is a daemon that ingests data from data source which is traffic data and sends it to InfluxDB. The data then is consumed by Kafka which uses publish-subscribe paradigm to intake data and sink it to various sources. A Kafka producer publishes a topic which is subscribed by CEP's Kafka consumer and the cognitive model's Kafka consumer. The CEP has rules written in SiddhiQL which helps in creation of complex events. The machine learning models in the CEP are pre-trained by the historical data. The Kafka consumer supplies real-time streaming data to the cognitive models and the machine learning models predictions are sent via MQTT (Message Queing Telemetry Transport Protocol) or Kafka back to the CEP. These predictions are forecasted values of intensity and velocity before their occurrence. The CEP uses its pattern matching techniques to form complex events and predict future complex "congestion" events.

The following Table 1 exhibits the experiments performed onto the traffic data. The best results were obtained when the input vector is taken as a tuple of 3 values which includes traffic velocity in the last three time periods and targeted to predict the upcoming traffic. The same experiments were executed to calculate the upcoming traffic intensity as well.

Table 1. Models used

S. No	Category	Inputs
1	Linear Regression	Hour, Minute
2	Decision Tree Regression	Hour, Minute
3	Random Forest Regression	Hour, Minute
4	Support Vector (RBF) Regression	Hour, Minute
5	AR Regressive Linear Regression	Hour, Minute, V(t-1)
6	AR Regressive Decision Tree Regression	Hour, Minute, V(t-1)
7	AR Regressive Random Forest Regression	Hour, Minute, V(t-1)
8	AR Support Vector (RBF) Regression	Hour, Minute, V(t-1)
9	ANN Regression with 3 inputs, $AR = -2$	Hour, Minute, V(t-2)
10	ANN Regression with 3 inputs, $AR = -3$	Hour, Minute, V(t-3)
11	ANN Regression with 4 inputs, $AR = -2$	Hour, Minute, V(t-2), TI(t-2)
12	ANN Regression with 4 inputs, $AR = -3$	Hour, Minute, V(t-3), TI(t-3)
13	Adaptive Moving Window Regression	Past window sized velocity values
14	Simple LSTM	v(t-1)
15	Window based LSTM with lookback = 3	v(t-1), v(t-2), v(t-3)

Each period is considered as an interval of 5 min while carrying out experiments. In algorithm 1, the heuristics are simple but works well under the circumstances. It proclaims that if the predicted value is wrong for last 1 h, than re-tunes the network for the data collected in last 3 days. It makes sure that if the data predicted is wrong for 1 h continuously or not within 10% of the actual value, it re tunes the network for the data collected in last 3 days. The parameters 3 days and 1 h are taken on experimental basis and holds a scope of research so that the retuning step can be optimized. Once the network is returned, it appeared to have adapted to situations and led to reduction in error in upcoming time period.

The models that have been deployed have to be validated by a standard. To verify its correctness, two commonly used metrics are used that can give us the validity of the deployed models, i.e., Mean Absolute Percentage Error (MAPE) and Root Mean Square Error (RMSE). They are defined as follows:

$$MAPE = \frac{1}{n} \sum_{i=1}^{n} \frac{|P_i - P_i'|}{P_i} \tag{12}$$

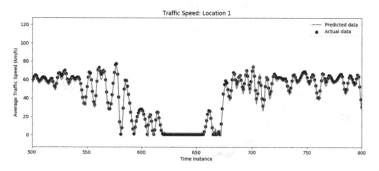

(a) $Location1(MAPEError = 15.59\%)$

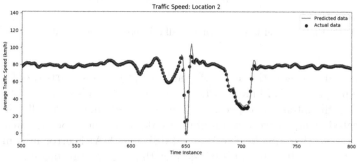

(b) $Location2(MAPEError = 8.64\%)$

Fig. 5. Prediction results on average traffic speed from locations 1 and 2

$$RMSE = \frac{1}{n}\sqrt{\left(\sum_{i=1}^{n}(|P_i - P_i'|)^2\right)} \qquad (13)$$

where P_i is the observed value and P_i' is the predicted value. Both the metrics (12) and (13) are the most generic metrics used in time series analysis and are useful in evaluation of the deployed models.

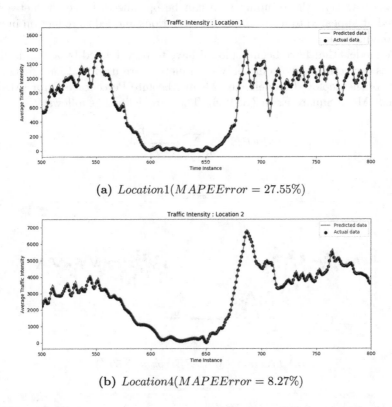

(a) $Location1(MAPEError = 27.55\%)$

(b) $Location4(MAPEError = 8.27\%)$

Fig. 6. Prediction results on average traffic intensity from locations 1 and 2

Figures 5 and 6 show the array of graphs or predictions carried out by comparing the proposed LSTM model with other conventional ML models over traffic speed and traffic intensity respectively.T he proposed model and all other models are created using native Python libraries like Tensorflow, Scikit-learn, Keras, etc. It is noticed that LSTM give lesser RMSE and MAPE error in comparison to its counterparts. It is because pf the fact that LSTM can process long term dependencies and can also remember past events to predict future events. This brings the proactivity feature in the proposed CEP. Figure 7 speaks about the MAPE values and RMSE values that has been computed from the all models as mentioned in Table 1. It includes an overall comparison with other regression

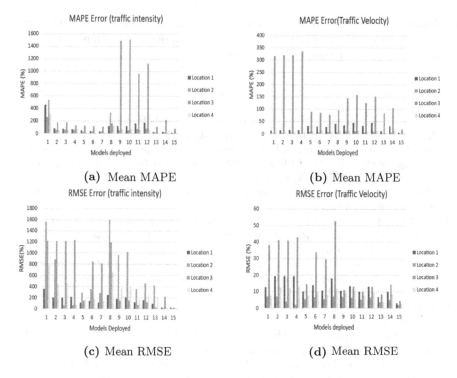

Fig. 7. Experimental results of RMSE and MAPE

models. As it can be seen the MAPE and RMSE error by the proposed model is least which adds to its suitability to be interfaced with CEP.

The experiments were executed on DELL Precision 3420 workstation containing an Intel Xeon E3 processor @3.7 GHz, NVIDIA Quaddro K620 GPU and 16 GB RAM, running on Ubuntu 16.04 operating system.

6 Conclusion and Future Work

The work done in this research proposes a generic framework that helps in the prediction of complex events. The nexus of CEP and Deep Learning methodology coupled with real time and historical data makes the implementation pro-active. The usage of Long Short Term Memory model had been a suitable choice in the time series analysis and had been a choice for sequence analysis in the past. This motivated us to use it on a real world traffic data in which it gave the least RMSE and MAPE errors proving its feasibility to be deployed in IoT architectures.

We aim to scale our architecture and in future apply it to other use-cases like smart energy, smart weather, etc. The framework is generic and modifiable because of the modularity in its architecture. Users can plug in and plug out modules from the concerned framework and update them with the modules of their own choice as and when required. This leaves room for scalability in future deployments. In future, we aim to develop faster predictions for other IoT applications that can add pro-activeness and more responsiveness thus, minimizing further human interaction with the devices.

References

1. Akbar, A., Khan, A., Carrez, F., Moessner, K.: Predictive analytics for complex IoT data streams. IEEE Internet Things J. **4**(5), 1571–1582 (2017)
2. Brenna, L., et al.: Cayuga: a high-performance event processing engine. In: Proceedings of the 2007 ACM SIGMOD International Conference on Management of Data, pp. 1100–1102. ACM (2007)
3. Cugola, G., Margara, A.: Complex event processing with T-REX. J. Syst. Softw. **85**(8), 1709–1728 (2012)
4. da Penha, O.S., Nakamura, E.F.: Fusing light and temperature data for fire detection. In: 2010 IEEE Symposium on Computers and Communications (ISCC), pp. 107–112. IEEE (2010)
5. Engel, Y., Etzion, O.: Towards proactive event-driven computing. In: Proceedings of the 5th ACM International Conference on Distributed Event-Based System, pp. 125–136. ACM (2011)
6. Etzion, O., Niblett, P., Luckham, D.C.: Event Processing in Action. Manning, Greenwich (2011)
7. Fülöp, L.J., Beszédes, Á., Tóth, G., Demeter, H., Vidács, L., Farkas, L.: Predictive complex event processing: a conceptual framework for combining complex event processing and predictive analytics. In: Proceedings of the Fifth Balkan Conference in Informatics, pp. 26–31. ACM (2012)
8. Gyllstrom, D., Wu, E., Chae, H.-J., Diao, Y., Stahlberg, P., Anderson, G.: SASE: complex event processing over streams. arXiv preprint cs/0612128 (2006)
9. Hofleitner, A., Herring, R., Abbeel, P., Bayen, A.: Learning the dynamics of arterial traffic from probe data using a dynamic Bayesian network. IEEE Trans. Intell. Transp. Syst. **13**(4), 1679–1693 (2012)
10. Jie, Y., Pei, J.Y., Jun, L., Yun, G., Wei, X.: Smart home system based on IoT technologies. In: 2013 Fifth International Conference on Computational and Information Sciences (ICCIS), pp. 1789–1791. IEEE (2013)
11. Jin, X., Lee, X., Kong, N., Yan, B.: Efficient complex event processing over RFID data stream. In: Seventh IEEE/ACIS International Conference on Computer and Information Science, ICIS 2008, pp. 75–81. IEEE (2008)
12. Kanungo, A., Sharma, A., Singla, C.: Smart traffic lights switching and traffic density calculation using video processing. In: 2014 Recent Advances in Engineering and computational sciences (RAECS), pp. 1–6. IEEE (2014)
13. Li, J.Z.: A logical agent-based environment monitoring and control system. Master in Engineering Project Report (2011)
14. Li, Y., Wang, J., Feng, L., Xue, W.: Accelerating sequence event detection through condensed composition. In: 2010 Proceedings of the 5th International Conference on Ubiquitous Information Technologies and Applications (CUTE), pp. 1–6. IEEE (2010)
15. JDK Oracle. Disponível em. http://www.oracle.com/technetwork/java/javase/downloads/index.html. Acessado em, 8, 2010
16. Schultz-Moeller, N.P., Migliavacca, M., Pietzuch, P.: Distributed complex event processing with query optimisation. In: International Conference on Distributed Event-Based Systems (DEBS 2009), Nashville, TN, USA. ACM (2009)

17. Serra, J., Pubill, D., Antonopoulos, A., Verikoukis, C.: Smart HVAC control in IoT: energy consumption minimization with user comfort constraints. Sci. World J. **2014**, 1–11 (2014)
18. Tommasini, R., Bonte, P., Della Valle, E., Mannens, E., De Turck, F., Ongenae, F.: Towards ontology-based event processing. In: Dragoni, M., Poveda-Villalón, M., Jimenez-Ruiz, E. (eds.) OWLED/ORE -2016. LNCS, vol. 10161, pp. 115–127. Springer, Cham (2017). https://doi.org/10.1007/978-3-319-54627-8_9
19. TongKe, F.: Smart agriculture based on cloud computing and IoT. J. Converg. Inf. Technol. **8**(2), 210–216 (2013)
20. Tóth, G., Fülöp, L.J., Vidács, L., Beszédes, Á., Demeter, H., Farkas, L.: Complex event processing synergies with predictive analytics. In: Proceedings of the Fourth ACM International Conference on Distributed Event-Based Systems, pp. 95–96. ACM (2010)
21. Wang, D., Rundensteiner, E.A., Wang, H., Ellison III, R.T.: Active complex event processing: applications in real-time health care. Proc. VLDB Endow. **3**(1–2), 1545–1548 (2010)

An Energy-Efficient Greedy MapReduce Scheduler for Heterogeneous Hadoop YARN Cluster

Vaibhav Pandey[✉] and Poonam Saini

Department of CSE, Punjab Engineering College (Deemed to be University),
Chandigarh, India
pandeyvaibhav5l@gmail.com, nit.sainipoonam@gmail.com

Abstract. Energy efficiency of a MapReduce system has become an essential part of infrastructure management in the field of big data analytics. Here, Hadoop scheduler plays a vital role in order to ensure the energy efficiency of the system. A handful of MapReduce scheduling algorithms have been proposed in the literature for slot-based Hadoop system (*i.e.*, Hadoop 0.*x* and Hadoop 1.*x*) to minimize the overall energy consumption. However, YARN-based Hadoop schedulers have not been discussed much in the literature. In this paper, we design a scheduling model for Hadoop YARN architecture and formulate the energy efficient scheduling problem as an Integer Program. To solve the problem, we propose a *Greedy* scheduler which selects the best job with minimum energy consumption in each iteration. We evaluate the performance of the proposed algorithm against the FAIR and Capacity schedulers and find out that our greedy scheduler shows better results for both CPU- and I/O intensive workloads.

Keywords: MapReduce · Scheduling · Energy-efficiency

1 Introduction

Hadoop MapReduce is the most prevalent distributed computing framework inspired by Google's MapReduce programming paradigm. The framework was initially used for long batch processing of production jobs which are executed periodically in data centers. Essentially Hadoop is an ecosystem and MapReduce is one of its core components. The other components include HDFS [1], Pig [2], Hive [3], Mahout [4] and ZooKeeper [5] etc. Four versions of Hadoop framework have been released since inception, namely, Hadoop 0.*x*, Hadoop 1.*x*, Hadoop 2.*x* and Hadoop 3.*x*. The slot-based resource management is used in Hadoop 0.*x* and Hadoop 1.*x*. On the other hand, Hadoop 2.*x* and Hadoop 3.*x* use a fine-grained container based resource management system known as YARN.

The performance of Hadoop is greatly influenced by its scheduler. Initially, the main purpose of Hadoop to run large batch jobs such as web indexing and log mining. Hence, users submit jobs to a queue, and the Hadoop execute the jobs in FIFO order. However, the amount of data has increased substantially in Hadoop clusters that require

A. Mondal et al. (Eds.): BDA 2018, LNCS 11297, pp. 282–291, 2018.
https://doi.org/10.1007/978-3-030-04780-1_19

various complex algorithms to be executed faster. In the view of this, a MapReduce cluster is being shared among multiple users for a variety of workload. In a shared environment, the scheduler first selects a user who generally submits multiple jobs of different characteristics for execution. Hence, job selection is to be done next. Once the job is selected, its map, reduce, or speculative tasks are scheduled for further execution. It is not always true that a single scheduler schedules all three entities. At times, the scheduler allocates resources to one or more entities depending upon the scheduling policy being used. In July 2008, the scheduler in Hadoop became a pluggable component and triggered the innovation in this domain.

A MapReduce scheduler has to cater various quality of service (QoS) requirements of two stakeholders, namely, *Hadoop user* and *Hadoop administrator*. The QoS parameter includes makespan, response time, availability, throughput, energy efficiency, security and resource utilization etc. Here, energy efficiency is important for Hadoop system administrator in today's era of *Green Computing*. The U.S. Department of Energy stated in its report that data centers consumed about 70 billion kilowatt-hours of electricity in 2014 [6]. There is an increase in data center electricity consumption by about 4% from 2010–2014, a large shift from the 24% percent increase estimated from 2005–2010 and the nearly 90% increase estimated from 2000–2005. The Hadoop MapReduce framework is widely used in data centers to analyze the huge amount of data. Thus, there is a need to consider the energy efficiency of MapReduce clusters by designing energy-aware scheduling techniques to minimize energy consumption in a Hadoop system.

In this paper, we formulate an energy efficient MapReduce scheduling problem for YARN-based Hadoop framework as an Integer Program (IP). As the formulated problem is NP-hard, we propose a greedy technique which allocates a task with minimum energy consumption on a particular node. The rest of the paper is organized as follows. Section 2 presents a short survey of related work in the area of energy efficient scheduling in Hadoop framework. In Sect. 3 we prepare a scheduling model for YARN architecture and formulated an energy efficient scheduling problem. We then propose a greedy scheduler for the formulated problem in Sect. 4 and evaluate the performance in Sect. 5. Lastly, Sect. 6 concludes the paper.

2 Related Work

Energy-efficient job scheduling techniques help to reduce the energy consumption in a Hadoop system which can be further classified into two categories. In the first category, the partial or whole cluster is usually kept on low power state whenever not in use in order to reduce energy consumption. In the second category, map and reduce tasks are placed on appropriate nodes to achieve better energy efficiency. Few of the techniques [7, 8], that combine DVFS technique with task scheduling yields better results. A brief discussion of energy efficient MapReduce scheduling techniques is as follows.

Leverich et al. [9] identified that cluster nodes might remain idle for 20–38% of the time and hence proposed a method for energy management of MapReduce jobs by selectively powering down nodes with low utilization. Further, using idle and low utilization periods of cluster nodes, Lang et al. [10] proposed workload energy aware

All-in-Strategy (AIS). AIS use batching for consistently low utilization periods. It powers down all nodes during low utilization periods, batches the jobs and powers on all nodes, performs all jobs and again power down all when all jobs are completed.

Chen et al. [11] designed an energy-efficient Map-Reduce workload manager called Berkeley Energy-Efficient MapReduce (BEEMR). In the scheme, the cluster is split into two zones i.e., a small interactive zone and a larger batch zone, each having different percentages of available capacity in terms of task slots, memory, disk and network. The interactive zone is always in full power state, while the batch zone oscillates between full and low power state. BEEMR segregates interactive and batch workloads into separate sub-clusters to improve energy efficiency.

Yigitbasi et al. [12] proposed an energy-efficient algorithm for scheduling hetero-geneous workload to the heterogeneous cluster consisting of high and low power machines. This provides an opportunity to save energy by intelligently placing jobs on its corresponding energy-efficient machine. In case the available node does not match the energy efficiency threshold, fairness and data locality criteria are used to select a job and its task respectively for scheduling.

Mashayekhy et al. [13] proposed a framework for improving the energy efficiency of MapReduce applications over heterogeneous machines while satisfying the service level agreement (SLA). The problem of energy-aware scheduling of a single MapReduce job has been modeled as an Integer Program called Energy-aware MapReduce Scheduling (EMRS-IP) with a deadline as a constraint. Two heuristic algorithms have been proposed to solve the EMRS-IP, called energy-aware MapRe-duce scheduling algorithms (EMRSA-I and EMRSA-II). Both of these heuristics take the energy efficiency differences of different machines into account and use a metric called energy consumption rate of the map and reduce slots that characterize the energy consumption of each machine and induces an order relation among the machines.

Bampis et al. [8] proposed energy-efficient scheduling algorithms for two different scenarios to minimize the weighted completion time of a set of n MapReduce job with a constraint of the energy budget. In the first scenario, the authors formulated the problem as a linear program assuming that the order of the job execution is not fixed. A polynomial time constant-factor approximation algorithm was derived to solve the formulated problem. In the second scenario, the scheduling problem has been formu-lated as a convex program with the order of jobs is given.

Cai et al. [7] proposed a YARN scheduler to minimize the energy consumption with a deadline as a constraint unlike [8]. The proposed scheduler works at both, job level and task level. At the job level, the scheduler in highly inspired by ARIA [14] and competes for the jobs within its deadline. At the task level, energy efficiency has been targeted through the user-space DVFS governor. While scheduling at the task level, energy consumption of tasks at a specific node has not been considered as in [13].

3 System Modeling and Problem Formulation

A MapReduce job comprises a specific number of map and reduce tasks that are executed on a cluster composed of multiple machines. Cluster may consist of a different generation of machines implying heterogeneous hardware. The execution of jobs consists of a map phase followed by a reduce phase. In YARN setting, a task (either map or reduce) may request a fixed amount of different resources available at a machine for its execution. Multiple types of resources on each machine are allocated to tasks in the form container. Each task request in YARN cluster is usually represented by a tuple $<p, r, n, l, b>$, where p represents the priority of a task, r gives the resource requirement vector of a task, n is the total number of tasks which have the same resource requirements r, l represents the location of a task's input data split, and b is a boolean value to indicate whether a task can be assigned to a *NodeManager* that does not have its input data split locally.

We consider a set of N MapReduce jobs $J = \{J_1, J_2, \ldots \ldots J_N\}$, which is submitted to a YARN cluster $S = \{S_1, S_2, \ldots \ldots S_M\}$ consisting of M machines. Each MapReduce job J_j consists of two distinct sets of the map and reduce tasks and precisely defined as $J_j = MT_j \cup RT_j$, where $MT_j = \{m_1^j, m_2^j, m_3^j, \ldots \ldots m_{m'}^j\}$ is a set of m' map tasks and $RT_j = \{r_1^j, r_2^j, r_3^j, \ldots \ldots r_{r'}^j\}$ is a set of r' reduce tasks. The task m_i^j represents the ith map task of jth job and similarly the task r_i^j represents the ith reduce task of jth job. Let the execution start time and processing time of map task m_i^j on k^{th} machine is sm_i^j and pm_{ik}^j respectively. Similarly, execution start time and processing time of reduce r_i^j on k^{th} machine is sr_i^j and pr_{ik}^j respectively. It is to be noted that execution start time of a map and reduce task is independent of machine on which it is scheduled.

Furthermore, we assume that K types of resources are available at each machine represented by $r_1, r_2, \ldots \ldots r_K$. A two-dimensional matrix A of size $M \times K$ is used to represent the current availability of resources at each node. $A[i, j]$ indicates the total amount of jth resource type r_j available at i^{th} machine S_i at a particular time instance $\emptyset$.

A matrix M of size $N \times K$ is also defined to store the amount of each resource type r_j required by map tasks of each job. The value $M[i, j]$ indicates the amount of resource type r_j requested by map tasks of ith job for its execution. Similarly, a matrix R of size $N \times K$ is also defined to store the amount of each resource type r_j required by reduce tasks of each job and value $R[i, j]$ indicates the amount of resource type r_j requested by reduce tasks of an ith job for its execution.

The YARN scheduler can assign a map task of ith job to a worker node S_j for execution as long as $M[i, p] \leq A[j, p], \forall p \in [1, K]$. Similarly, the scheduler can assign a map task of ith job to a worker node S_j for execution as long as $M[i, p] \leq A[j, p], \forall p \in [1, K]$.

We consider that the machines are heterogeneous and em_{ik}^j and er_{ik}^j represents the energy consumption of map task m_i^j and reduce task r_i^j respectively at machine S_k during the execution. We aim to minimize both, the energy consumption of whole YARN cluster and the completion time of the execution of all n jobs (i.e., makespan). We formulate the following energy-efficient MapReduce scheduling problem in YARN cluster as an Integer Program (IP).

$$minimize \sum_{i=1}^{m'} \sum_{k \in [1,M]} em_{ik}^j X_{ik}^j + \sum_{i=1}^{r'} \sum_{k \in [1,M]} er_{ik}^j Y_{ik}^j, \quad \forall j \in [1, N]$$

$$minimize \ max \sum_{i=1}^{r'} \sum_{k \in [1,s]} (sr_i^j + pr_{ik}^j) Y_{ik}^j, \forall j \in [1, n]$$

Subject to:

$$\max\{(sm_i^j + pm_{ik}^j)X_{ik}^j\} \geq sr_{i'}^j Y_{i'k}^j, \forall j \in [1,N], \forall i \in [1,m'], \forall i' \in [1,r'], \forall k \in [1,M] \quad (1)$$

$$\sum_{i,i' \in J(\emptyset)} M[i,p]X_{ik}^j + R[i',p]Y_{i'k}^j \leq A[j,p], p \in [1,K], j \in [1,M] \quad (2)$$

$$X_{ik}^j = \{0,1\}, \forall j \in [1,N], \forall i \in [1,m'], \forall k \in [1,M] \quad (3)$$

$$Y_{ik}^j = \{0,1\}, \forall j \in [1,N], \forall i \in [1,r'], \forall k \in [1,M] \quad (4)$$

$$sm_i^j, sr_i^j \geq 0 \quad (5)$$

where $J(\emptyset)$ is a set of active jobs at particular time instance $\emptyset$.

In this formulation, there are two objective functions. First objective function minimizes the energy consumption while executing the MapReduce application whereas, the second objective minimizes the completion time of all jobs. Constraint (1) ensures that for all MapReduce jobs, reduce tasks start only when all map tasks have been completed. Constraint (2) requires that the resources consumed by all active tasks at a particular worker node S_j cannot exceed its resource capacity. Constraints (3) and (4) represents the integrality requirement for the decision variable. The decision variable X_{ik}^j takes the value 1 if map task m_i^j is assigned to machine S_k and 0 otherwise. Similarly Y_{ik}^j takes the value 1 if reduce task r_i^j is assigned to machine S_k and 0 otherwise. Lastly, constraint (5) requires decision variables sm_i^j and sr_i^j to be non-negative.

4   Proposed Solution

The MapReduce scheduling problem formulated in the previous section is NP-hard Integer Programming (IP) problem. There are mainly three approaches to solve NP-hard IPs. First one is Heuristic approach which solves the IPs usually in polynomial time with a sub-optimal result without any guarantee on the sub-optimality. The second one is known as approximation algorithms which provides a sub-optimal result with an assurance on the quality of the sub-optimal result. Finally, the third approach solves the IPs optimally, however, takes exponential time e.g., Branch and Bound (B&B), Branch and Cut, and cutting plane methods. We opt the first approach and use the energy consumption of a task on a particular machine as a heuristic. We further use that heuristic to greedily select a task to assign to a particular machine. The energy efficient *greedy* approach minimizes the total energy consumption of the set of n jobs. We now explain our proposed approach in detail.

4.1   Energy Efficient *Greedy* Approach

We propose a greedy approach without aiming makespan to schedule map and reduce task over worker nodes. The scheduler is triggered whenever it receives the heartbeat message from any worker node *N* which is running the *NodeManager* demon. Upon receiving the heartbeat message, scheduler groups all jobs in two sets: NodeLocal_Jobs and RackLocal_Jobs. Jobs in the set NodeLocal_Jobs have a local copy of desired data split at node N, whereas, jobs in RackLocal_Jobs have a copy of data split at a different node which is in the same rack that of node N.

After grouping the jobs in two sets, the algorithm greedily selects a job first from the set NodeLocal_Jobs. Precisely, it picks a job which has the least energy consumption on node *N* and assigns as much map tasks as the available capacity of the node. If no local job exists then, the algorithm searches the set RackLocal_Jobs and greedily selects a job which has least energy consumption on node *N*. By this grouping of jobs, the proposed scheduling algorithm gives preference to NodeLocal_Jobs to achieve better data locality. This further improves energy efficiency because if a task gets its data from the local node, it does not need to perform energy consuming network I/O operation to move data split to its local node. The pseudo code of the proposed greedy approach for map tasks is given in Algorithm 1.

Algorithm 1:
```
Upon receiving a heartbeat from node N:
NodeLocal_Jobs ← GetJobsNodeLocal(J,N)
RackLocal_Jobs ← GetJobsNodeLocal(J,N)
while NodeLocal_Jobs ≠ NULL
     Best_Job ← SelectBestJob(NodeLocal_Jobs, N)
     NodeLocal_Jobs ← NodeLocal_Jobs - {Best_Job}
     TaskAssignment(Best_Job, N)
     If node N still has resource capacity to     a
     commodate new tasks
     then continue
while RackLocal_Jobs ≠ NULL
     Best_Job ← SelectBestJob(RackLocal_Jobs, N)
     RackLocal_Jobs ← RackLocal_Jobs - {Best_Job}
     TaskAssignment(Best_Job, N)
     If node N still has resource capacity to
      accommodate new tasks
     then continue
SelectBestJob(JobsList L, Node N)
```
$$\text{return } \mathrm{argmin}_{j \in L} \{em_{iN}^{j}\}$$
```
TaskAssignment(Job J, Node N)
     Assign as much map tasks of job J as the total
     remaining capacity of node N
```

In case of reduce tasks, the algorithm does not consider any data locality metric while assigning tasks. Data locality for reduce tasks is hard to achieve because reduce task receives the output of various map tasks running on many different nodes. For assigning the reduce task, the algorithm considers the whole job set and pics the best job which has the least energy consumption on the node from which heartbeat message is received. The pseudo code of the proposed greedy approach for reduce task is given in Algorithm 2.

Algorithm 2:

```
Upon receiving a heartbeat from node N:
Let R_Jobs be set of jobs whose map tasks are
finished
Best_Job ← SelectBestJob(R_Jobs, N)
TaskAssignment(Best_Job, N)
SelectBestJob(JobsList L, Node N)
return argmin_{j∈L} {em_{iN}^{j}}
TaskAssignment(Job J, Node N)
Assign as much map tasks of job J as the total
remaining capacity of node N
```

5 Experimental Results and Discussions

In this section, we will evaluate the performance of proposed energy efficient greedy scheduler on the basis of total energy consumption and completion time and compare the results with FAIR and Capacity schedulers. Before illustrating the final results, we mention the cluster configurations, workload and energy model used in our experiments.

5.1 Cluster Setup

We use a heterogeneous YARN cluster which is composed of four high-end machines. We use three different configurations of the machine to build the cluster. Two of the nodes have 12 cores of 2.6 GHz (Intel Xeon E5-2690 v3) with 32 GB of RAM and 1 TB of hard disk. The third node has 4 core of 3.5 GHz (Intel Xeon E3-1270 v3), with 16 GB RAM and 500 GB of HDD and the fourth one has 4 cores of 3.8 GHz (Intel Xeon E3-1270 v6), with 8 GB RAM, and 500 GB of HDD. Energy consumption of various components in these nodes are shown in Table 1.

Table 1. Energy consumption of various components in cluster nodes

Node type	CPU power (*CPU_POW*)	Disk read/write power (*DISK_POW*)	NIC I/O power (*NIC_POW*)
Intel Xeon E5-2690 v3	135 W	4.5 W	1.2 W
Intel Xeon E3-1270 v3	80 W	3 W	1.2 W
Intel Xeon E3-1270 v6	72 W	3 W	0.8 W

5.2   Workload Mix Used

We use two different workloads; (i) CPU- intensive and (ii) I/O-intensive. For CPU-intensive workload, we use WordCount job which counts the frequency of words in a text file. Whereas, for I/O intensive workload we use TeraSort job which sorts the terabytes of numbers.

5.3   Energy Model Used for Profiling

We assume that energy consumption of map/reduce tasks at every node in the cluster are known to the proposed algorithm in advance that can be easily measured by energy profiling. In order to profile WordCount and Terasort jobs, we run the single job on every node multiple time and take mean of all recorded values. We note that certain activity like CPU processing, Disk I/O and network I/O consumes energy during the execution of any map and reduce task. In view of this, we devise the Eq. 6 for calculating the energy consumption of a map task. Energy consumption of reduce task can be measured in the same manner.

$$em_{ik}^{j} = C_{ik}^{j} * CPU\_POW_k + D_{ik}^{j} * DISK\_POW_k + N_{ik}^{j} * NIC\_POW_k \qquad (6)$$

Where symbols have the following interpretations.

Symbol	Meaning
C_{ik}^{j}	CPU time in ms
CPU_POW_k	CPU power in watt
D_{ik}^{j}	Total disk input/output in bytes
$DISK\_POW_k$	Disk power consumption/byte read or written
N_{ik}^{j}	Number of shuffle bytes
NIC_POW_k	NIC power consumption/byte sent or received

5.4   Parameters Used for Evaluation

We use total energy consumption and completion time of all jobs (i.e., makespan) for comparing the greedy scheduler with FAIR and Capacity schedulers. We use both workloads for each of these evaluation parameters. During the experiments, we gradually increase the workload size for 5 jobs to 15 jobs. For each job, the input file size is kept 4 GB and block size at HDFS layer is kept 128 MB with a replication factor of 3. This creates 32 map tasks for each job and no of reducers explicitly set as 4.

5.5   Results

We perform two sets of experiments: one for measuring energy efficiency and another for measuring completion time. For energy efficiency experiments we perform a test on both workloads separately.

Energy Efficiency

We first evaluate the performance of the proposed scheme on the basis of total energy consumed during the execution of a set of jobs. We perform three sets of experiments with 5, 10 and 15 jobs individually for WordCount and TeraSort jobs. These experiments show (Figs. 1 and 2) that proposed Greedy scheme performs 15% and 17% better on average in comparison to FAIR and Capacity scheduling algorithm respectively for I/O intensive workload. In the case of CPU intensive workload, the proposed scheme performs up to 16% and 20% better in comparison to FAIR and Capacity algorithm respectively.

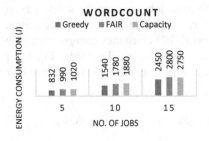

Fig. 1. Energy consumption of jobs for CPU-intensive workload

Fig. 2. Energy consumption of jobs for I/O-intensive workload

Completion Time

We further evaluate the performance of the proposed algorithm on the basis of completion time of all jobs. Here, in Figs. 3 and 4, we see that for both CPU- and I/O intensive workload, the greedy scheduler does not achieve better completion time than FAIR and Capacity scheduler. The reason for this degradation is due to consideration of only energy fitness metric during task assignment by the proposed greedy scheme. Hence we conclude that our greedy approach achieves better energy efficiency at the cost of job completion time. This motivates us to include the time parameter in our heuristic besides energy consumption.

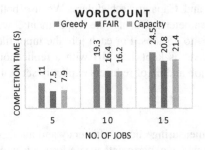
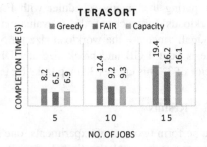

Fig. 3. Completion time of all jobs for I/O-intensive workload

Fig. 4. Completion time of all jobs for I/O-intensive workload

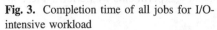

6   Conclusion

Energy efficient scheduler in YARN system is a critical component in the era of Green Computing. In this paper, we presented a scheduling model for YARN architecture and formulated an energy efficient scheduling problem as an Integer Program at task level of scheduling hierarchy. We proposed a greedy scheduler which selects the best job in terms of energy efficiency upon receiving the heartbeat message. Experimental results show better energy consumption performance of the proposed scheme in comparison to both FAIR and Capacity scheduler. In future work, we aim to develop a scheduler which minimizes the energy consumption and completion time of the last job.

Acknowledgment. Authors would like to thank Ministry of Electronics and IT, Govt. of India for providing financial support to perform this work under the Visvesvaraya Ph.D. scheme.

References

1. Shvachko, K., Kuang, H., Radia, S., Chansler, R.: The Hadoop distributed file system. In: 2010 IEEE 26th Symposium on Mass Storage Systems and Technologies, MSST2010 (2010)
2. Welcome to Apache Pig! https://pig.apache.org/. Accessed 25 June 2018
3. Apache Hive TM. https://hive.apache.org/. Accessed 25 June 2018
4. Apache Mahout: Scalable machine learning and data mining. http://mahout.apache.org/. Accessed 25 June 2018
5. ZooKeeper. https://zookeeper.apache.org/doc/trunk/zookeeperOver.html. Accessed 25 June 2018
6. Shehabi, A., et al.: United States Data Center Energy Usage Report, June 2016
7. Cai, X., Li, F., Li, P., Ju, L., Jia, Z.: SLA-aware energy-efficient scheduling scheme for Hadoop YARN. J. Supercomput. **73**(8), 3526–3546 (2017)
8. Bampis, E., Chau, V., Letsios, D., Lucarelli, G., Milis, I., Zois, G.: Energy efficient scheduling of MapReduce jobs. In: Silva, F., Dutra, I., Santos Costa, V. (eds.) Euro-Par 2014. LNCS, vol. 8632, pp. 198–209. Springer, Cham (2014). https://doi.org/10.1007/978-3-319-09873-9_17
9. Leverich, J., Kozyrakis, C.: On the energy (in)efficiency of Hadoop clusters. ACM SIGOPS Oper. Syst. Rev. **44**(1), 61 (2010)
10. Lang, W., Patel, J.M.: Energy management for MapReduce clusters. Proc. VLDB Endow. **3**(1–2), 129–139 (2010)
11. Chen, Y., Alspaugh, S., Borthakur, D., Katz, R.: Energy efficiency for large-scale MapReduce workloads with significant interactive analysis. In: Proceedings of the 7th ACM European Conference on Computer Systems – EuroSys 2012, p. 43 (2012)
12. Yigitbasi, N., Datta, K., Jain, N., Willke, T.: Energy efficient scheduling of MapReduce workloads on heterogeneous clusters. In: Green Computing Middleware on Proceedings of the 2nd International Workshop – GCM 2011, pp. 1–6 (2011)
13. Mashayekhy, L., Nejad, M.M., Grosu, D., Zhang, Q., Shi, W.: Energy-aware scheduling of MapReduce jobs for big data applications. IEEE Trans. Parallel Distrib. Syst. (1), 1 (2015)
14. Verma, A., Cherkasova, L., Campbell, R.H.: ARIA: automatic resource inference and allocation for MapReduce environments. In: Proceedings of the 8th ACM International Conference on Autonomic Computing - ICAC 2011, p. 235 (2011)

Predictive Analytics in Healthcare and Agricultural Domains

Analysis of Narcolepsy Based on Single-Channel EEG Signals

Jialin Wang[1,2], Yanchun Zhang[1,3(✉)], and Qinying Ma[4]

[1] Shanghai key Laboratory of Data Science, School of Computer Scinece,
Fudan University, Shanghai, China
yanchunzhang@fudan.edu.cn
[2] Cyberspace Institute of Advanced Technology, Guangzhou University,
Guangzhou, China
[3] Institute for Sustainable Industries and Liveable Cities, VU Research,
Victoria University, Melbourne, Australia
yanchun.zhang@vu.edu.au
[4] The First Hospital of Hebei Medical University, Shijiazhuang, China

Abstract. A normal person spends about third of his life in sleep. Healthy sleep is vital to people's normal lives. Sleep analysis can be used to diagnose certain physiological and neurological diseases such as insomnia and narcolepsy. This paper will introduce the sleep stage and the corresponding electroencephalogram (EEG) characteristics at each stage. We used the deep convolutional neural network (CNN) to classify original EEG data with narcolepsy. We use perturbations based on frequency to generate adversarial examples to analyze the characteristics of narcolepsy in different sleep stages. We find that perturbations at specific frequencies affect the classification results of deep learning.

Keywords: EEG data · Sleep analysis · Adversarial example

1 Introduction

Sleep is a process that is in the majority of organisms. There are five stages in human sleep process, each with characteristic and distinctive EEG features [13].

Sleep is the main way to eliminate physical fatigue. Sleep can drop blood pressure, reduce part of endocrine, and lower basal metabolic rate. Sleep can protect the brain and restore energy. Insufficient sleep can lead to irritability, lack of energy, distraction, memory loss, etc. Long-term lack of sleep can lead to hallucinations. The brain consumes a lot of oxygen during sleep, which is good for energy storage of brain cells. People with adequate sleep are energetic, quick-thinking and efficient. Therefore, sleep helps protect the brain and improve brain power [6].

In 1953, Nathaniel Kleitman and his student Eugene Aserinsky discovered the REM stage. In 1968, Rechtschaffen and Kales divided NREM into four stages.

Supported by the NSFC (No. 61332013 and No. 61672161).

In 2007, the American Academy of Sleep Medicine (AASM) changed NREM to three stages.

Narcolepsy is a debilitating lifelong rapid eye movement (REM) sleep disorder. It has a negative effect on the quality of life of its sufferers and can restrict them from certain careers and activities [1]. There are two different types of narcolepsy, which are narcolepsy with cataplexy and narcolepsy without cataplexy [5].

The current international classification of sleep disorders definition for narcolepsy includes: (1) excessive sleepiness during the day is almost at least 3 months; (2) the determined cataplexy does not exist; (3) the diagnosis must be confirmed by polysomnography or multiple sleep latency tests (MSLT); (4) The incubation period should be less than or equal to 8 min and greater than or equal to 2 sleep-onset rapid eye movement (SOREM); (5) Another disease or drug cannot better explain oversleeping [3].

EEG signals analysis is the most important dimensions to extract the features of human sleep. Extracting these features are a significant task with many real-world applications. The analysis of sleep is used in the diagnosis and treatment of certain physiological and neurological disorders such as apnea, insomnia, and narcolepsy [5]. A classic application to illustrate the usefulness of these features is sleep stage analysis.

Sleep research has made significant progress in different sleep stages. One of the remaining mysteries is how to represent and understand the way all night sleep relates to sleep disorders. In other words, how to detect sleep disorders. It has been proposed that the sleep EEG signals are correlated to the narcolepsy [1]. Unfortunately, although there is great interest in the features of narcolepsy in sleep EEG signals, little progress has been made towards developing methods to extract them.

The study of EEG frequency is an important area of research relating to the sleep disorder. EEG frequencies are used in the sleep analysis for sleeping stage scoring. A traditional EEG signals measurement used to analysis rhythm is low frequency (less than 200 Hz) extracellular neural oscillations that represent coordinated neural activity across distributed spatial and temporal scales [19].

EEG signals are useful for describing overall sleep stages since they reflect activity across all night sleep. We will examine sleep disorders under different sleep stages of the subject by recording EEG signals in multiple channels of the human brain.

Recently, deep learning based method has achieved good performance, and in some area is closed to export level. However, the high accuracy result from deep learning model cannot interpret the relationship between feature and class. Now, several studies have revealed that artificial perturbations on natural images can easily make deep learning model misclassify [9], even only attacking one pixel can fool deep learning model [17]. A common idea for generating adversarial examples are adding a data transfer, which is expected to find some knowledge to interpret the feature. Such perturbation can cause the classifier to label the modified input as a completely different class. Unfortunately, most of the previous attacks did not consider EEG signals for adversarial attacks, namely the adversarial mainly in image recognition. Additionally, investigating adversarial examples generated in EEG signals might give new insights about the EEG signals observe the result of classification by deep learning model.

The rest of the paper is organized as follows. Section 2 summaries the related works. Section 3 describes the rhythm of EEG signals in sleep stage. Section 4 proposes a generating adversarial example approach to analyze narcolepsy in a real-world database. Sections 5 reports on the experimental results. Sections 6 and 7 discuss and conclude the paper.

2   Related Works

In sleep disorders research, the analysis of EEG signals is largely centered around sleep stage scoring [17,18], where different sleep stages contains different features based on rhythm [7,10]. However, sleep stage scoring is not suitable for exploratory analysis of EEG data such as inferring unknown features. This work focuses on differences between the features in sleep stages suitable for exploratory analysis.

Perilously published sleep disorder studies distinguished between none rapid eye movement (NREM) and REM using proportions of lower frequency wave, k-complex wave, spindle wave and alpha wave in EEG signals [8,11,14]. Recent deep learning models can be applied to sleep stage scoring and sleep disorder detection. Discriminating sleep stages and sleep disorders may need more features by deep learning models.

The problem of detecting sleep stage has been studied in [12]. Some of the approaches based on neural network [4,16], while others consider traditional methods [12].

Although the results are accurate, the above methods do not interpret relationships between rhythm features and sleep disorder.

In this paper we try to overcome the shortcoming of previously published sleep stages scoring and sleep disorder detection methods by extracting a feature of EEG signals from a sleep stage. Therefore, we use a generating adversarial example method to analyze sleep EEG with narcolepsy [21,22]. Considered on these sleep stages, each EEG signal clip is assigned to an adversarial example. Each adversarial example is contained with a frequency band perturbation. Thus, all EEG signals clips as the adversarial example include a piece of the frequency band.

3   Sleep Stages of EEG Signals

In this section, we describe the sleep EEG signal and corresponding features.

In this paper, sleep EEG signal is a collection of brain activity obtained from sequential measurements oversleep. Kleitman et al. have discovered the REM stage in a sleep EEG signal and discussed the difference between REM and NREM stage [2]. According to the report of AASM, an example of different features are shown in Figs. 1, 2, 3, 4 and 5.

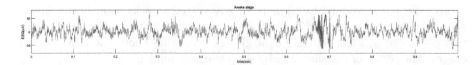

Fig. 1. Awake stage: The consciousness is clear, and the alpha wave appears when the subject closes his eyes. Low amplitude mixed frequency. The W stage is characterized by low amplitude and mixed frequency EEG; Alpha waves and high tonic EMG may also appear.

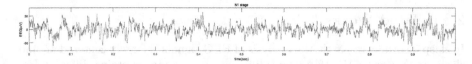

Fig. 2. N1 stage: The N1 stage is a light sleep and does not last long. The alpha wave will decrease and a vertex sharp will appear. Alpha waves. In stage N1, the EEG signal has the highest amplitude, a frequency range of 2–7 Hz, and the presence of Alpha waves in the EEG signal in less than half the epoch's duration.

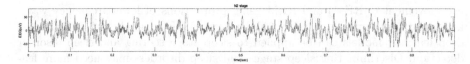

Fig. 3. N2 stage: K complex wave and spindle wave appear deeper than N1 sleep. Sleep spindle. Stage N2 is characterized by the presence of sleep spindles (12–14 Hz). K-complex. K complex wave with duration longer than 0.5 s.

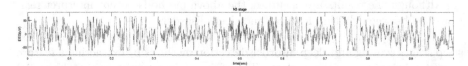

Fig. 4. N3 stage: Deep sleep is not easy to wake up and subjects will feel tired when subjects wake up. Waveforms are generally low-frequency waves. Delta waves. A low-frequency wave of 0.5–2 Hz will occur, with an amplitude greater than 75 uV accounting for 20%–50% in the N3 stage.

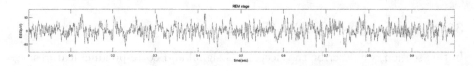

Fig. 5. REM stage: The EEG features are not obvious at this stage. The subjects can remember his dream when they are awakened in this stage. Sawtooth waves. The REM stage shows low voltage, mixed frequency EEG, sawtooth wave-like pattern, low amplitude EMG, and high-level EOG signal from both eyes.

4 Methodology

Due to the many channels of EEG data acquisition, the high sampling frequency and the long acquisition period, it is difficult to find the feature intuitively from EEG data. Therefore, we propose a method to extract EEG signal features.

In this paper, we study all of these approaches concerning the following topics:

1. EEG data of sleep stage: The EEG data is divided into different sleep stages by label.
2. EEG representation: The data is divided according to a certain length of time. The horizontal axis of the representation represents a sample point.
3. Convolution neural network: The model structure contains a convolutional layer, a pooled layer, and a fully connected layer.
4. Generating adversarial examples: We perturb specific EEG frequency to generate adversarial examples.
5. Features Extraction: The model is trained through original EEG data, and the adversarial examples are tested.

4.1 Framework and Architecture

For each subject $S = \{s_1, s_2, \ldots, s_n\}$, we have EEG signals T from S. These EEG signals T are split into m segments, $X = \{x_1, x_2, \ldots, x_m\}$, such that segments are same length. These segments are chosen to be non-overlapping, thereby isolating data points between consecutive segments. Each segment is considered as a single observation sample. The deep learning model aims to extract the feature in each of these samples.

This section describes the architecture of deep learning model that we used in the narcolepsy classification. We select VGG-19 as the network model [15]. The VGG-19 comprises 19 layers, not counting the input, all of which contain trainable parameters.

During training, the input to convolutional neural network is a fixed-size 1×1024 series. This is significantly different from the traditional VGG-19 model. The reason is that it uses one-dimensional convolutional (conv.) layers. The values of the input pixels are normalised. The preprocessing is segmenting the EEG signals into small equal clips. The kernel size of the filters is 1×3. The convolution stride is fixed to 1 pixel; the padding of conv. layer input is such that the spatial resolution is preserved after convolution, i.e. the padding is 1 pixel for 1×3 conv. layers. The pooling is carried out by five max-pooling layers, which follow some of the conv. layers (not all the conv. layers are followed by max-pooling). Max-pooling is performed over a 1×2 pixel window, with stride 2. All hidden layers are equipped with the rectification non-linearity.

4.2 Generate Adversarial Example

The adversarial examples are constructed by perturbed pixels in the image directly [9] while the frequency attack considered in this paper is the opposite which specifically focuses on the frequency domain.

In this paper, we proposed a frequency perturbation method. The method perturb frequency information between two classes of EEG signals. Perturbation includes single-direction perturbation and two-direction perturbation. The single-direction perturbation is to replace the EEG frequency of position n of the non-narcolepsy subject with the EEG frequency of the position n of the narcolepsy subject. Two-direction perturbation is the exchange of EEG frequencies at the position n of non-narcolepsy subjects and narcolepsy subjects. But the two-direction perturbation makes the results difficult to understand. In this paper, if there is no special explanation, we only use single-direction perturbation. This is illustrated in Fig. 7 for the case when position $n = 5\,\text{Hz}$.

We can infer from the results which frequency is associated with narcolepsy. The proposed generating adversarial example method has several advantages: 1 the perturbation does not need to know the deep learning network structure and the number of parameters; 2 the deep learning model only needs to train the original EEG data; 3 the deep learning model results are interpretative; 4 the perturbation can be applied to other time series data.

According to the AASM, each subject has five sleep stages in all night sleep process. For each sleep stages $A = \{W, N1, N2, N3, REM\}$, where W is the awake stage, N1 is the first non-rapid eye movement(NREM) stage, N2 is the second NREM stage, N3 is the third NREM stage and REM is the rapid eye movement stage. Each stage contains a different rhythm.

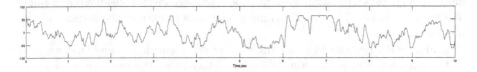

Fig. 6. Original EEG

EEG is divided into different sleep stages according to the sleep stage label. Experiments were performed on data from subjects with the same sleep stage. We use the Fourier transform to convert EEG data. We assessed differences in the spectrum of subjects with the same stage.

$$y_a(k) = \sum_{n=0}^{N-1} x(n)e^{-j\frac{2\pi}{N}nk} \tag{1}$$

We represent the input sample as the one-dimensional vectors in which each element represents one sample point.

We selected original EEG data with narcolepsy as shown in Fig. 6. We performed a Fourier transform on this data to obtain y_a from Eq. 1. At the same time, we performed the Fourier transform of the original EEG data with non-narcolepsy by the same method to obtain y_b. Then we take the value of position n from the EEG frequency of non-narcolepsy to replace the value of position n

from the EEG frequency of narcolepsy to get y_a'. We only perturb the portion of the frequency band length 1. Finally, the inverse Fourier transform is used to generate adversarial example x'.

$$y_a' = y_a + e(y_b) \tag{2}$$

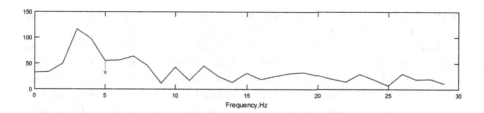

Fig. 7. Modify the original EEG 5 Hz spectrum to approach another class

The frequency perturbation can be seen as perturbing the consecutive data point of the frequency bands. The frequency bands perturbation is different than the pixel modification. The original EEG signals should be transferred into the frequency domain. Frequency bands perturbation changes the amplified of the frequency component (Fig. 8).

$$x'(n) = \frac{1}{N} \sum_{k=0}^{N-1} y_a'(k) e^{j\frac{2\pi}{N}nk} \tag{3}$$

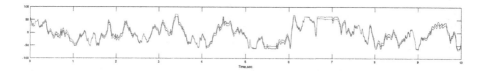

Fig. 8. Original EEG and modified EEG

We considered 50 frequency bands as candidate perturbation range, which is replaced by other class. One candidate perturbation range contains w frequencies and each perturbation is a tuple holding two elements: X-Y coordinates of the perturbation. The initial bandwidth is $w = 1$. Once generated, each candidate frequency band with their corresponding class construct an adversarial example feed into the deep learning model. The minimum result of accuracy is recorded. When the accuracy of adversarial example less than the original accuracy the frequency band is proven to be effective for distinguishing features.

We will replace the value of the corresponding frequency band in the second category with the specific frequency bands of all samples in the first category.

5　Experimental Results

In this section, we discuss the usefulness of EEG classification using a real-world database and we use deep learning model to extract time domain features.

5.1　Experimental Setup

The deep learning model is written in python and run on Ubuntu 14.04 machine having 16 GB of RAM. The details of the databases are as follows:

These datasets are sleep EEG signals. Subjects included different sleep stages while 6-channel EEG signals were recorded. In our experiment, we select a single channel of EEG for segmentation through the sleep stage label. Sleep stages are an imbalance in all night sleep. We segmented dataset that consists of over 12000 two-second EEG recordings, obtained from all-night sleep EEG samples, as described below. Each segment has a corresponding label. They are W, N3, and R.

Figure 9 shows the difference of frequency domain discovered in W and REM sleep stages, respectively. The X-axis represents the frequency used to discover the feature of EEG. The Y-axis represents the voltage amplitude. Each line in this figure represents the EEG signals of the subject. The range of frequency used in our experiments is 50 Hz. The reason for setting this range is to distinct low-frequency rhythm. The following observations can be drawn from this figure.

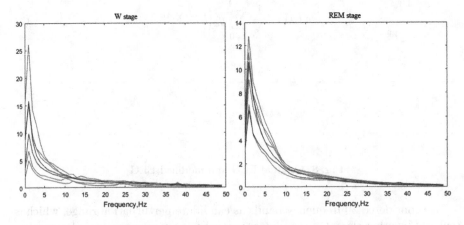

Fig. 9. Spectrum of different subjects in W and REM stages

We found that the frequency characteristics in the narcolepsy dataset are discriminative. As shown in Fig. 9, we found some interesting phenomena in the EEG signal at the 5 Hz position. In this frequency band, the EEG frequencies of narcolepsy and non-narcolepsy do not overlap.

5.2   Evaluation of Results

All night sleep EEG signals from the same brain regions were recorded from 7 subjects naturally transiting through different levels of sleep feature. We present only three sleep stages of sleep data from a single subject.

The evaluation of the proposed perturbation method is based on sleep EEG signals. We introduce two metrics to measure the effectiveness of the attacks.

Success rate (SR) is defined as the percentage of original EEG signals that were successfully classified by the deep learning model which tested by original EEG signals.

Attack rate (AR) is defined as the percentage of adversarial examples that were successfully classified by the deep learning model which tested by adversarial examples.

5.3   Results

The success rate (SR) and attack rates (AR) for frequency perturbations on deep learning models are shown in Table 1. In this section, by perturbing frequency band of EEG signals, we test a deep learning which trained by original EEG signals.

Table 1. Results of different spectrum perturbations.

Heading level	SR	AR1	AR5	AR10	AR20	AR30	AR50
W stage	0.9296	0.9290	0.9260	0.9270	0.9270	0.9270	0.9270
N3 stage	0.7830	0.7740	0.7825	0.7830	0.7830	0.7830	0.7830
REM stage	0.6300	0.6250	0.6250	0.6300	0.6300	0.6300	0.6300

According to Table 1, the accuracy of VGG19 is higher in the W stage than in other sleep stages. We found that perturbations based on frequency reduced the accuracy of the three sleep stage test results. In addition, by changing the perturbation position of the frequency band from 1 to 50 Hz, the attack rate will change correspondingly. The attack rate in 1 or 5 Hz frequency attack is higher than the attack rate in other frequency attack in three stages. Perturbing different frequencies has different effects on the results.

In this experiment, the perturbation at the 5 Hz position has the greatest impact on the results, especially in the W and REM stages. This conclusion is consistent with the conclusion of the [20].

6   Discussion

We use the deep learning model to extract features for sleep disorder analysis from a single-channel EEG signal. The results showed that the model could achieve high accuracy on real-world dataset with scoring standards (AASM). The results showed that the features extracted from the EEG signal helped improve the classification interpretably. These demonstrated that the deep learning model could extract features for sleep disorder analysis from a single-channel EEG signal.

The main reason that we evaluated narcolepsy with the W, N3 and REM stages from the dataset instead of all the sleep stage that rely on the feature in EEG signals. The reason is that some of the subjects do not have N1 stage and have very few N2 stages. Our results show that W, N3 and REM can extract feature from subjects, The N1 and N2 stage is not trained in our experiments.

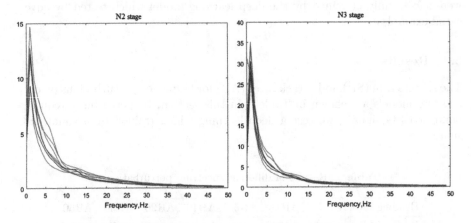

Fig. 10. Spectrum of different subjects in N2 and N3 stages

The deep learning model requires a lot of data training parameters. We tried input data of N2 stages for the narcolepsy classification and we found that the results were worse than others (Fig. 10). In the experiment, we only tested fewer subjects for analysis. In the following work we expect to analyze more subjects to verify our conclusions.

7   Conclusion

This paper presents a method to analyze sleep disorders. We explored the possibility of extracting narcolepsy features by generating adversarial examples. We used samples segmented from the EEG signal as input data for the one-dimensional VGG19 neural network. The experimental results of the real-world datasets demonstrate that deep learning can classify well. We found that perturbing the specific frequency of EEG with narcolepsy can mislead the trained deep learning model. We suspect that this specific EEG frequency is associated with narcolepsy. This inference is consistent with the conclusions of other studies. In future work, we expect to find the characteristics that accurately interpret sleep disorders through EEG signal adversarial examples.

Acknowledgment. This paper is supported by the National Science Foundation of China (No. 61332013 and No. 61672161).

References

1. Akintomide, G.S., Rickards, H.: Narcolepsy: a review. Neuropsychiatric Dis. Treat. **7**, 507 (2011)
2. Aserinsky, E., Kleitman, N.: Regularly occurring periods of eye motility, and concomitant phenomena, during sleep. Science **118**(3062), 273–274 (1953)
3. American Academy of Sleep Medicine. American Academy of Sleep Medicine: The International Classification of Sleep Disorders, Diagnostic and Coding Manual, 2nd edn. American Academy of Sleep Medicine, Westchester (2005)
4. Cecotti, H., Graeser, A.: Convolutional neural network with embedded Fourier transform for EEG classification. In: 19th International Conference on Pattern Recognition, ICPR 2008, pp. 1–4. IEEE (2008)
5. Dauvilliers, Y., Arnulf, I., Mignot, E.: Narcolepsy with cataplexy. Lancet **369**(9560), 499–511 (2007)
6. Hartmann, E.: The Functions of Sleep. Yale University Press, New Haven (1973)
7. Hor, H., Tafti, M.: How much sleep do we need? Science **325**(5942), 825–826 (2009)
8. Kuhn, A., Brodbeck, V., Tagliazucchi, E., et al.: Narcoleptic patients show fragmented EEG-microstructure during early NREM sleep. Brain Topogr. **28**(4), 619–635 (2015)
9. Kurakin, A., Goodfellow, I., Bengio, S.: Adversarial examples in the physical world. arXiv preprint arXiv:1607.02533 (2016)
10. Mahowald, M.W., Schenck, C.H.: Insights from studying human sleep disorders. Nature **437**(7063), 1279 (2005)
11. Mukai, J., Uchida, S., Miyazaki, S., et al.: Spectral analysis of all-night human sleep EEG in narcoleptic patients and normal subjects. J. Sleep Res. **12**(1), 63–71 (2003)
12. Olsen, A.V., Stephansen, J., Leary, E., et al.: Diagnostic value of sleep stage dissociation as visualized on a 2-dimensional sleep state space in human narcolepsy. J. Neurosci. Methods **282**, 9–19 (2017)
13. Shein-Idelson, M., Ondracek, J.M., Liaw, H.P., et al.: Slow waves, sharp waves, ripples, and REM in sleeping dragons. Science **352**(6285), 590–595 (2016)
14. Siddiqui, M.M., Srivastava, G., Saeed, S.H.: Diagnosis of narcolepsy sleep disorder for different stages of sleep using Short Time Frequency analysis of PSD approach applied on EEG signal. In: 2016 International Conference on Computational Techniques in Information and Communication Technologies (ICCTICT), pp. 500–508. IEEE (2016)
15. Simonyan, K., Zisserman, A.: Very deep convolutional networks for large-scale image recognition. arXiv preprint arXiv:1409.1556 (2014)
16. Sors, A., Bonnet, S., Mirek, S., et al.: A convolutional neural network for sleep stage scoring from raw single-channel EEG. Biomed. Signal Process. Control **42**, 107–114 (2018)
17. Su, J., Vargas, D.V., Kouichi, S.: One pixel attack for fooling deep neural networks. arXiv preprint arXiv:1710.08864 (2017)
18. Supratak, A., Dong, H., Wu, C., et al.: DeepSleepNet: a model for automatic sleep stage scoring based on raw single-channel EEG. IEEE Trans. Neural Syst. Rehabil. Eng. **25**(11), 1998–2008 (2017)

19. Ulrich, K.R., Carlson, D.E., Lian, W., et al.: Analysis of brain states from multi-region LFP time-series. In: Advances in Neural Information Processing Systems, pp. 2483–2491 (2014)
20. Vassalli, A., Dellepiane, J.M., Emmenegger, Y., et al.: Electroencephalogram paroxysmal theta characterizes cataplexy in mice and children. Brain **136**(5), 1592–1608 (2013)
21. Xiao, C., Li, B., Zhu, J.Y., et al.: Generating adversarial examples with adversarial networks. arXiv preprint arXiv:1801.02610 (2018)
22. Zhao, M., Yue, S., Katabi, D., et al.: Learning sleep stages from radio signals: a conditional adversarial architecture. In: International Conference on Machine Learning, pp. 4100–4109 (2017)

Formal Methods, Artificial Intelligence, Big-Data Analytics, and Knowledge Engineering in Medical Care to Reduce Disease Burden and Health Disparities

Sakthi Ganesh and Asoke K. Talukder[(⊠)]

Vibrant Health Sciences, Wanchai, Hong Kong
{sakthi,asoke}@vibranthealthsciences.com

Abstract. Medical errors and overtreatment combined with growing non-communicable disease population are responsible for increase in the **burden of disease** and **health disparity**. To control this burden and disparity, automation with zero defects must be introduced in evidence based medicine. In **safety critical** systems, zero defects are achieved through formal methods. A formal model is tested (proved) and the target system is generated through automation with the removal of error prone programming or construction phase. Inspired by similar ideas, we created **DocDx**, a novel formal method driven medical care framework without any programming phase involved. We convert **clinical pathways** into a **multipartite directed weighted graph** (MDWG) that embeds the medical intelligence. The autonomous interpreters in the server presents **natural language generator** (NLG) pathophysiology questions a doctor would normally ask a patient to understand the signs and symptoms of a disease. The biological terms and human understandable unstructured text entered in DocDx client is made machine understandable through AI NLP engine and translated into biomedical ontology concepts. A new medical condition or presentation of disease in DocDx will need a new clinical pathway translated into a MDWG without the need for any programming or application development process either at the client or at the server end.

Keywords: Formal methods · Artificial intelligence · Knowledge engineering
Clinical pathway · Evidence based medicine · Chatbot · Outcome assessments
Disease burden · Health disparity

1 Introduction

Health disparity refers to higher **disease burden** between populations. Many factors increase the disease burden in a population. At the center of this crisis are the medical errors caused by human factors. **Medical errors** are the third major cause of death in the United States of America [1]. Medical errors happen primarily due to systemic failures caused by information gap in patient and the doctor. Another factor is knowledge gap emanating from the gap between "truth" and "what one thinks one knows about the truth". These gaps comprise of patient's health knowledge gaps and the population

© Springer Nature Switzerland AG 2018
A. Mondal et al. (Eds.): BDA 2018, LNCS 11297, pp. 307–321, 2018.
https://doi.org/10.1007/978-3-030-04780-1_21

health knowledge gaps. Patient's health knowledge gaps relate to health professional's knowledge about the patient's health and patient's disease conditions. This will include the *current episode* and history of *all previous episodes* (interactions) and their outcomes. In addition, Patient's health knowledge will include patient's *medication history*, *drug toxicity* (side effect), *misdiagnosis* of the past, and patient's *family history*. In contrast, the population health knowledge will include both the *public health* knowledge and the *population health* knowledge.

The disease burden is further increased by avoidable over-treatments. This unnecessary care increases the cost of disease management and disparities. About one third of healthcare spending in the United States of America is unnecessary due to overuse or overtreatment – this includes unnecessary medical tests, excess medication, unnecessary procedures and unnecessary health care [2].

In United States of America about 18% of the GDP (Gross Domestic Product) is spent on healthcare, still, it ranked poorly in a survey on medical care (except specialist care). In other high income countries like Australia, Canada, France, Germany, the Netherlands, New Zealand, Norway, Sweden, Switzerland, and United Kingdom, shortfalls in care management were reported [3]. In low income countries disparities are due to low doctor density and poor health infrastructure. Available statistics show that over 45% of WHO Member States report to have less than 1 physician per 1000 people in the population [4].

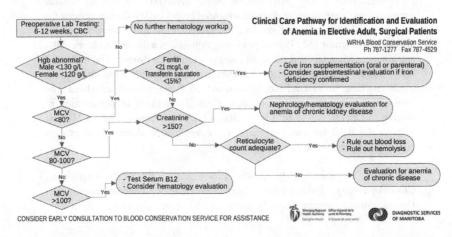

Fig. 1. The clinical pathway for anemia management at WRHA, Canada [18]. It can be noted that the algorithm (clinical pathway) for the anemia is similar to the flowcharts that are used in software design.

Communicable diseases are caused by external agents like bacteria, virus, fungus; whereas, non-communicable diseases are caused by pathogenic genes in the body. A communicable disease has acute (quick) onset and is diagnosed through morphologic signs or pathophysiologic symptoms at an early phase of the disease. Non-communicable diseases generally manifest at a late stage of the disease through complex co-occurring internal causes (including genomic and molecular causes) with confusing signs and symptoms. These confusing signs often lead to medical errors.

Researchers at Mayo Clinic in the United States found that 88% of patients who were referred to Mayo Clinic for a second opinion were misdiagnosed [5]. Also, there are errors in surgery known as Wrong-Site/Wrong Side, Wrong-Procedure, and Wrong-Patient (WSPE) adverse events [6]. Moreover, there are drug related casualties that include overdose, toxicity, and side effects. The age-adjusted rate of drug overdose deaths in the United States increased from 6.1 in 1999 to 19.8 per 100,000 standard population in 2016 (https://www.cdc.gov/nchs/products/databriefs/db294.htm). The other avoidable medical incidents are hospital acquired infections and hospital acquired sepsis [7].

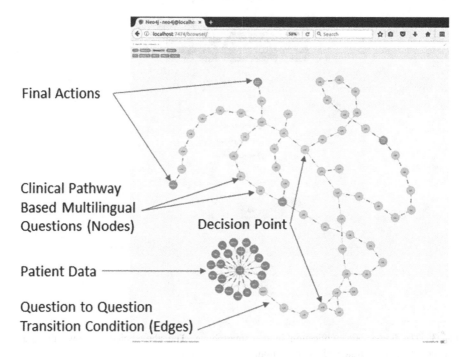

Fig. 2. A simplified directed multipartite probabilistic graph (MDWG) for chest pain. In this graph there are three independent sets of nodes. The orange nodes are pathophysiological questions related to the clinical pathway taken from the triage protocol [28]. The clinical pathway is based on population health and evidence based medicine. The pink nodes are the final actions. The grey nodes are the vital information about a patient. (Color figure online)

The disease burden is similar to the software crisis that was first debated in NATO Software Engineering Conference in 1968 [8]. The adhocism and human factors in the software development has highest propensity of introduction of software errors, bugs, and defects. In 1947 Goldstein and von Neumann [9] presented a formal technique of drawing the *flow diagram* of the algorithm to be implemented on a machine in anticipation of this propensity. In Safety critical systems formal methods are used to remove some of these error prone phases. In Paris, RATP (Régie Autonome des Transports Parisiens) has been using safety critical platform screen doors (PSD) for its driverless metro trains. This system used B formal method [10], where there was no programming phase at all in the software development lifecycle. This system used about 110,000 lines of B code and proved to be error free. This B code generated 86,000 lines of Ada code through automated tools [11] used in the target driverless trains. Our approach towards zero defects in medical care is based on similar philosophies. In Safety critical computer software and hardware designs defects are eliminated through the use of formal methods [12]. Formal methods are capable of eliminating both undeliberate and deliberate errors. We used similar techniques in medical care to mitigate the medical errors and medical overtreatment.

Fig. 3. The home care application in the smartphone. The above picture shows the multi selection check-box followed by the action.

Here in this paper we present a smartphone based medical care named *DocDx* that uses formal models and automation with a goal to reduce the disparities in health and healthcare due to lack in infrastructure and low physician density. We automated *clinical pathways*. Clinical pathways (Fig. 1) are formal procedures in medical decision making process used by medical practitioners. Clinical pathways are based on algorithms created with the help of *evidence based medicine* (EBM). We converted clinical pathways into *multipartite directed weighted graphs* (MDWG). DocDx uses various formal technology frameworks to eliminate errors and defects in medical care. This will promote timely, accurate, and actionable insight for patients and doctors alike to reduce the disparity.

2   Formal Methods in Evidence Based Medicine

Clinical pathways are peer-reviewed formal methods in medicine used in medical diagnosis and therapeutics. For example, the book "Detection, Prevention and Management of Urinary Tract Infections" by Kunin [13] was reduced from more than 300 pages of text representing more than thousands of pages of literature survey from more than 500 articles to one page with just three algorithms [14]. In different literature EBM is also referred as 'clinical pathways' [15]. Along with limiting errors, clinical pathways and algorithms reduce the variability and improves outcome [16, 17] in medical decision making.

Fig. 4. The outcome assessment application combined with medication alerts, and follow-ups.

A medical algorithm or a formal process in medicine is represented as a clinical pathway as shown in Fig. 1. This clinical pathway (Fig. 1) depicts the identification and evaluation process of anemia in elective adults for surgical patients standardized by Winnipeg Regional Health Authority (WRHA) [18]. Clinical pathways are in fact equivalent to flow diagrams proposed by Goldstein and von Neumann.in computer science [9]. Proper use of such algorithms can identify and eliminate many avoidable errors and unnecessary medical encounters [19, 20].

2.1   Benefits of Formal Methods in Medicine

Heckerman used formal methods in medical care in 1989 in his Pathfinder system. Heckerman used Bayesian network as part of a decision support system for the use of surgical pathologists for lymph node disease [21, 22]. Because of the inherent power of formal methods, the diagnostic accuracy of Pathfinder was higher than that using the routine methods by human experts [23].

Pathfinder was designed to address one specific disease though. In the latest release of ICD (International Classification of Disease) in 2018, there are more than 112,383 disease codes that have been catalogues in ICD11 [24]. Therefore, it is humanly impossible to follow the same strategy of Pathfinder today of having one application for one disease. Here in this paper we present a formal model in medical care named *DocDx* that is disease agnostic. Moreover, Pathfinder was designed to be used for expert physicians; however, our model is designed to be intuitive, to be used by patients, patients' families, advocacy groups, nurses, paramedics, pharmacists, and the physicians without any format training.

2.2   Machine Understandable Medical Notes Through AI and NLP

Public health is the science and art of preventing disease, prolonging life and promoting human health. Unless the disease data in the EMR is machine understandable, such epidemiological analysis will be impossible. As a first step towards reduction of health disparity, the health records and most importantly the medical notes must be *computer understandable*.

For centuries medical notes have been handwritten. The first documented medical notes were used during the Babylonian Civilization about 5000BCE. During Egyptian Civilization medical notes were handwritten notes on papyrus in 3000BCE. This millennium old tradition of handwritten medical notes is still in use across the globe. These handwritten notes are human readable and human understandable – a computer cannot read or understand the content of such medical text. We define handwritten human readable and human understandable medical notes as the *First Generation Health Care Systems*. During late 20th Century, EMR (Electronic Medical Records) or EHR (Electronic Health Records) systems were introduced. These medical records still need a medical expert to interpret. We define these healthcare systems as *Second Generation Health Care Systems* because they are machine readable but only human understandable.

Our DocDx formal methods transforms a 2nd Generation Health Care system into a *Third Generation (3G) Health Care Systems* using Artificial Intelligence (AI) and Natural Language Processing. We convert an unstructured medical note into a set of machine understandable disease codes in SNOMED CT (Systematized Nomenclature of Medicine – Clinical Terms) and ICD10 (International Classification of Disease version 10). For this function, we used UMLS (*Unified Medical Language System*) from National Library of Medicine, USA (https://www.nlm.nih.gov/research/umls/).

2.3   Knowledge, Artificial Intelligence, Reasoning and Big Data Analytics

It is estimated that a medical student is required to learn 130,000 bits of information whereas a practitioner needs more than 2 million bits of information for everyday work. It is humanly impossible today for unaided healthcare professionals to deliver timely medical care with the efficacy, consistency, safety, and accuracy that the full range of advanced knowledge could support. Therefore, all these knowledge need to be represented in formal methods such that they can be stored and consumed by computers as and when necessary.

A software defect is introduced during the construction phase of the intelligence into the application. Such errors result from the gaps between understanding of a business function (intelligence) by a domain expert and a software engineer. To realize zero defects, in DocDx, we isolated these two functions of software engineering and medical intelligence (knowledge) entirely. We constructed semantic networks that encapsulate both the medical intelligence and the medical knowledge.

The representational theory of mind (RTM) defines intentional mental states through relation of mental representations through two dimensional attributes where intentionality of the former is represented through the semantic property of the later. Semantic network with controlled vocabularies are used to organize knowledge in computers for subsequent retrieval. Such knowledge is used for reasoning and decision making by a computer accurately with the help of artificial intelligence (AI).

Various authors have defined big-data having four properties, which are Volume, Velocity, Variety, and Veracity. Biological big-data however needs 7 V viz., *Volume, Velocity, Variety, Veracity, Vexing, Variability*, and *Value* [25]. Databases like Gene Ontology (http://geneontology.org), KEGG (https://www.genome.jp/kegg), COSMIC (https://cancer.sanger.ac.uk/cosmic), OMIM (https://www.omim.org), Interactome (interactome.dfci.harvard.edu/), BioGRID (https://thebiogrid.org) etc. are all examples of biological big-data. In our system, most of these biological data are in the process of integration (future work) with the SNOMED CT clinical database.

3 The DocDx Formal Model for Medical Care

In DocDx, we developed three frameworks. These frameworks are, (a) Home care, (b) Primary care, and (c) Peer-to-peer surveillance, outcome assessment, and follow-ups. Keeping primary care and home care in mind, we used smartphone as the user device. Another advantage of using smartphone is to remove the challenges related to difficult patients and reserved or shy patients who will not feel intimidated by the presence of a doctor.

4 Implementation of the Formal Model

We used two graph databases in our system. They are (a) disease graph (for patient), and (b) knowledge graph (for population). The disease graph is a multipartite directed weighted graph (MDWG) to represent a clinical pathway into a computer. Disease graph is used for a single patient ($n = 1$); whereas, knowledge graph is a multipartite directed graph generated from population health and research ($n = N$). We used the open-source Neo4j graph database (https://neo4j.com/) for disease graph, clinical knowledge graph, and biological knowledge graph. Figure 5 shows a disease graph for chest pain medical condition.

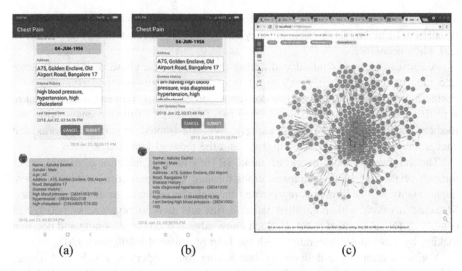

(a)   (b)   (c)

Fig. 5. (a) Shows the user input of "high blood pressure" and "hypertension" are interpreted by the machine as SNOMED CT code 38341003 and ICD10 code of I10. (b) Another user enters "I am having high blood pressure". In this case also the AI system is able to resolve it to the same SNOMED CT/ICD code of 38341003 and I10 respectively. (c) SNOMED CT semantic network for hypertension (code 38341003) with the diameter of 3. In this graph pink nodes are finding concepts, yellow nodes are description concepts, and grey nodes are attribute concepts. (Color figure online)

The machine understandable clinical knowledge body is created from SNOMED CT database (https://www.snomed.org/). The SNOMED CT knowledge base is converted into a semantic graph in Neo4j (https://neo4j.com/) graph database [26]. Figure 5 shows a multi-partite graph of hypertension in Neo4j. We defined such systems as the *3rd Generation Health Care System*.

Our system is multilingual that supports seven languages, viz., English, Hindi, Tamil, Kannada, Bangla, Chinese, and Arabic. Such systems are very useful in a real life situation where the patient may be conversant with Arabic and the nurse is comfortable with Hindi. We mapped clinical pathways into triage systems that generate multilingual questions using *natural language generator* (NLG). The original concepts of triage were primarily focused on mass casualty situations. Triage controls and prioritize patient treatment efficiently when resources are insufficient [27]. Figure 2 shows a MDWG for Chest Pain. The questions for this graph (Fig. 2) are taken from a telephone triage protocol [28]. The graphs prioritized the patient cases into four actions groups like, immediate (call the ambulance), urgent (refer to the hospital appointment system for outpatient examination within 24 h), non-critical categories (schedule appointment within a week), and non-urgent (schedule appointment within a fortnight).

One of the AI examples of DocDx is shown in Fig. 5. In this example the user enters three disease names in the smartphone App as the disease history. These three diseases are "high blood pressure", "high cholesterol", and "hypertension". These are three words, two words, and one word disease names respectively. The three words "high blood pressure" disease name is converted into machine understandable ontology concepts in SNOMED CT and ICD as 38341003 and I10 respectively. Two words "high cholesterol" disease name is converted into SNOMED CT code 13644009 and ICD code E78.00. It may be noted that the one word disease name "hypertension" is also converted into SNOMED CT code 38341003 and ICD code of I10. It is interesting to note that in Fig. 5(b), another user has written six words "I am having high blood pressure" in disease history. The artificial intelligence within the system is able to understand like an expert doctor that when somebody is saying "I am having high blood pressure" the person in fact is referring to "hypertension". Therefore, in this case also the AI engine in DocDx has resolves the six words disease name into SNOMED CT code of 38341003 and ICD code I10.

As part of the implementation, we start navigating the graph at the ***root*** node of the disease graph in the Neo4j database. We use the Cypher query language of Neo4j to navigate through the graph. Let us take an example to explain the function of the graph. The question "Do you Smoke?" is associated with a node "CHP010030". This node is a Binary node that has two nodes in the directed graph as NEXT nodes. The path with "Yes" relation will be chosen if the user types "Yes". The "Yes" path has the question "CHP010040 How many cigarettes a day?" In case the user types "No", the "No" path will be chosen, which is "CHP010050 Are you Diabetic?"

4.1 Home Care

In the directed multipartite disease graph there are three independent types of nodes (Fig. 2) to realize home care function (Fig. 3). First set of nodes represents the questions in the clinical pathway. Second set of nodes represent the final action. The third set of nodes are the instance of the patient details that will normally come from the EHR (Electronic Health Record) or the PHR (Personal Health Record) databases maintained by a hospital, clinic, or HIE (Health Information Exchange) [29].

4.2 Primary Care

The medical condition and the questions in ***primary care*** are more exhaustive. Here questions are often generated based on the answer of the previous answer using NLG. When we want to use a different medical condition, we define a new graph from the clinical pathway of the disease, get it verified by a certified medical doctor or surgeon and deploy.

This Primary Care interface will be used as 24 x 7 telemedicine as well. Especially in low and middle income countries where the physician density is less, DocDx will be of tremendous help. A multilingual DocDx will not only be able to play the role of a GP (General Physician), but also will function as a registry for public health. The chat records thus captured in the DocDx log will be analyzed to help understand the disease demography in a population. It will also help understand symptoms of a disease and the effectiveness of an intervention or therapeutics.

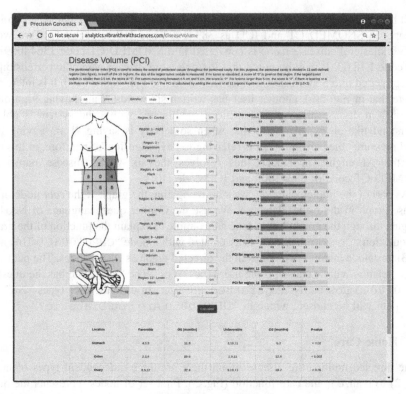

Fig. 6. The disease volume calculator for peritoneal cancer along with the prognosis and overall survivability.

4.3   Outcome Assessments, Medication Reminders, and Follow-Ups

There are many challenges in medication compliances [30]. Many patients do not take medicines as prescribed by the doctors. This is a common problem across the world independent of country, culture, or socioeconomic conditions. Patients often do not communicate with their doctors about their medicine-taking behavior and the outcome of the medical encounter. Figure 4 depicts the DocRx (outcome assessment) application. This framework has the same graph structure and the navigational framework at its core like the Home care and Primary care. However, for this application we use a simple directed graph with three main calendar databases at the server end.

These calendar databases relate to the medication schedules (pill-card), signs and symptoms of a patient, and the follow-up schedules. The pill-card calendar contains the medication schedule and the dosages. The alert and notification system of our framework sends the alert for medication to the concerned parties. The signs and symptom calendar includes questions about the patients' health and the efficacy of the medication. The follow-up schedules include all types of follow-ups starting from lab visits, visit to the doctor or visit to the dialysis center or an injection by nurses etc.

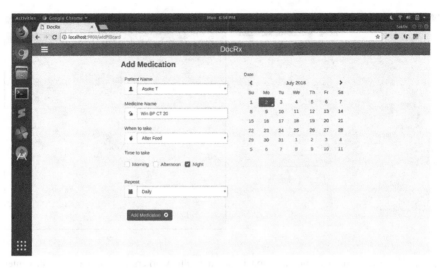

Fig. 7. The setting up of medicine alerts through DocRx for the Medication Alert in the peer-to-peer protocol

4.4   Peer-to-Peer Communication (P2P) and Surveillance

In standard Web based EMR/EHR application messages are paired and synchronized. In a hospital EMR/EHR a doctor or a hospital staff always initiates a conversation and receive a response back from the application. Patient-provider communication often need peer-to-peer communication protocol for emergency services and disease surveillance. In such protocol any node in the network should be able to initiate a conversation – a patient or a doctor or a hospital support staff will be able to send one or more messages in sequence. On the other hand, a hospital will also be able to send one or more messages in an asynchronous manner to a patient or another doctor. The hospital will in addition will be able to send alerts and notifications to a patient or a doctor in duty. Figure 7 shows the screenshot of setting up of the medication alert. The alert is set using the Medication Alert as shown in this figure. Everyday evening around 8:00PM the user (Asoke T in this case) will receive a notification that "Win BP CT 20" needs to be taken.

Unlike non-communicable diseases that has late onset, communicable disease cases are always acute in nature and needs immediate attention. Also, communicable or infectious diseases can penetrate in the population very fast causing pandemics. Therefore, they need peer-to-peer surveillance protocol (P2P). Also, the control room user can communicate with the health worker as a requester any time through the P2P protocol. There are other types of emergency requirements as well that will be serviced by this application like road accidents, natural disasters etc. A regional hospital or a primary care will also report the cases of birth and deaths using this application.

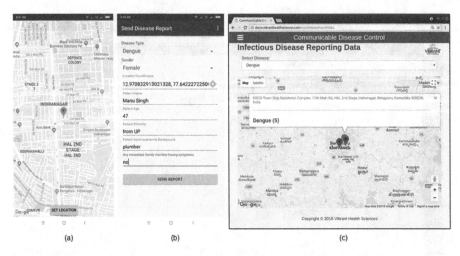

Fig. 8. The peer-peer to communication between health workers and primary health centers with the disease control center

In Fig. 8 we show the sequence of events in a use case of dengue fever in a neighborhood of Bangalore, India. Authorized healthcare personnel enter the infectious disease cases using their smart phones as shown in Fig. 8(a). In this use case, the GPS (Global Positioning System) application in the smartphone client is used along with Google Maps. The healthcare staff uses the mobile App to identify the location of the dengue case. Once the location is identified, the healthcare staff enters the detail of the patients and sends the same to the server (Fig. 8(b)). The server in the cloud receives the detail of the dengue case through the P2P protocol and enters into the database. Figure 8(c) shows the detail of the dengue fever cases on the Google maps at the control room. Health controllers at the disease control room will click the pin of interest in the map to know the spread of the disease. Further statistical details can be retrieved from the database as formal reports through standard database queries. The infectious disease data will be used to counter pandemic. This data will be used for geospatial diffusion analysis [31]. Diffusion analysis will be able to predict the vector of the disease penetration. This knowledge will be used to control the epidemic quickly and ensure public health.

In case of other emergency services like road accidents or disasters, a medical team will be rushed to the accident or disaster site without delay. Because the data has been entered using the GPS, the precise location of the accident is known. The data captured in case of child birth and deaths will help preventive actions to reduce the burden of disease as well.

5 Predictive and Precision Medicine for NCD

Our medical analytic system deals with big-data analytics engine for various communicable and non-communicable diseases starting from cholera to cancer. Figure 6 shows a case of volume of disease for peritoneal cancer calculated and visually represented by our analytic system. At the bottom of Fig. 6, the likely survivability of the cancer patient is also shown. This is for a case of late stage ovarian cancer that has metastasized.

Our big-data analytic system in evidence based medicine and *precision medicine* has already been published [29]. It uses *exploratory data analysis* (EDA) techniques and *statistical data mining* to create hypothesis from the medical data. The data sources are anonymized hospital EMR, Lab, radiology, data of patients. The analytic system described in [29] analyzes these data to obtain the population specific spatial and temporal health related knowledge. This health knowledge will be used to determine the weights in the MDWG. The comorbidity and genomic knowledge and the knowledge about biology and disease on contrast will determine the disease interaction and its probabilities. This will define the probability of one or multiple disease both in spatial and temporal coordinates. With this disease relationship knowledge out MDWG will become probabilistic.

In our future work the multipartite directed weighted graph will become *multipartite directed probabilistic weighted graph* (MDPWG), the weights for edges will be computed from the population clinical and medical data gathered from hospital data. The probability of edges in contrast will be assigned from the literature and molecular biology knowledge of disease. For example the *minor allele frequency* (MAF) helps determine the probability of a pathogenic mutation at which the second most common allele occurs in a given population. In our next phase, this TDPWG will be used in Bayesian statistics and Hidden Markov Model to make it a *predictive formal model*. This predictive model will soon be used by the autonomous formal model in our future work. This will help the non-communicable diseases to be diagnosed at early stage of the onset and reduce the disease burden. This predictive analysis will also help to determine whether an intervention or a medical interaction is really necessary or not and decrease avoidable encounters. This will also be used for lifestyle change and improve the health equity.

6 Conclusion

Medical errors and overuse are some of the main causes of high disease burden and poor outcomes. In addition, there is changing disease demography. Moreover, as medicine is reactive and fully manual, the cost of expert hour is pushing the medical cost up. High burden causes high disparity as well. Only way to mitigate this burden and disparity is to bring in automation. However, medicine involves human life; therefore, the automated system must be zero defects. Software industry taught us how to design zero defects safety critical system. It also taught us how to overcome software crisis caused by human errors and adhocism through formal methods and mathematical models. We borrowed these techniques and used in our medical care system.

In this paper, we presented DocDx and DocRx – a set of formal autonomous method driven systems. Our formal methods automate evidence based clinical pathways. In our system we do not write any code for implementation of any specific medical condition. We translate an evidence based clinical pathway into a directed graph. In our system the domain intelligence is not coded by human into the software; but, it is embedded in the semantic network within the graph. Our framework navigates through these graphs and does intelligent decisions at runtime. This increases the accuracy of the medical decision and reduces the medical error. It reduces the interaction time between the doctor and the patient, which increases the productivity.

Out autonomous model will reduce the testing and maintenance cost of clinical medical system at the point of care. Greatest advantage of this model is that the variability in diagnosis will be limited and will always be reproducible. We have presented a set of use-cases for Home care, and outcome assessment in medical care and showed how they contribute towards reducing disease burden and arrest the medical crisis.

Our autonomous model uses Internet and AI to reduce the dependence on human, infrastructure, and geographical constraints. Patients can be anywhere, doctors can be anywhere – they are connected through Internet. These systems offer medical helps to anybody anywhere anytime to reduce healthcare disparity which will finally reduce the disease burden and the health disparity.

References

1. Makary, M.A., Daniel, M.: Medical error-the third leading cause of death in the US. BMJ **3** (353), i2139 (2016)
2. Gawande, A.: Overkill. Analysis of Health Care, The Newyorker, May 11, 20 issue (2015). (https://www.newyorker.com/magazine/2015/05/11/overkill-atul-gawande)
3. Osborn, R., Squires, D., Doty, M.M., Sarnak, D.O., Schneider, E.C.: In new survey of eleven countries, US Adults still struggle with access to and affordability of health care. Health Aff (Millwood) **35**(12), 2327–2336 (2016). Epub 16 Nov 2016
4. WHO: Density of Physicians. (http://www.who.int/gho/health_workforce/physicians_density/en/)
5. Van Such, M., Lohr, R., Beckman, T., Naessens, J.M.: Extent of diagnostic agreement among medical referrals. J. Eval. Clin. Pract. **4**, 870–874 (2017)
6. Seiden, S.C., Barach, P.: Wrong-side/wrong-site, wrong-procedure, and wrong-patient adverse events: Are they preventable? Arch. Surg. **141**(9), 931–939 (2006)
7. Mehta, Y., et al.: Guidelines for prevention of hospital acquired infections. Indian J. Crit. Care Med. **18**(3), 149–163 (2014)
8. Software crisis. (https://en.wikipedia.org/wiki/Software_crisis)
9. Goldstein, H.H., von Neumann, J.: Planning and coding problems for an electronic computing instrument (1947). Part II, vol I. Rep. prepared for the U.S. Army Ordinance Dept., 1947. Reprinted in von Neumann, J. Collected Works, Vol. V, A.H. Taub, ed., McMillan, New York, pp. 80–151 (1947)
10. Abrial, J.R.: The B tool (Abstract). In: Bloomfield, R.E., Marshall, L.S., Jones, R.B. (eds.) VDM 1988. LNCS, vol. 328, pp. 86–87. Springer, Heidelberg (1988). https://doi.org/10.1007/3-540-50214-9_8

11. Lecomte, T., Servat, T., Pouzancre, G.: Formal methods in safety-critical railway systems. In: Proceedings of Brazilian Symposium on Formal Methods: SMBF 2007 (2007)
12. Talukder, A.K., Chaitanya, M.: Architecting Secure Software Systems. Auerbach Publications, Boston (2008)
13. Kunin, C.M.: Detection, Prevention and Management of Urinary Tract Infections. 2nd Revised edn. Lea & Febiger, U.S. (1974)
14. Djulbegovic, B.: Reasoning and Decision Making in Hematology. Churchill Living-stone (1992)
15. Kinsman, L., Rotter, T., James, E., Snow, P., Willis, J.: What is a clinical pathway? Development of a definition to inform the debate. BMC Med. **8**, 31 (2010). https://doi.org/ 10.1186/1741-7015-8-31
16. Panella, M., Marchisio, S., Di Stanislao, F.: Reducing clinical variations with clinical pathways: do pathways work? Int. J. Qual. Health Care **15**(6), 509–521 (2003)
17. Rotter, T., et al.: The effects of clinical pathways on professional practice, patient outcomes, length of stay, and hospital costs: Cochrane systematic review and meta-analysis. Eval. Health Prof. **35**(1), 3–27 (2012). Epub 24 May 2011
18. WRHA Clinical Care Pathway for the identification and evaluation of anemia in elective adults, surgical patients. (https://bestbloodmanitoba.ca/wp-content/uploads/2014/08/Anemia-Care-Pathway.pdf)
19. Schubart, J.R., Fowler, C.E., Donowitz, G.R., Connors Jr., A.F.: Algorithm-based Decision Rules to Safely Reduce Laboratory Test Ordering. MEDINFO 2001, IOS Press (2001)
20. Eaton, K.P., et al.: Evidence-based guidelines to eliminate repetitive laboratory testing. JAMA Int. Med. **177**(12),1833–1839 (2017)
21. Heckerman, D.E., Horvitz, E.J., Nathwani, B.N.: Toward normative expert systems: part I The Pathfinder project. Methods Inf. Med. **31**(2), 90–105 (1992)
22. Heckerman, D.E., Nathwani, B.N.: Toward normative expert systems: Part II. Probability-based representations for efficient knowledge acquisition and inference. Methods Inf. Med. **31**(2), 106–116 (1992)
23. Nathwani, B.N., et al.: Evaluation of an expert system on lymph node pathology. Hum. Pathol. **28**(9), 1097–1110 (1997)
24. ICD-11 Update. (http://www.who.int/classifications/2017_10_ICD11_Newsletter.pdf?ua=1)
25. Talukder, A.K.: Genomics 3.0: big-data in precision medicine. In: Kumar, N., Bhatnagar, V. (eds.) BDA 2015. LNCS, vol. 9498, pp. 201–215. Springer, Cham (2015). https://doi.org/10. 1007/978-3-319-27057-9_14
26. Campbell, W.S., et al.: An alternative database approach for management of SNOMED CT and improved patient data queries. J. Biomed. Inform. **57**, 350–357 (2015)
27. Robertson-Steel, I.: Evolution of triage systems. Emerg. Med. J. **23**(2), 154–155 (2016)
28. Briggs, J.K. (ed.) Telephone Triage Protocols for Nurses. Lippincott, Philadelphia (2002)
29. Talukder, A.K.: Big data analytics advances in health intelligence, public health, and evidence-based precision medicine. In: Reddy, P.K., Sureka, A., Chakravarthy, S., Bhalla, S. (eds.) BDA 2017. LNCS, vol. 10721, pp. 243–253. Springer, Cham (2017). https://doi.org/ 10.1007/978-3-319-72413-3_17
30. Bezreh, T., Laws, M.B., Taubin, T., Rifkin, D.E., Wilson, I.B.: Challenges to physician–patient communication about medication use: a window into the skeptical patient's world. Patient Prefer. Adherence **2012**(6), 11–18 (2011)
31. Congdon, P.: Spatiotemporal Frameworks for Infectious Disease Diffusion and Epidemiology. Int. J. Environ. Res. Public Health **13**(12), 1261 (2016)

Adaboost.RT Based Soil N-P-K Prediction Model for Soil and Crop Specific Data: A Predictive Modelling Approach

Rashmi Priya and Dharavath Ramesh[(⊠)] [iD]

Department of Computer Science and Engineering,
Indian Institute of Technology (Indian School of Mines), Dhanbad, India
rashmipriya.aec@gmail.com, ramesh.d.in@ieee.org

Abstract. In relation to the evaluation of the soil breeding status of a region or realm, the soil characteristics are an important aspect in terms of agricultural production. Nitrogen, phosphorus, potassium, and sulfur are important elements of soil that regulate its fertility and yield of crops. Due to the low efficiency of other inputs or due to the use of unbalanced and inadequate fertilizer, the reaction of chemical fertilizer nutrients (production) efficiency in recent years has reduced considerably under intensive agriculture. Stability in crop productivity cannot be extended without the judicial use of macro and micro nutrients to overcome existing deficiencies. The information on the availability of macro nutrients in the study area is low. Therefore, the current study has been done to know the condition of soil nutrients. Use of advanced agricultural technology can help in predicting soil nutrient content and can help farmers to decide the amount of fertilizers to use on a particular land. The proposed study focuses on the accurate prediction of N-P-K content in the given land by utilizing the predication method using Adaboost.RT method. A comparison is also made in between the nutrient utilized using traditional methods and the proposed method. Experimental results show that the proposed stream outperforms with other existing methodologies.

Keywords: Prediction · Soil nutrients content · Adaboost.RT algorithm

1 Introduction

India is the land of agriculture. About 70% of the population is involved in crop production and related activities. Agriculture sector helps the country's economy because it contributes about 18% of GDP. According to the Economic Survey (2015–16), it has been said that the Green Revolution, which has played an important role in the growth of crop productivity in India, had left its dark signs. Undoubtedly, with the introduction of high-yielding seed variety, new methods of farming techniques and fertilizer have increased manifold in crop production, but there were also negative environmental effects. Negative effects include water table, greenhouse emissions, ground and surface water pollution. Also, as soon as the population is growing rapidly, crop productivity remains the same, therefore, there is a need to meet the growing crop demand.

Soil is an important component of the earth system, not only for the production of food, fodder and fibres but also for the maintenance of local, regional and global

© Springer Nature Switzerland AG 2018
A. Mondal et al. (Eds.): BDA 2018, LNCS 11297, pp. 322–331, 2018.
https://doi.org/10.1007/978-3-030-04780-1_22

environmental quality. Farmers have been in Asia for centuries and practiced a cultural system that ensures moderate but stable yield, still maintains the desired level of fertility in the soil. This equilibrium was disturbed by the need to increase production through the use of high yielding seed varieties, chemical fertilizers and pesticides and the extensive use of sustainable farming [1, 2].

After the Green Revolution, there was an increase in crop production in the country, but more such revolution is required. Since the population has grown rapidly but crop production is not enough to feed the current population. The importance of soil in the supply of plant nutrients has been known since the beginning of agriculture. The concept was more around 1840, when several attempts were made to estimate the potential of soil nutrient supply and to get the exact basis for predicting the fertilizer requirements of crops. By the middle of the twentieth century, agriculture in South Asia, however, was largely dependent on the inherent nutrient reserve of soil and organic compost. With the introduction and expansion of modern high-yielding varieties as well as the development of irrigation facilities in India, the use of fertilizers has been increasing for more than sixty years. However, in recent reports of South Asia suggested the trend of decline in the production of rice and wheat. Among the suggested major causes, due to the decline in improper fertilizer applications and soil organic material content, there is a gradual decline in the supply of nutrients in soil due to the nutrient (macro and micro) imbalance. With decline in yield, partial factor productivity is going down with the applicable nutrients, resulting in cost of production being increased. In case of loss of nutrients in gaseous form and leaching, it is linked to environmental costs at the same time.

As indicated by the report by ICAR [3], farmers in India are making the use of the soil at its maximum, growing at least two products every year, rather than one without appropriate soil management. This unplanned intensity is increasing nutrient deficit and the chemical structure of the soil is changing. Levels of natural carbon in soil are dropping the nation over, making soils more susceptible to erosion and perhaps bringing about the quantity of worms falling [4, 5]. Therefore, it is necessary to ensure continuous monitoring of soil nutrient supply capacity to ensure and improve the sustainability of agriculture. For keeping up of soil quality and achievable product yield, it is required to include a legitimate measure of manures and limit the misuse of soil assets which is conceivable by knowing the absolute soil condition through perception, examination and soil testing. A Soil test is a basic means of surveying the compost or excrement necessity for maintainable creation of yields and for supporting soil fertility [6].

Because the growing smart machines and sensors of the crop manifest on the quantity and scope of the crop on farmland and farm data, the agricultural process will turn out to be progressively information driven and information empowered. Fast improvements in the Internet of Things and Cloud Computing are pushing the marvel of what is called Smart Farming [7]. Precision agriculture is an essential part of the durable intensity of agriculture, where information and communication technologies and other technologies are essential, but not enough for sustainable agricultural systems [8]. Technology should be fit into the practice of farmers and should be handled by their experienced-based, valuable knowledge to contribute to the increasing sustainability in their farming. Precision agriculture utilizes a proximal and remote sensor reports to depict and analyse regional variations within soil and crop characteristics, directing variable rate control of data sources, so that soil monitoring can be effective, e.g. use of key nitrogen manure application to site-particular field conditions. It can possibly enhance productivity and nutrient use, guaranteeing that nutrients do not filter out or aggregate in a particular part of the field, which creates ecological issues. With the advent of affordable GPS (Global Positioning System), discipline emerged in the 1980s; and has additionally created with access to a variety of reasonable soil and crop sensors, enhanced PC power and programming, and hardware with accurate application control, e.g. convertible rate of compost and irrigation system frameworks. Precision farming focuses on enhancing nutrients use proficiency at the proper scale aiming at (i) correct decision support system (e.g. computerized maps), and (ii) tools capable of making separate applications on these varying scales (e.g. the impression of [a] one water system sprinkler or [b] a compost spraying drones) [9, 19].

The rest of the paper is organized as follows. First, the related work in the field of precision agriculture is discussed in Sect. 2. The advantages and limitation of different models are presented and also the areas to focus on are discussed. Section 3 discusses about the proposed methodology for the N-P-K prediction for the particular crop. Section 4 describes the outcome of the proposed method and also compares it with the results of traditional practices. Finally, we conclude the entire work in Sect. 5 with possible future enhancements.

2 Related Work

In literature, many agricultural advisory systems have been proposed to make agriculture environment more suitable for related agricultural crops [10]. By increasing the high quality seeds genetically, crop productivity can be increased. But the problem of handling such seeds is that while predicting, one has to look at the reaction of the genotype and the phenotype as well as the reaction to the environmental variable [11]. Other factors affecting productivity include soil, weed and nutrients and water content in their management [12].

Machine learning algorithms are playing an important role towards the development of precision agriculture [13]. An ANN approach was used to predict the yields of potatoes in Iran by forecasting the Greenhouse gas emission [14]. The data were collected by meeting 260 farmers in person, hence the output from the ANN model developed could be cross checked by the experts and was concluded that the models were efficient. The factors influencing the output were electricity supplied amount of fertilizers used and the seed variety. Out of various ANN models it was concluded, 12-8-2 model performed best with an efficiency of 98% for potatoes output energy prediction and 99% for GHG emission.

In [15], the authors showed the use of wireless sensor for soil moisture detection and automated irrigation system. The sensor was deployed to collect information like soil moisture, soil pH, so that these variables can be analyzed to maximize crop production. It provides farmers with information related to agriculture such as seeds, moisture levels, and types of soil requirement, weather forecasting, fertilizers and pesticides.

A methodology using linear regression with neural network has been used to predict the occurrence of rainfall in the relevant geographical area. For this, data was collected from Ahmednagar, India's weather station and daily maximum and minimum temperature, humidity and rainfall of the last 10 years were studied. Linear regression and ANN helps to predict future rainfall more accurately [16]. For building analytical models various data mining techniques have been applied on datasets. In a work presented in [17], problem of soil clustering was addressed using k-mean and fuzzy k-mean algorithms. A database of soil sample characteristics monitored in Montengro was used for comparative analysis of the applied algorithms. The results obtained using K-mean are presented on the static Google map and dynamic open street maps. The authors' claim that the soil data and the data mining results presented on the map is a good way to present data to scientists and land users. In [18], a smart irrigation system is proposed using two machine learning algorithm, PLSR and ANFIS. The irrigation system helps in estimating the weekly water requirement of plant on the basis of weather and soil condition as well as plant characteristics.

3 Proposed Methodology

In many cases it is quite difficult to apply regression to real world data to predict the exact values as seen in the classification. Due to non-uniform distribution of real value data, there is a huge difference between the predicted and observed values which is most often unavoidable. Use of algorithms like SVM, to draw margins between the predicted and observed values is one way to overcome this limitation, but is mainly used in classification. The proposed work uses Adaboost.RT [20] to predict the N-P-K content in the soil and the suggest amount good for yield enhancement. The dataset used for our experimentation is shown in Table 1.

Table 1. Dataset used

Crop type	Soil type	Available			Required			Predicted			Yield before	Yield after
		N	P	K	N	P	K	N	P	K		
Paddy	Sandy loam	316	119	56	237	43	30	233	41	29	1850	2140
Cotton	Sandy loam	92	106	28	32	51	44	29	50	45	1680	2146
Ground-nut	Sandy clay loam	283	70	99	173	51	55	171	52	55	2750	3613
Maize	Sandy loam	182	81	53	109	59	42	111	56	39	3280	4421
Gram	Maize	20	14	2	12	6	2	11	9	3	3165	4092
Ragi	Red Laterite	109	132	66	75	54	35	78	55	34	1940	3257
Soybean	Loamy	28	73	16	46	111	13	43	115	11	1350	1980
Sunflower	Loamy	119	79	0	96	62	0	93	59	0	2030	3549

Adaboost.RT is a boosting algorithm for regression analysis by using user defined threshold value. By comparing the errors between the values and the threshold values φ it divides the training dataset into two classes, i.e. good prediction and poor prediction. It uses a different weight updating method to consider strong training data, where the error is considerably low. It also gives its user flexibility to make any number of iterations even if the error rate is greater than a predefined value as compared to other algorithms. At the time of final prediction, all the outputs of different subsets are combined using the weighted mean, hence the output is more précised. The performance of the algorithm after each iteration is evaluated by calculating the error rate. If the absolute relative error for any test data sample is greater than the defined threshold φ value, then the prediction is incorrect, otherwise it is correct. In the weight updating phase the weight w_t is updated according to the power raised to error rate, where power varies in between 1 to 3, this helps in evaluating the tough subsamples where data sets are more dispersed. The number of iterations is determined based on the cross-validation error Er. The algorithm stops if the Er starts to increase, hence the algorithm is not dependent on error rate and does not terminate after a few iterations resulting in more accurate output.

Algorithm: *Adaboost.RT*

Input: n no. of input sample for $S_n = (X_1, Y_1), (X_2, Y_2), ..., (X_n, Y_n)$ where

$Y \in R$.   *(WL = The WeakLearner, I = No. of iterations)*

(φ = Threshold to distinguish between correct and incorrect prediction

$\varphi \in (0,1)$*)*

Output : $O_{fin(x)} = \dfrac{\sum\limits_{k} (\log \dfrac{1}{Wt}) O_k(x)}{\sum\limits_{k} (\log \dfrac{1}{Wt})}$

Data distribution $Dist_k(j) = \dfrac{1}{n} \forall j = [1, 2, ..., n]$, Error rate $Er = 0$

Begin: for $k = 1$ to I call WL

for $l = 1, ..., k$

pass training set through WL_l to obtain $Y_j^l(X_j)$ where $j = 1, ..., n$

$Abs\_rel\_err_j = \dfrac{|Y_j^l(X_j) - Y_j|}{Y_j}$ where $j = 1, ..., n$

$Er = \sum\limits_{j:Abs\_rel\_err>\varphi} Dist_k(j)$ *end for*

Update weight $Wt = Er^m$

if $Abs\_rel\_err_j <= \varphi$ then $Dist_k(j) = \dfrac{Dist_k(j)}{Z_t} \times Wt$

else $Dist_k(j) = \dfrac{Dist_k(j)}{Z_t}$ and $k = k+1$

end for

4   Result Analysis

In this section, the amount of nutrients required by the crop and applied in the past 10 years has been discussed. Figure 1 shows the amount of n-p-k applied to the cotton plant from the year 1999 to 2011, in comparison to amount which could have been applied predicted by the Adaboost.RT algorithm. From the figure it can be inferred that the amount of nitrogen applied is almost double compared to the amount actually required by the plant in particular soil. Again for phosphorous and potassium also the amount applied is more than the actual requirement.

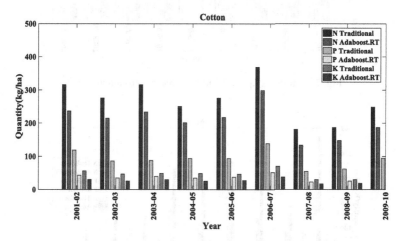

Fig. 1. Nutrients applied and required by cotton plant

From Fig. 2, the amount of n-p-k applied to groundnut in kg/ha, from year 1991 to 2011 can be inferred. The graph shows the amount of these nutrients applied using traditional and manual calculation as compared to the proposed method that must have been applied for higher yield.

Figure 3 shows the nutrient comparison graph for maize crop. The graph shows the nitrogen applied to the crop is twice the exact amount needed for a good yield. Also the amount of phosphorous and potassium is twice as compared to actually needed.

The graph for paddy shown in Fig. 4 illustrates the ratio of nitrogen applied to crop as compared to other two nutrient is very high. Again the quantity of these nutrient applied is high compared to actual amount of nutrient required by the rice crop.

From the Fig. 5(a) for paddy-gram, it can be inferred that the quantity of phosphorous and nitrogen applied for crop is high. For potassium, the value applied to crop is also more but for the year 2006-07 and 2008-09 the amount applied is almost same as predicted. For the crop Ragi, phosphorous applied to the crop is twice or even more in amount as shown in Fig. 5(b). The nitrogen is given to the crop is through out double is quantity in past 10 years. For potassium, the amount applied in the year 2007-08 is same as compared to the prediction by the algorithm.

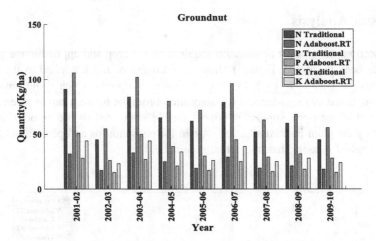

Fig. 2. Nutrients applied and required by groundnut plant

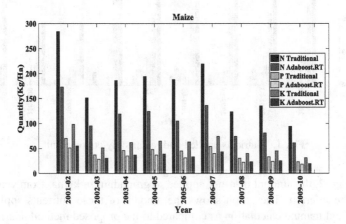

Fig. 3. Nutrients applied and required by maize plant

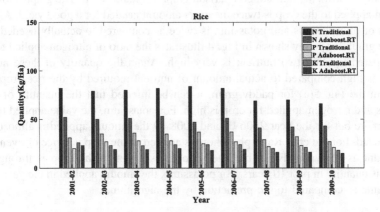

Fig. 4. Nutrients applied and required by paddy plant (rice)

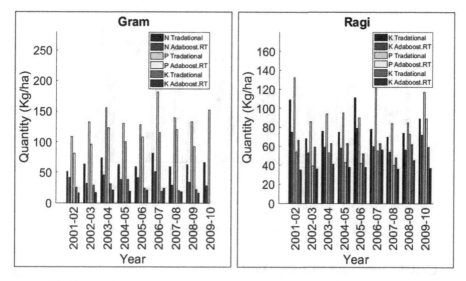

Fig. 5. Nutrients applied and required by (a) paddy-gram plant (b) Ragi crop

Figure 6 shows the amount of potassium applied to soybean crop is near to the amount predicted by the methodology, but for the other nutrient amount applied is again not according to the actual requirement.

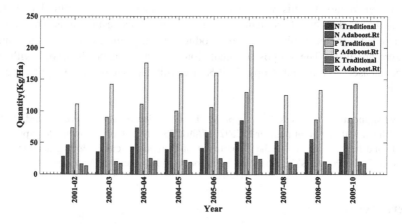

Fig. 6. Nutrients applied and required by soybean plant

After the conduct of experiment a general increase in production was calculated and it was found that if the right amount of nutrient is supplied the production of crop would increase. It is also observed that the yield of crop obtained by applying N-P-K to soil based on traditional methods or manual calculation as compared to the proposed work. It can be concluded, the yield increases for every crop under study if the nutrient supplied is according to the proposed methodology. Figure 7 shows the average increase in production in past 10 years if the applied methodology is used.

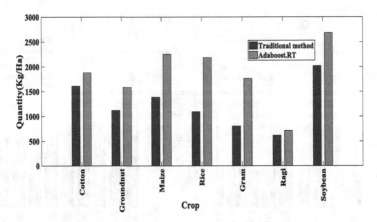

Fig. 7. Average increase in production in last 10 years

5 Conclusions and Future Work

The proposed work introduces adaboost.rt based N-P-K prediction method. The comparison with the traditional method of agriculture and the proposed work shows, the method works outstanding in determining the nutrient requirement of the crops. The proposed methodology gives a superior ability to decide required soil N-P-K content by using one time soil testing information like accessible soil N-P-K content, type of soil, crop and yield target. This will encourage the farmers and agricultural experts to evaluate the required N-P-K content without utilizing manual estimations. This can also be concluded from the Table 1, the crop production can be increased if right amount of fertilizers are used. The proposed calculation can likewise be displayed with high volume dataset of different crops, N-P-K accessibility, yield targets, soil type and location.

Acknowledgments. This work was supported by the Science and Research Board (SERB), Govt. of India with the grant number **ECR/2017/001273**. The authors also wish to express their gratitude and heartiest thanks to the Department of Computer Science & Engineering, Indian Institute of Technology (ISM), Dhanbad, India for providing their research support.

References

1. Lundström, C., Lindblom, J.: Considering farmers' situated knowledge of using agricultural decision support systems (AgriDSS) to Foster farming practices: the case of CropSAT. Agric. Syst. **159**, 9–20 (2018)
2. Arunima, G., Antaryam, M.: Evaluation of STCR targeted yield approach on pumpkin (Cucurbita moschata) under rice-pumpkin cropping system. Int. J. Tropic. Agric. **33**(2 (Part IV)), 1583–1586 (2015)
3. https://icar.org.in/files/Degraded-and-Wastelands.pdf
4. Agriculture in India - of Planning Commission. http://www.planningcommission.nic.in/reports/sereport/ser/vision2025/agricul.doc

5. STCR Crop Wise Recommendations. http://www.iiss.nic.in/downloads/stcrCropwiseRecommendations
6. Rao, N.H.: Big Data and Climate Smart Agriculture-Review of Current Status and Implications for Agricultural Research and Innovation in India (2017)
7. De Mauro, A., Greco, M., Grimaldi, M.: A formal definition of Big Data based on its essential features. Libr. Rev. **65**(3), 122–135 (2016)
8. Friess, P.: Digitising the Industry-Internet of Things Connecting the Physical, Digital and Virtual Worlds. River Publishers (2016)
9. Wolfert, S., Ge, L., Verdouw, C., Bogaardt, M.J.: Big Data in smart farming–a review. Agric. Syst. **153**, 69–80 (2017)
10. Morshed, A., Dutta, R., Aryal, J.: Recommending environmental knowledge as linked open data cloud using semantic machine learning. In: 2013 IEEE 29th International Conference on Data Engineering Workshops (ICDEW), pp. 27–28. IEEE (2013)
11. Parent, B., Tardieu, F.: Can current crop models be used in the phenotyping era for predicting the genetic variability of yield of plants subjected to drought or high temperature? J. Exp. Bot. **65**(21), 6179–6189 (2014)
12. Rosenzweig, C.E., Antle, J., Elliott, J.: Assessing impacts of climate change on food security worldwide (2015)
13. Mishra, S., Mishra, D., Santra, G.H.: Applications of machine learning techniques in agricultural crop production: a review paper. Indian J. Sci. Technol. **9**(38) (2016)
14. Khoshnevisan, B., Rafiee, S., Omid, M., Mousazadeh, H., Rajaeifar, M.A.: Application of artificial neural networks for prediction of output energy and GHG emissions in potato production in Iran. Agric. Syst. **123**, 120–127 (2014)
15. Patil, A., Beldar, M., Naik, A., Deshpande, S.: Smart farming using Arduino and data mining. In: 2016 3rd International Conference on Computing for Sustainable Global Development (INDIACom), pp. 1913–1917. IEEE (2016)
16. Bendre, M.R., Thool, R.C., Thool, V.R.: Big data in precision agriculture: Weather forecasting for future farming. In: 2015 1st International Conference on Next Generation Computing Technologies (NGCT), pp. 744–750. IEEE, September 2015
17. Hot, E., Popović-Bugarin, V.: Soil data clustering by using K-means and fuzzy K-means algorithm. In: 2015 23rd Telecommunications Forum Telfor (TELFOR), pp. 890–893. IEEE, November 2015
18. Navarro-Hellín, H., Martínez-del-Rincon, J., Domingo-Miguel, R., Soto-Valles, F., Torres-Sánchez, R.: A decision support system for managing irrigation in agriculture. Comput. Electron. Agric. **124**, 121–131 (2016)
19. Villarrubia, G., Paz, J.F.D., Iglesia, D.H., Bajo, J.: Combining multi-agent systems and wireless sensor networks for monitoring crop irrigation. Sensors **17**(8), 1775 (2017)
20. Shrestha, D.L., Solomatine, D.P.: Experiments with AdaBoost.RT, an improved boosting scheme for regression. Neural Comput. **18**(7), 1678–1710 (2006)

Machine Learning and Pattern Mining

Deep Neural Network Based Image Captioning

Anurag Tripathi[1], Siddharth Srivastava[1], and Ravi Kothari[2]($\boxtimes$)

[1] Indian Institute of Technology Delhi, New Delhi, India
{eez128368,eez127506}@ee.iitd.ac.in
[2] Ashoka University, Sonepat, India
ravi.kothari@gmail.com

Abstract. Generating a concise natural language description of an image enables a number of applications including fast keyword based search of large image collections. Primarily inspired by deep learning, recent times have witnessed a substantially increased focus on machine based image caption generation. In this paper, we provide a brief review of deep learning based image caption generation along with a brief overview of the datasets and metrics used to evaluate the captioning algorithms. We conclude the paper with some discussion on promising directions for future research.

Keywords: Image captioning · Deep neural networks
Natural language generation

1 Introduction

Image captioning refers to the generation of a concise natural language description of an image. Amongst other applications, the natural language description (often spanning a sentence or two) allows for text based search and retrieval of images.

Machine based image caption generation is typically comprised of 3 parts, *viz.*,

- The first part focuses on the detection and identification of the objects and their contexts in the image. This has historically been an exceptionally difficult problem for machines to solve and has only been partially addressed by classical image processing in highly constrained environments and with a limited number of objects. Indeed, one of the more promising approaches was based on deep learning and proposed as recently as 2012 [14].
- The second part relates to natural language generation of words and sentences that describe the objects in the image and possibly their contexts (e.g. "tree", "grass", "bench on the grass"). The description needs to be grammatically correct for it to be further consumed and assimilated. At this stage, the text snippets correspond more to individual objects (or potentially simple relationships between the objects e.g. "bench on the grass") as opposed to the most concise or the most relevant description of the overall image [15,19].

© Springer Nature Switzerland AG 2018
A. Mondal et al. (Eds.): BDA 2018, LNCS 11297, pp. 335–347, 2018.
https://doi.org/10.1007/978-3-030-04780-1_23

- The third part removes extraneous (or conditionally irrelevant) information, produces generalizations and appropriate contextualization. As an example, it is possible that a lot of "trees", "grass", and "benches" are detected in the first two parts and captioned as a "park" in this part. It is possible for this part to be subsumed within one or both of the above sub-problems [15].

To highlight the challenges in image captioning, Fig. 1 shows multiple captions of the same image. The captions are accurate at each level though they progressively capture the essence of the scene and the ones that a human being would generate. Thus, while "There is an animal in the jungle" is the simplest, it lacks in detail. It does not inform the reader of the type of animal or what the animal is doing in the jungle. The last caption in contrast is the most descriptive and resolves the type of animals as well as the activity they are performing in the jungle. Similar observation can be made for Fig. 2, where the first caption describes the scene but with limited details while the last caption describes the men along with their activity and their surroundings.

Fig. 1. Possible captions for this image include (i) "There is an animal in the image", (ii) "A giraffe is standing in front of a tree", and (iii) There are two giraffe's standing next to trees in a jungle. One of the giraffe's is eating leaves. Image from the MSCOCO dataset [2]

Fig. 2. Possible captions for this image include (i) "Two men are sitting on a bench", (ii) "Two men are sitting on a bench and eating", and (iii) "Two men are sitting on a bench facing each other and having lunch". Image from the MSCOCO dataset [2]

These examples serve to bring out the complexity of image captioning. Even more, it also highlights the problem of determining the quality of the generated captions. While observations were made above, they were human observations. How does one algorithmically determine the quality of a caption? To an extent, if one could determine that, it would provide an excellent objective function that could be optimized to arrive at an optimal caption (in the context of that specific objective function).

Not surprisingly then, despite some initial success based on deep learning, caption generation is still in a nascent stage. The generated captions often fail to capture the essence of the image or end up sounding mechanical as opposed to

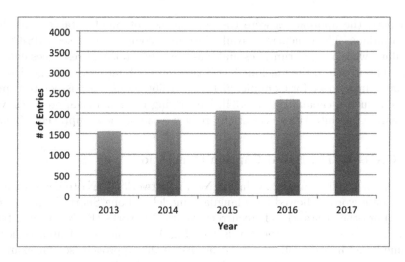

Fig. 3. Year-wise number of contributions retrieved by Google Scholar [1] in response to an "image captioning" query

a vivid summarization. Consequently, the focus on caption generation has been steadily increasing (see Fig. 3).

We have laid out the rest of the paper as follows. In Sect. 2, we provide a brief review of some state-of-the-art techniques. These techniques depend largely on deep learning. Given the empirical nature of deep learning, we devote a section (Sect. 3) to datasets that are available for constructing these empirical models and another section (Sect. 4) on evaluating the performance of these empirical models of caption generation. We discuss potential applications of image captioning in Sect. 5 and conclude the paper with some observations and directions for future research.

2 Deep Learning Approaches for Image Caption Generation

A lot of the initial development of machine based image annotation relied on supervised or unsupervised learning. Note that annotation required associating keywords with the image without the deeper descriptive semantics that must also resolve how the keywords interact. The granularity of the keywords in the initial development remained coarse - for example, "outdoor scene", "indoor scene", "landscape" and so on. The task could be approached using supervised learning, unsupervised learning, or template matching. In the supervised learning case, one collected a labeled dataset with enough instances of the entities of interest and induced a classifier to classify a new image as having or not having that entity. The unsupervised learning relied on inducing a joint density of semantic labels and image features. For this earlier work, we refer the reader to [7, 22, 23, 33] and references therein.

In part, the approaches were the consequence of small datasets with limited variations. The concurrent availability of a complex data set (MS-COCO [18]) along with parallel runtimes and massive computational resources enabled a frontal approach to the machine based generation of natural language descriptions of an image (see for example, [6,11,27,31,36]) along with associated applications of such a capability [4,25,34]. Underlying these newer approaches were deep learning architectures. We briefly review some of these approaches below.

2.1 CNN+LSTM Based Caption Generation

This approach relies on a Convolution Neural Network (CNN) to generate a set of candidate words followed by a language model or Long Short Term Memory (LSTM) or other instances of Recurrent Neural Networks (RNNs) to construct a coherent sentence from the words (see Fig. 4). A number of approaches to machine based image caption generation utilize this approach (see for example, [9,37]) and the availability of pre-trained CNN's makes it even easier to use this approach).

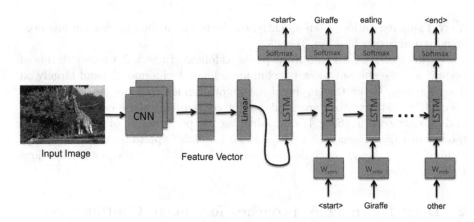

Fig. 4. CNN+LSTM based image captioning

A variation to the above uses a bottom-up approach by discovering important regions in an image (for example, through the use of attentional models to identify regions where a human is most likely to fixate his or her gaze). It is anticipated that such a bottom-up approach is computationally more efficient and accurate. In [35], an attention based model is introduced. The model learns to attend to selective regions while generating the captions for given images. The works explores "soft" deterministic attention and "hard" stochastic attention. Figure 5 shows the this attention based framework for image captioning.

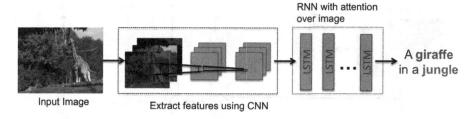

Fig. 5. CNN+Attention based LSTM for image captioning

2.2 CNN+CNN Based Image Captioning

LSTM/RNN are commonly used for image caption generation though suffer from an inherent sequentiality, and despite the improvements introduced in LSTM, remain susceptible to the vanishing gradient problem. A CNN+CNN model [32] has thus been used for machine based image caption generation (see Fig. 6).

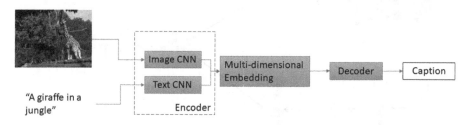

Fig. 6. CNN+CNN based image captioning

2.3 Generative Adversarial Networks

Generative Adversarial Networks (GANs) are the newest type of deep networks to be used for machine based image caption generation. GANs are motivated by a zero sum game framework [20]. However, unlike image synthesis tasks, sequence (i.e. caption) generation based on GAN encounters gradient back-propagation problems because discrete sequences are non-differentiable. Yu et al. [39] used a policy gradient method borrowed from reinforcement learning [29], to accomplish unconditional sequence generation.

A GAN is composed of two networks, a Generator and a Discriminator. The task of Generator is to produce a distribution which belongs to training data and the task of the Discriminator is to distinguish between real and fake distributions. In [8], the authors propose a conditional GANs. They condition the image and sentences and introduce early feedback using reinforcement learning (see Fig. 7).

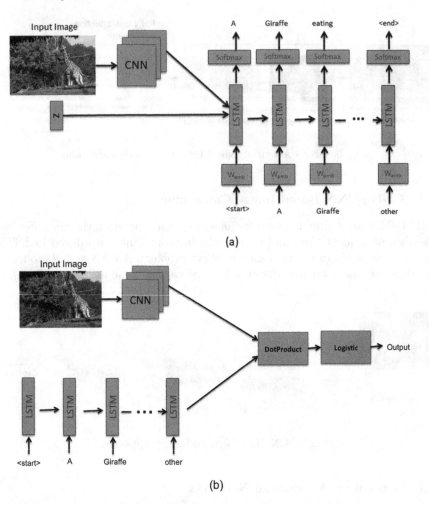

Fig. 7. GAN based Image Captioning (a) the generator for paragraph and sentence generation (b) the evaluator for single sentences

Another issue in image captioning is discrimination which implies that the image caption is not only accurate but also diversified so that it can be discriminative across different images. To achieve that, a novel conditional generative adversarial network for generating diverse captions has been proposed [16]. Instead of estimating the quality of a caption solely on one image, they propose a comparative adversarial learning framework to better assess the quality of captions by comparing a set of captions within the joint image-caption space. By contrasting with human-written captions and image-mismatched captions, the caption generator exploits the characteristics of human languages to generate more discriminative captions.

3 Datasets

Given the centrality of data in empirical model construction, we provide a brief review of some of the publicly available datasets.

Flickr8K: The Flickr8K dataset [12] has 8,000 images each with 5 captions.

Flickr30K: The Flickr30K dataset [38] is an extension of the Flicker8k dataset. It has 31,783 images each with 5 captions.

MSCOCO: The MSCOCO dataset [18] is probably the most widely used and standardized dataset. It has 82,783 images for training and 40,504 images for validation and testing. Each image has 5 captions.

Visual Genome Dataset: The Visual Genome Dataset [13] consists of 94000 image and 41,00,000 captions for different regions in the images. All the captions are human generated on Amazon Mechanical Turk and the descriptions are provided with a bounding box with the corresponding region. This image set is taken from the intersection of MSCOCO and YFCC100M datasets. Each image has an average of 21 objects, 18 attributes and 18 pairwise relationships between objects. This relationships is provided by the metadata of the dataset.

SBU: The SBU dataset [21] contains 1 million images and their description collected by querying Flickr for specific objects and actions.

Pascal VOC Dataset: The Pascal VOC dataset [10] is built around the Pascal VOC 2008 dataset which is a dataset for object classification. There are 50 images from each of the 20 classes. Amazon's mechanical turk is then used to generate 5 captions for each image. Each image is also labeled with a triplet which consists of the main object in the image, the main action and the main place.

Pascal-50S: The Pascal-50S dataset [30] is based on the popular UIUC Pascal sentence dataset [26]. It has 50 human generated descriptions for each of the 1,000 images.

Abstract-50S: The Abstract-50S dataset [30] has 50 human generated captions for each image and has 2,54,721 such images.

4 Metrics

The evaluation of machine generated captions is not straightforward. Indeed, one could argue that if there were a good mechanism that could evaluate a caption, then the same mechanism would itself be the perfect caption generator. Indeed, that mechanism could be used for credit assignment and parameter adjustment of the empirical model for progressively better caption generation.

The most appropriate caption for an image is not unique and a good metric should be able to similarly score equally meaningful descriptions. The metric should also be able to empirically weigh the grammatical structure, the contextual relevance, and the completeness of a caption. Ultimately however, natural language is subjective and it is quite likely to find differing opinions on the quality of a piece of text.

In this section, we provide a brief review of various metrics used for evaluating the quality of image captioning. As may be expected, these metrics originated in machine translation where they were used to measure the agreement of the translation to the reference. Here, they are adopted to measure the agreement between the generated caption (candidate) and the ideal caption(s) (reference).

4.1 BLEU: Bilingual Evaluation Understudy

BLEU [24] was originally proposed to evaluate the quality of a machine translation. At its core, BLEU relies on the degree of match between the n-grams of the translation with the n-grams of the source. BLEU recognizes that there be equally meaningful translations and measures the match between a candidate and each of the possible reference (acceptable) translations. However, since it is possible for a candidate translation to repeatedly use an n-gram and obtain a higher score, BLEU clips the contribution of an n-gram to the maximum number of occurrences of that n-gram in any of the reference translations.

Of course, n can assume various values with lower values favoring fidelity and higher values favoring fluency.

Succinctly,

$$p_n = \frac{\sum_{C \in \{C\}} \sum_{n\text{-}gram \in C} Count_{clip}(n\text{-}gram)}{\sum_{C' \in \{C\}} \sum_{n\text{-}gram' \in C'} Count_{clip}(n\text{-}gram')} \tag{1}$$

where, $\{C\}$ is the set of candidates, $count_{clip}(n\text{-}gram)$ is the clipped n-gram count, p_n is the precision. The BLEU score is then based on a geometric mean of the modified n-gram precisions,

$$\log B = \min(1 - \frac{r}{c}, 0) + \sum_{n=1}^{N} w_n \log p_n \tag{2}$$

where, B is the BLEU score and w_n is the weight given to the different precisions (often taken to be $1/N$).

The BLEU score can be used to compare the machine generated caption (candidate) with oracle supplied caption(s) (reference).

4.2 ROUGE-N: Recall-Oriented Understudy for Gisting Evaluation

ROUGE [17] is a set of metrics for evaluating a machine translation against a set of references. ROUGE-1, ROUGE-2, ... ROUGE-N computes the overlap of n-grams between a candidate translation and the references and can be computed as,

$$ROUGE\text{-}N = \frac{\sum_{R \in \{R\}} \sum_{n\text{-}gram \in S} Count_{match}(n\text{-}gram)}{\sum_{R \in \{R\}} \sum_{n\text{-}gram \in S} Count(n\text{-}gram)} \tag{3}$$

where, $Count_{match}$ is the maximum number of n-grams co-occurring in a candidate and set of references, $\{R\}$ is the set of references. Note that while BLEU is precision oriented, ROUGE is more recall oriented.

4.3 Meteor: Metric for Evaluation of Translation with Explicit ORdering

METEOR [5] is another metric for evaluating the output of machine translation which tries to incorporate both precision and recall. It computes precision as the ratio of the number of unigrams in the candidate that are aligned with unigrams in the reference to the total number of unigrams in the candidate. It computes recall as the ratio of the number of unigrams in the candidate that are aligned with unigrams in the reference to the total number of unigrams in the reference. Precision (P) and Recall (R) so obtained are combined using a geometric mean that is weighed heavily in favor of recall.

To begin with, an alignment is created between the candidate and reference. Each unigram in the candidate matches to one or none of the unigrams in the reference. The best alignment is one with the maximum number of mapped unigrams and least number of crossed lines (i.e. ordered unigrams are mapped). Then, the precision is computed as, $P = \frac{m}{w_t}$ where, m is the number of common unigrams between the reference and the candidate w_t is the number of unigrams in the candidate. The recall is computed as, $R = \frac{m}{w_r}$ where, m is the same as above and w_r is the number of unigrams in the reference sentence.

The precision and recall are combined as,

$$F_{mean} = \frac{10PR}{R + 9p} \tag{4}$$

To account for the length of the match, unigrams in the candidate that matched the reference are grouped into the minimum number of *chunks*. A penalty term is then computed as,

$$PE = 0.5 * \frac{Count_{chunks}}{Count_{unigrams\ matched}} \tag{5}$$

and finally, the METEOR score is computed as,

$$Score = F_{mean}(1 - PE) \tag{6}$$

4.4 CIDEr: Consensus-Based Image Description Evaluation

CIDEr [30] incorporates human consensus for evaluating machine generated text. Often, a large number of individual human opinions are required to arrive at a consensus. Our discussion below follows the one given in [30].

To compute CIDEr, a TF-IDF (Term Frequency-Inverse Document Frequency) weighting for each *n-gram* is computed. The motivation is similar to that of TF-IDF i.e. an *n*-gram that occurs for a few images is more discriminating than the one that appears for all images.

Let an *n*-gram, ω_k, occur $h_k(R_{ij})$ times in a reference sentence R_{ij}. S_i is the consensus opinion about image i. Likewise, let the *n*-gram occur in a candidate sentence $h_k(C_i)$ times. The TF-IDF weighting for ω_k is then,

$$g_k(R_{ij}) = \frac{h_k(R_{ij})}{\Sigma_{\omega_l \in \Omega} h_l(R_{ij})} \log \frac{|I|}{\Sigma_{I_p \in I} \min(1, \Sigma_q h_k(R_{pq}))} \tag{7}$$

where ω is the vocabulary of all the n-grams and I is the set of all images in the dataset. The first term measures the TF and second measures the IDF.

The average cosine similarity for n-grams of length, say n, among generated and reference sentence gives the CIDErn score, i.e.,

$$CIDEr_n(C_i, R_i) = \frac{1}{m} \sum_j \frac{g^n(C_i) \cdot g^n(R_{ij})}{\|g^n(C_i)\| \, \|g^n(R_{ij})\|} \tag{8}$$

where $g_n(C_i)$ is a vector formed by $g_k(C_i)$ corresponding to all the n-grams of length n. Similarly for $g^n R_{ij}$

Finally, the scores of various n-grams can be combined as follows,

$$CIDEr_n(C_i, R_i) = \sum_{n=1}^{N} w_n CIDEr_n(C_i, R_i) \tag{9}$$

where, w_n is the weight given to an n-gram of length n.

The above metrics relies on n-grams in some shape or form. There are several other metrics including some that do not rely on the notion of n-grams to determine agreement between a candidate and a reference. SPICE: Semantic Propositional Image Caption Evaluation [3] is based on the premise that two sentences may have the same meaning while having minimal n-gram overlap. SPICE thus postulates that semantic propositional content is important for caption evaluation. It borrows from the notion of scene graphs to define agreement between a candidate and a reference piece of text.

5   Real World Applications

Image Captioning has many practical applications. Perhaps the biggest and most direct application lies in being able to do fast keyword based search in large image collections. More importantly, it allows easier incorporation of a user's intent based on the context of the search.

Caption generation also helps in generating a description of the image which can be useful to the visually impaired [28] and can also be used to emphasize specific situations which may require special attention (e.g. "a person is crossing the road ahead" being spoken in a car using text-to-speech technology to alert the user to what lies ahead) [2].

Multiple other use cases including personalized captioning [25], virtual assistants, textual summarization of a video, organizing photos in an album based on caption and the like are enabled by machine based image caption generation.

We also conjecture that caption generation and image analysis and understanding have much to gain from a symbiotic relationship. For example, one of the authors (RK) has proposed the use of language models (n-grams) as an attentional mechanism in images. We also anticipate that inference would be more fluid across modalities and the present work represents a small but purposeful step towards that goal.

6   Conclusion

The paper provided a brief overview of deep learning based image captioning. We discussed the algorithms, the datasets available to train such deep networks and some metrics that can be used to evaluate the quality of machine generated captions.

We also discussed some potential applications of caption generation and some possible advanced and symbiotic existence of caption generation and image analysis and understanding.

References

1. Google Scholar. http://scholar.google.com
2. MSCOCO Image Captioning Challenge (2015). http://cocodataset.org/#captions-leaderboard. Accessed 05 Oct 2018
3. Anderson, P., Fernando, B., Johnson, M., Gould, S.: SPICE: semantic propositional image caption evaluation. In: Leibe, B., Matas, J., Sebe, N., Welling, M. (eds.) ECCV 2016. LNCS, vol. 9909, pp. 382–398. Springer, Cham (2016). https://doi.org/10.1007/978-3-319-46454-1_24
4. Aneja, J., Deshpande, A., Schwing, A.: Convolutional image captioning. In: IEEE Conference on Computer Vision and Pattern Recognition (CVPR), pp. 5561–5570 (2018)
5. Banerjee, S., Lavie, A.: Meteor: An automatic metric for MT evaluation with improved correlation with human judgments. In: ACL Workshop on Intrinsic and Extrinsic Evaluation Measures for Machine Tanslation and/or Summarization, pp. 65–72 (2005)
6. Baraldi, L., Grana, C., Cucchiara, R.: Hierarchical boundary-aware neural encoder for video captioning. In: IEEE Conference on Computer Vision and Pattern Recognition (CVPR), pp. 3185–3194. IEEE (2017)
7. Carneiro, G., Chan, A.B., Moreno, P.J., Vasconcelos, N.: Supervised learning of semantic classes for image annotation and retrieval. IEEE Trans. Pattern Anal. Mach. Intell. 29(3), 394–410 (2007)
8. Dai, B., Fidler, S., Urtasun, R., Lin, D.: Towards diverse and natural image descriptions via a conditional GAN. arXiv preprint arXiv:1703.06029 (2017)
9. Devlin, J., et al.: Language models for image captioning: the quirks and what works. arXiv preprint arXiv:1505.01809 (2015)
10. Everingham, M., Van Gool, L., Williams, C.K., Winn, J., Zisserman, A.: The PASCAL visual object classes (VOC) challenge. Int. J. Comput. Vis. 88(2), 303–338 (2010)
11. Guo, Y., Liu, Y., Oerlemans, A., Lao, S., Wu, S., Lew, M.S.: Deep learning for visual understanding: a review. Neurocomputing 187, 27–48 (2016)
12. Hodosh, M., Young, P., Hockenmaier, J.: Framing image description as a ranking task: data, models and evaluation metrics. J. Artif. Intell. Res. 47, 853–899 (2013)
13. Krishna, R., et al.: Visual genome: connecting language and vision using crowd-sourced dense image annotations. Int. J. Comput. Vis. 123(1), 32–73 (2017)
14. Krizhevsky, A., Sutskever, I., Hinton, G.E.: Imagenet classification with deep convolutional neural networks. In: Advances in Neural Information Processing Systems, pp. 1097–1105 (2012)

15. Kulkarni, G., et al.: Baby talk: understanding and generating image descriptions. In: Proceedings of the 24th CVPR. Citeseer (2011)
16. Li, D., He, X., Huang, Q., Sun, M.T., Zhang, L.: Generating diverse and accurate visual captions by comparative adversarial learning. arXiv preprint arXiv:1804.00861 (2018)
17. Lin, C.Y.: Rouge: a package for automatic evaluation of summaries. In: ACL Text Summarization Branches Out (2004)
18. Lin, T.-Y., et al.: Microsoft COCO: common objects in context. In: Fleet, D., Pajdla, T., Schiele, B., Tuytelaars, T. (eds.) ECCV 2014. LNCS, vol. 8693, pp. 740–755. Springer, Cham (2014). https://doi.org/10.1007/978-3-319-10602-1_48
19. Lu, S., Zhu, Y., Zhang, W., Wang, J., Yu, Y.: Neural text generation: past, present and beyond. arXiv preprint arXiv:1803.07133 (2018)
20. Mirza, M., Osindero, S.: Conditional generative adversarial nets. arXiv preprint arXiv:1411.1784 (2014)
21. Ordonez, V., Kulkarni, G., Berg, T.L.: Im2Text: describing images using 1 million captioned photographs. In: Advances in Neural Information Processing Systems, pp. 1143–1151 (2011)
22. Pan, J.Y., Yang, H.J., Duygulu, P., Faloutsos, C.: Automatic image captioning. In: IEEE International Conference on Multimedia and Expo, ICME 2004, vol. 3, pp. 1987–1990. IEEE (2004)
23. Pan, J.Y., Yang, H.J., Faloutsos, C., Duygulu, P.: GCap: graph-based automatic image captioning. In: Conference on Computer Vision and Pattern Recognition Workshop, CVPRW 2004, p. 146. IEEE (2004)
24. Papineni, K., Roukos, S., Ward, T., Zhu, W.J.: Bleu: a method for automatic evaluation of machine translation. In: 40th Annual Meeting on Association for Computational Linguistics, pp. 311–318. Association for Computational Linguistics (2002)
25. Park, C.C., Kim, B., Kim, G.: Towards personalized image captioning via multi-modal memory networks. IEEE Trans. Pattern Anal. Mach. Intell. (2018)
26. Rashtchian, C., Young, P., Hodosh, M., Hockenmaier, J.: Collecting image annotations using Amazon's mechanical turk. In: NAACL HLT Workshop on Creating Speech and Language Data with Amazon's Mechanical Turk, pp. 139–147. Association for Computational Linguistics (2010)
27. Rennie, S.J., Marcheret, E., Mroueh, Y., Ross, J., Goel, V.: Self-critical sequence training for image captioning. In: IEEE Conference on Computer Vision and Pattern Recognition (CVPR), vol. 1, p. 3 (2017)
28. Rohrbach, A., Rohrbach, M., Schiele, B.: The long-short story of movie description. In: Gall, J., Gehler, P., Leibe, B. (eds.) GCPR 2015. LNCS, vol. 9358, pp. 209–221. Springer, Cham (2015). https://doi.org/10.1007/978-3-319-24947-6_17
29. Sutton, R.S., McAllester, D.A., Singh, S.P., Mansour, Y.: Policy gradient methods for reinforcement learning with function approximation. In: Advances in Neural Information Processing Systems, pp. 1057–1063 (2000)
30. Vedantam, R., Lawrence Zitnick, C., Parikh, D.: CIDEr: consensus-based image description evaluation. In: IEEE Conference on Computer Vision and Pattern Recognition (CVPR), pp. 4566–4575 (2015)
31. Vinyals, O., Toshev, A., Bengio, S., Erhan, D.: Show and tell: lessons learned from the 2015 MSCOCO image captioning challenge. IEEE Trans. Pattern Anal. Mach. Intell. **39**(4), 652–663 (2017)
32. Wang, Q., Chan, A.B.: CNN+ CNN: convolutional decoders for image captioning. arXiv preprint arXiv:1805.09019 (2018)

33. Wang, X.J., Zhang, L., Jing, F., Ma, W.Y.: AnnoSearch: image auto-annotation by search. In: IEEE Computer Society Conference on Computer Vision and Pattern Recognition, vol. 2, pp. 1483–1490. IEEE (2006)
34. Wu, Q., Shen, C., Wang, P., Dick, A., van den Hengel, A.: Image captioning and visual question answering based on attributes and external knowledge. IEEE Trans. Pattern Anal. Mach. Intell. **40**(6), 1367–1381 (2018)
35. Xu, K., et al.: Show, attend and tell: neural image caption generation with visual attention. In: International Conference on Machine Learning, pp. 2048–2057 (2015)
36. Yao, T., Pan, Y., Li, Y., Mei, T.: Exploring visual relationship for image captioning. In: European Conference on Computer Vision (ECCV), pp. 684–699 (2018)
37. Yao, T., Pan, Y., Li, Y., Qiu, Z., Mei, T.: Boosting image captioning with attributes. In: International Conference on Computer Vision (ICCV), pp. 22–29 (2017)
38. Young, P., Lai, A., Hodosh, M., Hockenmaier, J.: From image descriptions to visual denotations: new similarity metrics for semantic inference over event descriptions. Trans. Assoc. Comput. Linguist. **2**, 67–78 (2014)
39. Yu, L., Zhang, W., Wang, J., Yu, Y.: SeqGAN: sequence generative adversarial nets with policy gradient. arXiv preprint arXiv:1609.05473 (2017)

Oversample Based Large Scale Support Vector Machine for Online Class Imbalance Problem

D. Himaja[1](✉), T. Maruthi Padmaja[1], and P. Radha Krishna[2]

[1] Department of Computer Science and Engineering, VFSTR University, Guntur, Andhra Pradesh, India
himajadirsumilli@gmail.com, tmp_cse@vignanuniversity.org
[2] Department of Computer Science and Engineering, National Institute of Technology, Warangal, Telangana, India
prkrishna@nitw.ac.in

Abstract. Dealing with online class imbalance from evolving stream is a critical issue than the conventional class imbalance problem. Usually, the class imbalance problem occurs when one class of data severely outnumbers the other classes of data, thus leads to skewed class boundaries. In the case of online class imbalance problem, the degree of class imbalance changes over time and the present state of imbalance is not known a prior to the learner. To address such problem, in this paper, we present an Oversampling based Online Large Scale Support Vector Machine (OOLASVM) algorithm which is a hybrid of active sample selection and over sampling of Support Vectors and thereby both oversampling and under sampling coexists while learning the new boundary. Further, OOLASVM maintains the balanced boundary throughout the learning process. Results on simulated and real world datasets demonstrate that proposed OOLASVM yields better performance than existing approaches such as Generalized Oversampling based Online Imbalanced Learners and Over Online Bagging.

Keywords: Online learning · Dynamic class imbalance
Active learning · Support Vector Machines · Oversampling

1 Introduction

Learning from imbalanced evolving streams possesses several challenges than the conventional class imbalance problem. The class imbalance problem occurs when one class of data severely outnumbers the other classes of data such that the performance of the underlined classification algorithm biased towards the majorly represented data. This problem is widely studied and several algorithms are proposed at both data level and algorithm level [1,2]. In addition to these methods, several hybrids of data and algorithm levels and their ensembles are proposed in the literature [3].

© Springer Nature Switzerland AG 2018
A. Mondal et al. (Eds.): BDA 2018, LNCS 11297, pp. 348–362, 2018.
https://doi.org/10.1007/978-3-030-04780-1_24

In case of online learning, the data is learned sample by sample, and there is no way to the learner to prior know the status of class distributions i.e., degree of imbalance $d = \frac{maj}{min} > 1$, provided $maj > min$, where maj is number of majority samples and min is number of minority samples. Further, unlike the conventional class imbalance problem, the degree of class imbalance is not constant and it changes dynamically over time. Therefore, it is mandatory for the online learners to adapt to these dynamic changes of the degree in class imbalance.

Usually, the approaches for conventional class imbalance problem are guided by the degree of imbalance of the data, which can be directly counted from the standalone imbalance training data. However, this is not possible with evolving class Imbalanced stream, and the models should adaptively guide by the dynamic degree of class imbalance which could be computed on-the-fly. There are very few works in the literature that address this problem [4].

Motivated from the works on Support Vector Machines (SVMs) in handling conventional class imbalance problem [5,6] in this paper, we propose an active learning based adaptive oversampling algorithm Over Online Large Scale Support Vector Machine (OOLASVM) to handle online class imbalance problem. The proposed work is based on LASVM active learning algorithm [7].

The rest of the paper is organized as follows. Section 2 presents the related work pertaining to the online dynamic imbalance problem. Section 3 describes the necessary background techniques and motivation for our work. The proposed algorithm is presented in Sect. 4. Section 5 presents the discussion on obtained results and finally, the paper concludes in Sect. 6.

2   Related Work

Conventional class imbalance problem can be handled at (i) data level (ii) algorithm level and (iii) hybrid of both data level and algorithm level [1,2].

(i) Data Level: The solutions at data level preprocess the data before the classification task. Here, the main focus is to balance the class distribution either by oversample the minority class or by under sample the majority class [5]. This over and under sampling is carried either by random [5] or by selective procedure [8].

(ii) Algorithm Level: These solutions modifies the learning algorithm in such a way that the priority is given in predicting the minority class. These methods includes adjusting the decision threshold [9], ensembles of the base classifiers such as bagging and boosting [3] and novelty detection methods [10].

(iii) Hybrid Methods: These methods includes the hybrids of data and algorithm level solutions such as bagging and sampling, boosting and sampling, boosting and cost sensitive learning method [3].

For dynamic class imbalance problem, Wang et al. [11] proposed a Resampling-Based ensemble method namely weighted Oversampling based

Online Bagging and Under sampling based Online Bagging. Here, the oversampling and under sampling rates are guided by an adaptive weighing strategy based on minority class *Recall* and a time decay function. There are single classifier variants such as Recursive Least Square Adaptive Cost Perceptron (RLSACP) [12] and Online Neural Network (ONN) [13]. Here, the class imbalance is handled with adaptive weighing of Perceptron error either with the classification rate or with the degree of imbalance. Barua et al. [14] proposed a Generalized Over-Sampling based online Imbalanced Learning Framework (GOS-IL) for online learners to handle dynamic class imbalance change. Recently, Yan et al. [15] proposed a unified cost sensitive online learning framework for multi class unbalanced data. Proposed framework trains multiple classifiers simultaneously with different costs and prediction made out on a random classifier. The selection of the random classifier is based on the performance of the individual learners. But, determining the costs to learn the classifiers is brute force on pool of generated multiple costs. There are works that addressed the combined problem of class imbalance and concept drift in non-stationary environments [4]. However, the proposed work focus is on handling dynamic changes of class imbalance status in evolving streams.

3   Motivation and Background

This section presents the motivation behind carrying this work and necessary background required for the work presented in this paper.

3.1   Motivation

In case of class imbalance problem, due to the concept of maximum margin hyperplane SVM has been exhibited better generalization abilities compared to other standalone learners [5, 16]. Due to its ability of being directly computing the distance from a data point to the boundary, it is widely studied in active learning [17]. Rather learning on entire training data, during the process of learning, these methods select the informative training instances that are required. This scenario is possible with incremental learning of training data. Hence, these methods are widely applied in semi supervised learning where there is a limited labeled data and abundant unlabeled data is available for training [18]. In addition to these, the boundary has been altered for each selection of new data point from training. Thus, the learner is even adaptable to change in environment. Based on the aforementioned advantages, in this paper, active learning of SVM is studied for online class imbalance learning problem.

The single sample based query active learning methods that better fits for online learning settings are computationally expensive. Therefore this work extends a fast Large Scale Support Vector Machine (LASVM) active Learning algorithm for online class imbalance problem.

3.2   Support Vector Machine

Given a Training set $T(x_i, y_i)$ where $x_i \epsilon X$, $y_i \epsilon Y\{-1, 1\}$ of N number of points, the SVM [19] describes an optimal hyperplane (Eq. 1), in feature space:-

$$\min_{w, b, \xi_i} \frac{1}{2} w.w^T + C \sum_{i=1}^{N} \xi_i \tag{1}$$

Subject to

$$\begin{cases} \forall i \ y_i(w^T x^i + b) \geqslant 1 - \xi_i \\ \forall i \ \xi_i \geqslant 0 \end{cases}$$

where C is the tuning parameter for corresponding loss function for misclassification cost, ξ is slack variable for handling nonlinearity and w is the norm of the hyperplane, b is the intercept of hyperplane from the origin, x_i is the input vector to feature space, y_i is the corresponding class label. Generally, Eq. 1 is solved by convex Quadratic Programming (QP). The QP in turn converts the Eq. 1 into set of linear equations with linear constraints by substituting Lagrangian multipliers α_i, where i = 1 to N. The Sequential Minimum Optimization (SMO) [20] is widely used for solving this QP, which works based on the principle of identifying a τ-violating pairs. Here τ is the tolerance on the gradients of the pair (i, j), $g_i - g_j > \tau$. The SMO terminates when there is no such τ-violating pair and the final equations after solving QP are

$$w = \sum_{i=1}^{N} \alpha_i y_i \Phi(x_i) \tag{2}$$

and the test instances are classified with

$$y(x) = sign[\sum_{i=1}^{N} \alpha_i y_i k(x_i, x) + b] \tag{3}$$

Here the α's with non zero formulates the boundary and referred as Support Vectors (SVs) and $k(\ ,\ )$ is the non-linear kernel function such as Polynomial and Gaussian.

LASVM. Unlike the random sampling from training set, active learning methods selects the informative samples while learning. The traditional SVM active learning methods [17] selects the informative samples for the current learning based on requesting the sample by query that is nearest to the boundary. For each new sample by query, the Lagrangian multipliers α's of SVM QP solver are computed by solving the QP problem from the beginning. Unlike these methods, LASVM *process* the new sample by trying to add it to the existing support vectors set S. Next, it *reprocess* the S by removing some blatant non support vectors from S. Thus, for each new sample, at the end of *process* (and *reprocess*) a new SVM boundary is learned. The QP problem of LASVM simply extends the optimization procedure of Sequential Minimum Optimization (SMO), instead, for *process* and *reprocess* step it starts the computations of gradient from the previous α's and S. Hence, the learning becomes faster and the boundary adaptively changes with respect to the change in data. At the end, the *reprocess* step of the boundary continues until the gradient $\delta = (g_i - g_j)$ of the most τ-violating pair $\delta < \tau$, i.e., no τ-violating pair.

The LASVM tailored to process random samples in offline setting [21]. It is modified to support the computation of the *Recall* and *G-Mean* evaluation Prequential in online fashion. Algorithm 1 shows the LASVM algorithm. The working procedure for LASVM is depicted in Fig. 1; the final boundary with LASVM is skewed towards the majority class (see Fig. 1(e)).

Result: *Recall* and *G-Mean*

Input: Initial training set with a set of samples from each class;

Initialize $\alpha - > 0$ and compute the initial gradient g;

NewModel=process(Initial TrainingSet);

Initialize Confusion Matrix;

while *samples left in the stream* **do**

> Get *NewSample* from the stream;
>
> Predict the *NewSample* on *NewModel*;
>
> Update Confusion Matrix;
>
> *NewModel=process(NewSample)*;
>
> do *NewModel=reprocess(NewModel)* until $\delta \leq \tau$;

end

Algorithm 1: LASVM algorithm for streaming data with Evaluation Prequential

Evaluation Prequential [22] of individual class *Recall* and class *G-Mean* are used as performance indicators. The Evaluation Prequential is an interleaved test-and-train procedure to evaluate the data stream by testing each evolving sample on the learned model and then used as training.

4   Proposed Approach

In this section, we present the proposed Online Oversample based LASVM algorithm that maintains the stable boundary in the context of dynamic degree of imbalance. We assume that a single minority sample at initial training is existing for the high degree of imbalanced case, i.e., 1:9. Based on the state of imbalance, at the initial training, the minority class can be positive or negative class. The main objective of OOLASVM is to maintain the balanced boundary all through the learning process. This is achieved through oversampling in two stages:

i. At initial training the minority class boundary is pushed forward by generating synthetic samples between half the path way from minority class data to majority data.

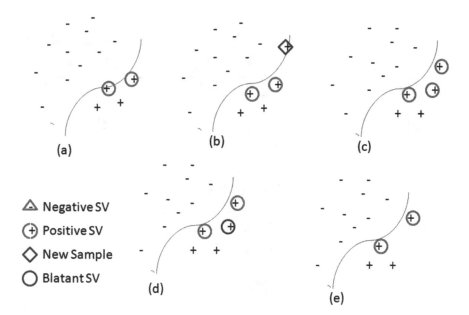

Fig. 1. Example scenario with LASVM (a) Imbalanced Training set, (b) New Test sample (c) *process* (d) *reprocess* (e) Final boundary.

ii. At each new model, the dynamic change in the state of imbalance is measured by computing the difference between number of positive and negative class SV's: nsv_+, nsv_-, the synthetic SV's are generated to bridge the difference.

Algorithm 2 and Fig. 2 show the OOLASVM algorithm. Synthetic Minority Oversampling Technique (SMOTE) [8] is used to generate synthetic samples for generate_synthetic($<x_{min}, y_{min}>$, $<x_{kmaj}, y_{kmaj}>$, d) function with current degree of imbalance d. The balanced initial training set is processed to formulate new boundary/model. Every incoming sample is first tested with the newly formulated boundary and later added to the support vector set S by process step. The new boundary is refined using reprocess step by removing some blatant non support vectors. Now the dynamic change in state of imbalance is handled by oversampling around the under represented SV's class. If positive class SV's less than the negative class SV's, positive synthetic samples are generated around the positive class, and vice versa.

The generated synthetic SV's are added to the boundary by *process* step. Thus, each new sample is always tested with the balanced boundary. Though, the boundary is pushed at initial training, it gradually adjusted with respect to evolving data. This methodology includes, both oversampling (adding synthetic SV's to boundary) under sampling (removing non SV's from the learned model), thus forming a hybrid sampling approach. Algorithm 2 and Fig. 2 show the OOLASVM algorithm and illustration of a sample scenario respectively. The working model of the OOLASVM is described below:

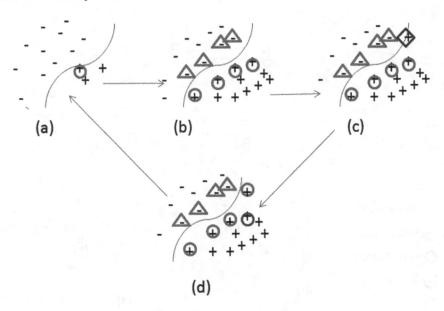

Fig. 2. Example scenario with OOLASVM

Step 1: To avoid the skewed boundary (like Fig. 2(a)) at initial training, the plane boundary is adjusted by generating synthetic minority samples across the minority and majority samples and added to the training set. A New model is formulated with the new training set through *process (NewTrainingset)* (shown in Fig. 2(b)). The new training set labels are predicted over the new model and the performance evaluation indicators namely *Recall* and *G-Mean* are calculated.

Step 2: The new sample (see Fig. 2(c)) from evolving stream is at first predicted over the new model. The *Recall* and *G-Mean* are updated accordingly. The *New sample* is added to the SV's set through process(*New sample*), which leads to a next new model.

Step 3: The New model trained in the previous step is updated by the *reprocess(NewModel)* to remove the blatant non support vectors according to the boundary (see Fig. 2(d)).

Step 4: After the removal of the blatant non support vectors, again the new boundary may go imbalance. As it is an online learning, initially there may not be enough samples. Also, removing a single point from minority has significant impact on the performance. Thus, the boundary formulated can be imbalanced (see Fig. 2(a)).

Step 5: As the stream evolves with dynamic degree of imbalance, at time t the positive class may be majority or minority, this is handled by generating d synthetic SV's around minority class SV's (either positive or negative). At the end of each *reprocess*, the positive class and negative class SV's from updated boundary, (nsv_+, nsv_-) are updated by $[nsv_+, nsv_-] = \text{get\_sv}(NewModel)$.

If $(nsv_+ < nsv_-)$ then synthetic samples are generated across positive class, otherwise the synthetic samples are generated across negative class.

Step 6: A new boundary is formulated with the newly generated synthetic samples by $NewModel = process(\text{synth\_data})$, that is, adding the synthetic SV's to the SV's set S. This procedure is repeated from Step 2 to step 6 until the last sample.

    Result: *Recall* and *G-Mean*

    Input: $< x_i, y_i >$ initial training set with degree of imbalance d

    $< x_t, y_t >$, *NewSample* at current time t;

    for each minority sample $< x_{min}, y_{min} >$ in $< x_i, y_i >$;

    synth_ data=generate_ synthetic ($< x_{min}, y_{min} >$, $< x_{kmaj}, y_{kmaj} >$, d);

    NewTrainingSet= [$< x_{kmaj}, y_{kmaj} >$, all $< x_{min}, y_{min} >$, $<$ synth _ data $>$] ;

    Initialize $\alpha - > 0$ and compute the initial gradient g;

    NewModel=*process*(*NewTrainingSet*);

    Predict the label of *NewTrainingSet* on *NewModel*;

    Compute *Recall* and *GMean* ;

    [nsv_+ , nsv_-] = get_ sv(*NewModel*);

    while *samples left in the stream* **do**

        Get $< x_t, y_t >$ from the stream;

        Predict the label of $< x_t, y_t >$ on *NewModel*;

        Update *Recall* and *GMean*;

        NewModel=*process*(*NewSample*);

        do *New Model*=*reprocess*(*New Model*)until $\delta \leq \tau$;

        [nsv_+ , nsv_-] =get_ sv(*NewModel*);

        if $nsv_+ < nsv_-$ **then**

            $d=\frac{nsv_-}{nsv_+}$;

            synth_ data=generate_ synthetic ($< x_{sv+}, y_{sv+} >$, $< x_{ksv+}$, $y_{ksv+} >$, d);

            NewModel=*process*(synth_ data);

        else

            $d=\frac{nsv_+}{nsv_-}$;

            synth_ data=generate_ synthetic ($< x_{sv-}, y_{sv-} >$, $< x_{ksv-}$, $y_{ksv-} >$, d);

            NewModel=*process*(synth_ data);

        end

    end

Algorithm 2: OOLASVM algorithm for dynamic Imbalanced streams

5   Experimental Results

This section presents the effectiveness of OOLASVM when compared to LASVM, GOS-IL and OOB algorithms. We used synthetic datasets and real world data sets for the experimental study. The *Recall* and *G-Mean* are used as performance indicators to evaluate the OOLASVM.

5.1   Data Sets

To generate synthetic data streams, CIRCLE and HYPERPLANE functions [23] are used from the stream generating environment. Table 1 shows the settings for artificial stream generators. From these stream generators two levels of imbalances are generated STATIC and DYNAMIC. In STATIC imbalance, the degree of imbalance remains static for entire stream whereas in DYNAMIC imbalance case the prior probabilities p(y) of the classes change dynamically. Here, for the DYNAMIC datastream the degree of imbalance change is considered exactly from the middle of the stream i.e., the Change Point (CP) is 501th timestep (See Table 2).

For each of these STATIC and DYNAMIC level of imbalance, five streams with degrees of imbalance (1:9, 2:8, 3:7, 4:6 and 5:5) are generated. Table 2 depicts the 9:1 scenario for dynamic degree of imbalance generation, it is same for rest of the degree of imbalance cases. Hence in total twenty synthetic streams, ten for STATIC imbalance streams and ten for DYNAMIC imbalance streams are generated using stream generators. Three real world datasets namely, Pima, Yeast, and Glass are considered from UCI machine learning repository [24]. Some of these are multiclass datasets which are converted into two class datasets [25].

Table 1. Settings for concept drift generators.

Problem	Fixed values
CIRCLE	$a = b = 0.5\ r = 0.2- > 0.2$
HYPERPLANE	$a1 = 0.1\ a0 = -0.25- > -0.25$

Table 2. Type of imbalance before and after the drift for 1:9 case.

Type of imbalance	Before CP	After CP
STATIC	1:9	1:9
DYNAMIC	1:9	9:1

5.2   Discussion

Figures 3, 4 and 5 shows the results obtained for OOLASVM on synthetic data sets. To show the effectiveness of OOLASVM, initially we compared it with

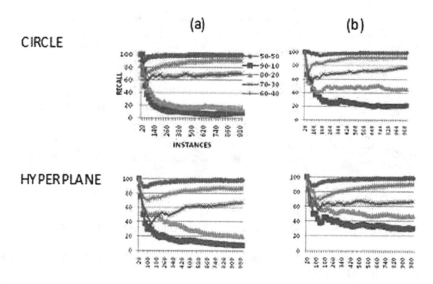

Fig. 3. Minority class Prequential Recall values before and after OOLASVM. (STATIC IMBALANCE) (a) LASVM (b) OOLASVM

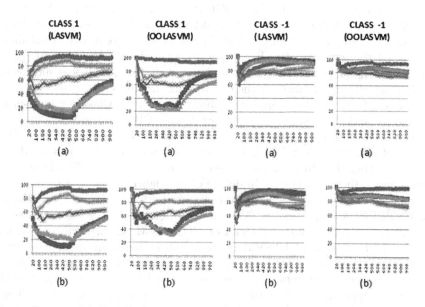

Fig. 4. Prequential Recall for minority and majority classes before and after OOLASVM. (DYNAMIC IMBALANCE) (a) CIRCLE (b) HYPERPLANE

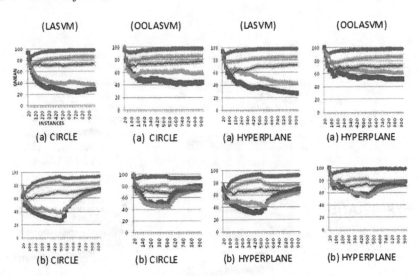

Fig. 5. Prequential G-Mean before and after OOLASVM. (a) STATIC IMBALANCE (b) DYNAMIC IMBALANCE

LASVM, which is not tailored to handle class imbalance (both STATIC and DYNAMIC cases) for evolving streams. Figure 3 presents the minority class Recall values for CIRCLE and HYPERPLANE in STATIC imbalance case. It is observed from the results that, OOLASVM outperforms the LASVM. At high degree of imbalance cases such as 9:1 and 8:2 the minority class Recall, OOLASVM is 20% more than LASVM. In case of moderate degree of imbalance such as 7:3 the minority class Recall performance of OOLASVM is 10% more than LASVM. Finally, in case of low and zero degree of imbalances such as 6:4 and 5:5, algorithms saturates at the end to same *Recall* value.

In case of DYNAMIC degree of imbalance, initially Class +1 is assumed as minority and −1 is assumed as majority, after CP, the degree of imbalance is assumed as in Table 2. From Fig. 4, on the minority class (+1), OOLASVM performed better than LASVM in all cases in terms of Recall. It is noticed that in LASVM, there is a drop/increase in performance (drift) from the middle of the stream where there is a change in degree of imbalance. Whereas this impact is observed as marginal with OOLASVM compared with LASVM on both CIRCLE and HYPERPLANE generators. At high degree of imbalances cases such as 9:1 and 8:2 the minority class +1 recall of OOLASVM on an average 20% and 10% better than that of LASVM on both the streams. At moderate degree of imbalance such as 7:3 OOLASVM yields 5% improvement on LASVM. At low and zero degree of imbalance cases such as 6:4 and 5:5, the impact of the drift due to dynamic degree of imbalance is low compared to LASVM (see Fig. 4 LASVM (a), (b) and OOLASVM (a), (b)). In case of 6:4 degree of imbalance, the performance seems to be low when compared with LASVM, after the drift both saturates to equal performance. However, in case of 5:5 there is clear domination of OOLASVM from the beginning.

Concerned with DYNAMIC degree of imbalance at majority class (-1), both algorithms yielded better *Recall* values compared with the degrees of imbalance of the minority class $(+1)$. This is due to enough number of instances that are learned at the first half, though this class becomes minority after the first half. Eventually, for this class also the OOLASVM exhibited better performance compared with the LASVM due to the over sampling of minority SV's at that time t. The class -1 becomes the minority after CP.

Figure 5 shows the comparison of *G-Mean* in STATIC and DYNAMIC imbalance cases for both CIRCLE and HYPERPLANE stream generation functions. In STATIC imbalance (STATIC) case, there is 20% improvement at high degree of imbalance such as 9:1 and 8:2, with OOLASVM when compared with LASVM. At moderate, low degree of imbalance ratios such as 7:3, 6:4 and at zero degree of imbalance, the LASVM alone exhibits good performance on both [CIRCLE:72%–96%] and [HYPERPLANE:69%–97%]. Whereas with OOLASVM the obtained performances on both CIRCLE and HYPERPLANE: are of [75%–96%] [71%–97%]. (see Fig. 5(a)).

In case of dynamic change of imbalance (DYNAMIC level – see Fig. 5(b)), for the degree of imbalances such as 9:1 and 8:2 the performance improvement is observed as 20% and 10% with OOLASVM. Further, the impact of the drift is low compared to LASVM on both CIRCLE and HYPERPLANE streams. For moderate, low and zero degree of imbalance cases such as 7:3, 6:4 and 5:5 the performance of OOLASVM is stable from the beginning (in terms of drift) compared with LASVM, and there is performance improvement from [69%–89%] to [72%–95%] on CIRCLE and from [67%–89%] to [72%–97%] on HYPERPLANE.

Tables 3 and 4 show the mean values of the minority class *Recall* and classifier *GMean* for real world data sets. Here, the comparison is carried out with GOS-IL and OOB algorithms which are intended to solve online class imbalance problem. Those values are taken at different time steps such as 25%, 50%, 75%, and 100% of the total number of instances. The best values are bold-faced. From the comparison it can be identified that OOLASVM performance is better than other two algorithms at each time step of the individual dataset. The superior performance of OOLASVM over LASVM, GOS-IL and OOB is due to:

(1) The standalone SVM is not much sensitive to conventional class imbalance problem [5, 15].
(2) The SVM boundary learned by dynamic oversampling and active learning approaches is adaptable to the dynamic change in the context of online learning.

Table 3. Mean values of Prequential minority recall for real world data sets.

Prequential minority recall					
Dataset	Method	t = 25%	t = 50%	t = 75%	t = 100%
Pima	OOB	32.86	41.89	40.38	42.54
	GOS-IL	38.57	43.92	45.19	44.78
	OOLASVM	**68.55**	**63.28**	**61.15**	**60.52**
Yeast	OOB	42.31	43.55	44.91	46.71
	GOS-IL	39.74	40.86	41.10	45.07
	OOLASVM	**70.69**	**72.30**	**71.09**	**70.70**
Glass	OOB	0.00	0.00	0.00	82.35
	GOS-IL	0.00	0.00	0.00	92.16
	OOLASVM	**100**	**100**	**100**	**95.83**

Table 4. Mean values of Prequential G-Mean for real world data sets.

Prequential G-Mean					
Dataset	Method	t = 25%	t = 50%	t = 75%	t = 100%
Pima	OOB	49.23	55.89	55.53	57.75
	GOS-IL	50.29	54.74	56.18	56.70
	OOLASVM	**68.86**	**65.22**	**64.03**	**63.97**
Yeast	OOB	58.93	59.89	60.35	63.48
	GOS-IL	58.75	59.24	61.19	62.92
	OOLASVM	**79.12**	**80.33**	**79.65**	**79.59**
Glass	OOB	0.00	0.00	0.00	89.06
	GOS-IL	0.00	0.00	0.00	93.91
	OOLASVM	**99.28**	**98.92**	**98.65**	**96.46**

6   Conclusion

The performance of the learners is hampered by the class imbalance problem, which occurs when one class of data severely outnumbers the other class. In case of dynamic imbalance in evolving streams, the degree of class imbalance changes dynamically and the distribution statistics are not prior known. To address this problem, this paper presents a SVM active learning based adaptive online learner, namely OOLASVM to cope with dynamic class imbalance problem. The proposed OOLASVM dynamically balances the SVM boundary throughout the stream by using active learning and oversample around minority class Support Vectors using Synthetic Oversampling Technique. The initial boundary is balanced by pushing the boundary from minority side by oversampling. As this method involves both dynamic oversampling of SV's and dynamic active sample selection/reduction to represent the current training set, thus formulates an hybrid oversample active learning algorithm. Experimental results on synthetic as well as real world streams demonstrated better performance in terms of minority class *Recall* and *G-Mean* when compared to LASVM, GOS-IL and OOB.

Acknowledgement. This work is supported by the Defense Research and Development Organization (DRDO), India, under the sanction code: ERIPR/GIA/17-18/038. Center For Artificial Intelligence and Robotics (CAIR) is acting as the reviewing lab for the work is concerned.

References

1. He, H., Garcia, E.A.: Learning from imbalanced data. IEEE Trans. Knowl. Data Eng. **21**(9), 1263–1284 (2009)
2. Sun, Y., Wong, A., Kamel, M.: Classification of imbalanced data. Int. J. Pattern Recognit. Artif. Intell. **23**(4), 687–719 (2009)
3. Galar, M., Fernandez, A., Barrenechea, E., Bustince, H., Herrera, F.: A review on ensembles for the class imbalance problem: bagging-, boosting-, and hybrid-based approaches. IEEE Trans. Syst. Man Cybern. Part C (Appl. Rev.) **42**(4), 463–484 (2012)
4. Wang, S., Minku, L.L., Yao, X.: Systematic study of online class imbalance learning with concept drift. IEEE Trans. Neural Netw. Learn. Syst. **29**(10), 1–20 (2018)
5. Nathalie, J., Stephen, S.: The class imbalance problem: a systematic study. Intell. Data Anal. **6**(5), 429–449 (2002)
6. Wu, G., Chang, E.: Class-boundary alignment for imbalanced dataset Learning. In: ICML 2003, Santa Barbara, California (2003)
7. Bordes, A., Ertekin, S., Weston, J., Bottou, L.: Fast kernel classifiers with online and active learning. J. Mach. Learn. Res. **6**, 1579–1619 (2005)
8. Chawla, N.V., Bowyer, K.W., Hall, L.O., Kegelmeyer, W.P.: SMOTE: synthetic minority over-sampling technique. J. Artif. Intell. Res. **16**, 321–357 (2002)
9. Morik, K., Brockhausen, P., Joachims, T.: Combining statistical learning with a knowledge-based approach-a case study in intensive care monitoring. In: ICML (1999)
10. Lee, H., Cho, S.: The novelty detection approach for different degrees of class imbalance. In: King, I., Wang, J., Chan, L.-W., Wang, D.L. (eds.) ICONIP 2006. LNCS, vol. 4233, pp. 21–30. Springer, Heidelberg (2006). https://doi.org/10.1007/11893257_3
11. Wang, S., Minku, L.L., Yao, X.: Resampling-based ensemble methods for online class imbalance learning. IEEE Trans. Knowl. Data Eng. **275**, 1356–1368 (2014)
12. Ghazikhani, A., Monsefi, R., Yazdi, H.S.: Recursive least square perceptron model for non-stationary and imbalanced data streams classification. Evol. Syst. **42**, 119–131 (2013)
13. Ghazikhani, A., Monsefi, R., Yazdi, H.S.: Online neural network model for non-stationary and imbalanced data stream. Int. J. Mach. Learn. Cybern. **51**, 51–62 (2013)
14. Barua, S., Islam, M.M., Murase, K.: GOS-IL: a generalized over-sampling based online imbalanced learning framework. In: Arik, S., Huang, T., Lai, W.K., Liu, Q. (eds.) ICONIP 2015. LNCS, vol. 9489, pp. 680–687. Springer, Cham (2015). https://doi.org/10.1007/978-3-319-26532-2_75

15. Yan, Y., Yang, T., Chen, J.: A framework of online learning with imbalanced streaming data. In: Sing, S.P., Markovitch, S. (eds.) Conference on Artificial Intelligence 2017, San Francisco, pp. 2817–2823. AAAI Press (2017)
16. Tang, Y., Zhang, Q.-Y., Chawla, N.V., Krasser, S.: SVMs modeling for highly imbalanced classification. IEEE Trans. Syst. Man Cybern. Part B **39**(1), 281–288 (2009)
17. Kremer, J., Steenstrup Pedersen, K., Igel, C.: Active Learning with support vector machines. Wires Data Min. Knowl. Discov. **4**(4), 313–326 (2014)
18. Calma, A., Reitmaier, T., Sick, B.: Semi-supervised active learning for support vector machines: a novel approach that exploits structure information in data. Inf. Sci. **456**, 3–33 (2018)
19. Cortes, C., Vapnik, V.: Support-vector networks. Mach. Learn. **203**, 273–297 (1995)
20. Platt, J.C.: Sequential minimal optimization: a fast algorithm for training support vector machines. A technical report MSR-TR-98-14 (1998)
21. Ertekin, S., Huang, J., Lee Giles, C.: Active learning class imbalance Problem. In: Kraaij, W., de Vries, AP., Clarke, L.A.C., Fuhr, N., Kando, N. (eds.) Conference on Research and Development in Information Retrieval 2007, Netherlands, pp. 823–824 (2007). https://doi.org/10.1145/1277741
22. Dawid, A.P., Vovk, V.G.: Prequential probability: principles and properties. Bernoulli **51**, 125–162 (1999)
23. Minku, L., White, A., Yao, X.: The impact of diversity on online ensemble learning in the presence of concept drift. IEEE Trans. Knowl. Data Eng. **225**, 730–742 (2010)
24. UCI Machine Learning Repository. http://archive.ics.uci.edu/ml/. Accessed 26 June 2018
25. Barua, S., Islam, M.M., Yao, X., Murase, K.: MWMOTE-majority weighted minority oversampling technique for imbalanced data set learning. IEEE Trans. Knowl. Data Eng. **262**, 405–425 (2014)

Using Crowd Sourced Data for Music Mood Classification

Ashish Kumar Patel[1], Satyendra Singh Chouhan[1(✉)], and Rajdeep Niyogi[2]

[1] Shri G.S.I.T.S, Indore 452001, India
patelashish53@gmail.com, schouhan@sgsits.ac.in
[2] Indian Institute of Technology (IIT), Roorkee, Roorkee 247667, India
rajdpfec@iitr.ac.in

Abstract. Music has been part of human lives since ancient times. We have hundreds of millions of songs representing different cultures, mood and genres. These songs are readily accessible using Internet and streaming services. However, the discovery of the right music piece to listen is hard and an automated assistance to find the right song among the millions is always desired. There have been several attempts to classify music on the basis of their genres but their efforts have not been much fruitful because of lack of good and large datasets. Moreover, identifying the set of features to represent the music in a summarized way is also a challenging task. In this work, we present an automated music mood classification approach that uses crowd-sourced platforms to label the songs. It eliminates the subjectivity of one's perception of mood on a song. We have confined our work to two classes of mood: happy and sad. The proposed approach is tested with three machine learning models: artificial neural networks (ANN), Decision Tree (DT) and Support Vector Machine (SVM). The experimental results show that ANN performs better than DT and SVM.

Keywords: Music Classification · Mood classification
Supervised learning

1 Introduction

Music, a euphony, forms an integral part of a humans harmonic experience. It represents culture, mood and expression and is considered an art. Music can convey feelings of individuals, thus it has a very important role in human life. Despite listening to such varying songs, we can relate to them on the basis of emotions they induce [1]. There is a lot of diverse research in music that has demonstrated that there exist some common psychological and emotional characteristics in music that can exceed the restrictions of language and accomplish cultural infiltration together [1,10].

Emotion modelling in psychology and music psychology research normally relies on explicit-textual or scale based-evaluations of emotion term relationships [23,25] and their application to music [32]. Based on these evaluations,

© Springer Nature Switzerland AG 2018
A. Mondal et al. (Eds.): BDA 2018, LNCS 11297, pp. 363–375, 2018.
https://doi.org/10.1007/978-3-030-04780-1_25

dimensional [23] and categorical [8] models of emotions have been proposed. Categorical emotion models either emphasize the existence of a limited set of universal basic emotions [8], or the variance between moods by means of a few underlying affect dimensions [27] or a larger number of emotion dimensions based on factor analyses [32]. In this paper, we considered categorical based emotion modelling.

In categorical based emotion modelling, the music moods can be divided into various categories: sadness, happiness, anger, calmness etc [23]. In this research work, we considered two categories of music: happy and sad [3]. There have been number of studies on automated music mood classification, which assume different kinds of songs representations such as (1) Symbolic (MIDI) [2] (2) Low-Level and Acoustic Audio features (content-based [3,16,18] (3) Lyrics [14,15]. Moreover, some of the works combine features from more than one representation (i.e., hybrid representation) [17,19,29].

Music mood classification has several issues. The first one is the subjectivity of one's perception while labelling a song. The second one is ambiguous mood taxonomy, for example, depressed and sad. Thus, mood categories must be very distinct and non-overlapping like happy and sad. The third one is computing content-based audio features and presence on noise and distortion in audio content [18,30]. In addition, lack of large and unbiased datasets in the public domain is also a major challenge in music mood classification.

In this work, we present an automated music mood classification technique that categories the music in two non-ambiguous categories i.e., happy and sad. The important features of the proposed technique are: (1) it uses crowd-sourced labelled tracks from *Last.Fm* to avoid the individual's perception towards the mood of a song; (2) unlike existing literature, it utilizes semantically close higher level audio features such as *danceability, speechiness*, etc [20]. In the AI component, the proposed classification technique evaluates Artificial Neural Network (ANN), Decision Tree (DT), and Support Vector Machine (SVM) models for learning.

The remainder of this paper is organized as follows. Section 2 presents the related work. Section 3 presents the proposed music mood classification approach. Performance evaluation is given in Sect. 4. Section 5 presents the conclusion and future directions.

2   Related Work

In this section, we first discuss the different music mood recognition (MMR) method used in the literature. Next, we discuss the works that are closely related to the work presented in this paper. Finally, we discuss the popular datasets used in MMR community.

The state-of-the-art MMR methods use different types of ground truth data about the music emotions. The ground truth data can be numeric or label type thus both regression and classification methods have been used for MMR [30]. MMR classification methods can be categorized into single-label classification [5], multi-label classification [11], and a special case of multi-label classification called fuzzy classification [12].

Single-label classification expresses the music emotion as a certain single emotion label that is mostly an adjective. These adjectives are simple to use and well known. But, the restriction to use a single label may fail to reflect one's perception when the music piece has a dynamic or ambiguous mood. Multi-label classification classifies the emotion of the music segment into a number of emotion categories, a set of labels are used as an annotation. These sets can accommodate the vagueness and inexactness of human perception. Fuzzy classification expresses the emotion of the music segment as the discrete possibility distribution of a number of emotion categories. Thus, each annotation is a fuzzy set with labels and their possibility measures. Since, we present a single label classification method, therefore now we discuss the works that are closely related to our work.

In [18], an approach for Mood Detection and Tracking of Music Audio Signals is proposed. It divides the music track into frames and uses the frame-based intensity features, timbre features, and rhythm features as feature set. It uses a hierarchical framework consisting of a Gaussian Mixture Model (GMM) with Expectation Maximization (EM) algorithm. It uses four target classes in hierarchical fashion Group 1 (Contentment and Depression), Group 2 (Exuberance and Anxious/Frantic). It achieves an accuracy of 86.3% in a dataset of 800 (75% : 25%) using 10 fold cross-validation.

In [2] an unsupervised agglomerative clustering algorithm applied on MIDI based dataset of songs. It divides a song into segments and calculates pitch, duration and velocity as features for each segment. The final feature vector consists of the music scale, accuracy, sound intensity, basic sound, interval, direction, velocity and duration of notes. It applies clustering for 5, 12, and 20 number of clusters. Finally, a SOM neural network has been shown for visualization of song segments. The system achieved an accuracy of 80% on a dataset of 104/70.

In [3] a system for detecting emotion in music, based on a deep Gaussian Process (GP), is proposed. In this approach, a song is divided into frames. For each frame features such as rhythm, dynamics, timbre, pitch and tonality are calculated. Next, statistical values, such as mean and standard deviation, of frame-based features are calculated to generate a 38-dimensions feature vector. For emotion classification, a deep GP is utilized. It treats the classification problem from the perspective of regression. Finally, 9 classes of emotion are categorized by 9 one-versus-all classifiers. The system achieved an accuracy of 71.3% on a dataset of 1080 songs.

In [21], It uses a dataset of 60 songs. It extracts a total of 45 feature using two free toolkits PsySound and Marsyas. It compares W-D-KNN with KNN and SVM. The highest recognition rate is 96.7% with W-D-KNN classifier weighted by the Fibonacci series scheme. The accuracy of other classifiers ranges from 78.3% to 90.0%.

Emotion annotation is a difficult task and is a primary issue that hinders research in MMR. Researchers have contributed and have made some datasets public. These datasets enable establishing a comparison between various MMR systems. Thus, here we discuss some of the popular MMR datasets used in the community. CAL500 dataset [28] consists of 500 popular music songs, each of which has been labelled using adjectives by at least 3 subjects. MER60 [31] consists of 60 pieces of 30-s clips manually selected from the chorus parts of English pop songs. It is annotated by 40 non-experts. MTurk [26] dataset consist of 240, 15-s clips labelled using crowd-sourcing platform Mechanical Turk (http:// mturk.com) and AMC [7] is a dataset of 600 tracks having five mood categories. AMC is built using metadata analysis and human assessments. Table 1 shows the summarized comparison of proposed work with state-of-the-art music mood classification approaches.

Table 1. Comparison of proposed work with state-of-the-art

Paper-ID\Characteristics		Ref. [18]	Ref. [2]	Ref. [3]	Ref. [21]	Proposed work
Crowd sourced data		No	No	No	No	**Yes**
Features type	High level	No	No	No	No	**Yes**
	Low level	Yes	Yes	Yes	Yes	**Yes**
Size of dataset		800	174	1080	60	**16527**
Classification technique		Gaussian Mixture Model	Agglomerative clustering	Deep Gaussian process	W-D-KNN, KNN, SVM	ANN, SVM, DT

3 Automated Music Mood Classification

The overall architecture of the proposed technique is shown in Fig. 1. The architecture consists of three phases; the first phase combines the data aggregation and data cleaning. The output of the first phase is evaluated using feature selection algorithms–the second phase. In the third phase, we apply three machine learning models ANN, DT, and SVM for performance evaluation.

Data Acquisition and Data Cleaning. For an unbiased dataset, each category of the songs is needed. First, a list of songs is required and then each song needs to be classified into a category. Traditionally, a group of experts or non-experts (subjects) may be asked to classify these songs manually. This approach is not scalable as building large datasets is tedious. Moreover, the categories may be biased with respect to the perceived emotion. Existing online music communities on the Internet contain valuable meta-data about music tracks like emotion tags [24]. We fetch a list of songs using emotion tags from *Last.fm*, that is a music website, which has large corpus of music meta-data and user listening habits. The procedure is shown in Fig. 2(a).

The dataset contains a total of 18 features;
track_mode_confidence, track_key_confidence, Track_time_signature_confidence, track_time_signature, track_tempo_confidence, track_end_of_fade_in, duration_ms,

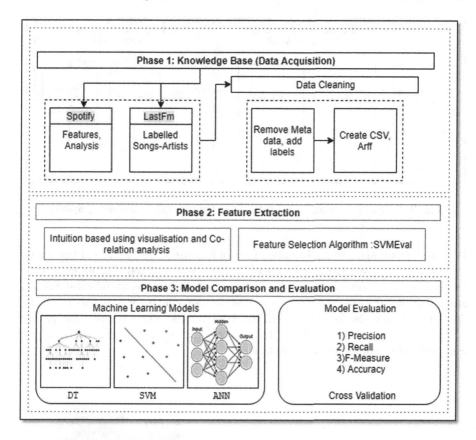

Fig. 1. An overview of Model Learning Process

tempo, valence, liveness, instrumentalness, acousticness, speechiness, mode, loudness, key, energy, danceability, and *mood.* The description of features is given in the Experiments section.

Once the list of songs has been compiled, we need a set of features to classify them. Music is a time-series signal, sampled at a frequency of 44 Khz (for a CD quality mp3). Using music in its raw form requires a large storage space and heavy computation. Thus, features are needed that can summarize music without losing much information. However, coming up with the right feature set is difficult and is a challenging task. We get features for a track from *Spotify*, which is a digital music streaming service and provides music features through its API. The significance of the features from *Spotify* is that they contain high-level semantic information like *danceability* and *accoustiness*. The high-level semantic features are closer to human perception of music. The procedure is shown in Fig. 2(b).

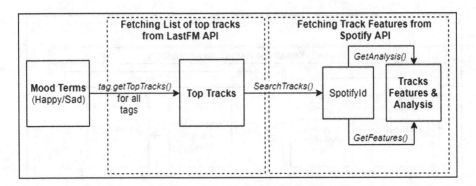

Fig. 2. (a) Song fetching from LastFM (b) Feature extraction using Spotify

Feature Analysis. After obtaining a set of feature, to get a sense of the distribution of data, some statistics for each feature is calculated. For numeric type features, we calculate the minimum value, maximum value, standard deviation and mean. We then plot the frequency distribution charts for each feature, using different colour codes for the two classes. For example, Fig. 3 is a plot for feature energy and mode respectively. In these plots, the red colour shows the happy class and blue represents the sad class.

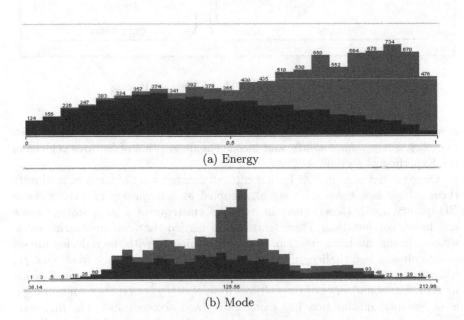

Fig. 3. Plots for feature: (a) Energy (b) Mode

In next step, we determine the usefulness of all the 18 features. We utilize the frequency distribution plots for this purpose. For example, the feature energy (Fig. 3(a)) most of the Happy (red) songs have values greater than 0.5, while most Sad (blue) songs have values less than 0.5. We use this approach to intuitively evaluate features usefulness. Features like mode (Fig. 3(b)) seems less useful as the distribution of both the categories of songs looks similar for it.

After primary analysis, to come up with the right set of features, we apply SVM recursive feature elimination (SVM-RFE) algorithm. It ranks the features according to their merit of distinguishing the target class. It proceeds in a recursive manner, we stop the procedure after no feature remains to be eliminated below a certain threshold merit value. To validate the process, we use 10 fold cross-validation. The results of the SVM-RFE elimination step is reported in Sect. 4.

AI Classifiers. In this subsection, we start with the discussion of how we minimize the sampling biases using k-fold cross-validation. Then we briefly discuss the classification algorithms use for the learning purpose. Finally, we discuss metrics used to evaluate and compare the algorithms.

In k-fold cross-validation, The data set is randomly partitioned into k subsets of approximately equal size. Then, the model is evaluated k-times i.e. k-folds. In each fold, one of the k subsets is held back from the training process for validating the model. Finally averaged result of the k-folds is reported. We use a variation of k-fold cross validation called stratified k-fold cross-validation, In which each subset contain roughly the same proportions of class labels. To keep a balance between performance and computational time, we choose 10 number of folds [6].

Artificial Neural Networks(ANN) are biologically inspired computational network. They have been widely employed to solve pattern recognition, classification, and regression problems. They have been used previously for Music moods classification [9]. In this paper, we use the Sigmoid activation function for all the layers. We have also employed Multi-Layer Perceptron(MLP) learning model that uses backpropagation algorithm.

Decision trees are also widely used classification algorithms. They are popular for their simplicity and efficiency. The dataset is partitioned at the root node, by applying splitting criteria on a single feature. The resultant smaller datasets are further partitioned recursively until maximal pure subsets are obtained. Some of the popular Decision Tree algorithms include ID3, C4.5, and CART. We use J48, which is an implementation of C4.5 in Java.

The third classification algorithm we use is SVM. A number of Music Classification works have shown their effectiveness in Music Moods Classification [4]. To classify, The data vectors are mapped to a high-dimensional space, followed by finding a hyperplane that separates the classes with maximum margin. We use a method of SVM implementation called SMO that avoid quadratic programming problem with SVM [22].

To compare the different models and their respective performance we use (1) Accuracy (2) F-Score (3) Recall. The selected metrics are suitable for binary classification problems and are widely used in other machine learning literature as well. For details of their calculation and their implications, readers are advised to go throw relevant reading sources (Ref. [13]).

4   Results and Discussion

In this section, the proposed approach is evaluated on a dataset of 16527 songs using the method discussed in Sect. 3. The dataset is randomized. The Table 2 shows the description of the feature used in the data set.

The experiments are performed on a Linux machine, with 8 GB memory and Intel i5 processor. We used Weka, a Java based machine learning package to carry out the experiments. The full dataset is available at https://github. com/ashish3805/AMMC/blob/master/Datasets/combined.csv. From an expert systems perspective, the music mood classification problem can be divided into two components: (1) what information and features are needed to be tracked as a part of our "knowledge base"; and (2) what machine learning models can be used for effective classification.

For the first component, we selected the significant features using some attributes selection algorithms such as SVM-RFE. The goal of this step is to find a subset of features that can result in accurate predictions without overfitting. The features are ranked and the less important ones are sequentially eliminated. Table 3 shows the results obtained by SVM-RFE. The results are averaged for 10-folds of the cross-validation.

Based on the output of various attribute selection algorithms, we divided the attribute set in three cases: The first case includes all the features. The second case is on the basis of the merit obtained through SVM-RFE, it includes features with merit greater than 10.0. In the third case, we used the subset features obtained from the algorithm. Table 4 depicts the three different cases that are used for the empirical study.

For the second component, we compare the performance of ANN, SVM and DT over the three cases. In Table 5, we report the Precision, Recall and F-Measure, for the three cases. Figure 4 shows the Accuracy comparison of these three models. From the results, we observe that

- ANN performs better than Decision tree and SVM in all the given cases.
- for some cases, precision and recall have the same value. This is due to the fact that, number of happy songs and sad songs are almost equal in the data set.
- Accuracy of ANN with 8 features (Case 2) is slightly better than the Case 1 and Case 3.

Table 2. Music mood classification Data set Description

Features	Description	Type and range
Acousticness	A confidence measure that represents degree of confidence that the particular track is acoustic	Numerical (0.0 to 1.0)
Danceability	It describes how suitable a track is for dancing based on a combination of musical elements including tempo, rhythm stability, beat strength, and overall regularity	Numerical (0.0 to 1.0)
Energy	It represents a perceptual measure of intensity and activity of a track	Numerical (0.0 to 1.0)
instrumentalness	Predicts whether a track contains no vocals	Numerical (0.0 to 1.0)
	Value closer to 1.0 represents the greater likelihood that the track contains no vocal content	(0.0 to 1.0)
mode	Mode indicates the modality (major or minor) of a track, the type of scale from which its melodic content is derived	Binary (0 and 1)
Liveness	Higher liveness values represent an increased probability that the track was performed live	Numerical (0.0 to 1.0)
Loudness	The overall loudness of a track in decibels (dB)	Numerical (-60 to 0 dB)
Key	The key the track is in. Integers map to pitches using standard Pitch Class notation E.g. 0 = C, 1 = C?/D?, 2 = D, and so on	numerical
speechness	Speechiness detects the presence of spoken words in a track. Values above 0.66 describe tracks that are probably made entirely of spoken words	Numerical (-60 to 0 dB)
track_tempo_confidence	Confidence Values: Many elements at the track and lower levels of analysis include confidence values, a floating-point number ranging from 0.0 to 1.0. Confidence indicates the reliability of its certainty	Numerical (0.0 to 1.0)
Track_time_sign_confidence		
track_mode_confidence		
track_key_confidence		
track_time_signature	An estimated overall time signature of a track (No. of beats per bar)	Numerical
track_end_of_fade_in	the end of the fade-in introduction to a track in seconds	Numerical
duration_ms	Total duration of the track in millisecond	Numerical
Tempo	The overall estimated tempo of a track in beats per minute (BPM)	Numerical (0.0 to 1.0)
Valence	Represents positive-ness in the Tracks. Tracks with high valence sound more positive (e.g. happy, cheerful, euphoric)	Numerical (0.0 to 1.0)
mood	Target class: happy or sad	Happy and sad

Table 3. SVM RFE results

Attribute	Average rank	Average merit
Speechiness	1 +− 0	18 +− 0
Energy	2 +− 0	17 +− 0
Danceability	3 +− 0	16 +− 0
loudness	4 +− 0	15 +− 0
valence	5 +− 0	14 +− 0
duration_ms	6 +− 0	13 +− 0
acousticness	7.2 +− 0.6	11.8 +− 0.6
track_mode_confidence	8.6 +− 0.92	10.4 +− 0.917
instrumentalism	9.1 +− 1.14	9.9 +− 1.136
track_key_confidence	9.6 +− 0.92	9.4 +− 0.917
track_time_signature	10.9 +− 1.22	8.1 +− 1.221
liveness	11.7 +− 0.46	7.3 +− 0.458
track_tempo_confidence	13.1 +− 0.54	5.9 +− 0.539
track_end_of_fade_in	14.1 +− 0.7	4.9 +− 0.7
tempo	14.7 +− 0.46	4.3 +− 0.458
track_time_signature_confidence	16.5 +− 0.67	2.5 +− 0.671
key	16.8 +− 0.75	2.2 +− 0.748
mode	17.7 +− 0.46	1.3 +− 0.458

Table 4. Experiment cases

Cases	Features		
One	Speechiness	acousticness	track_tempo_confidence
	Energy	track_mode_confidence	track_end_of_fade_in
	Danceability	instrumentalism	tempo
	loudness	track_key_confidence	track_time_signature_confidence
	valence	track_time_signature	key
	duration_ms	liveness	mode
Two	Speechiness	loudness	acousticness
	Energy	valence	track_mode_confidence
	Danceability	duration_ms	
Three	Speechiness	loudness	acousticness
	Energy	valence	Danceability
	duration_ms		

Table 5. Experimental results

Case I

Algorithm	Precision	Recall	F measure
ANN	0.868	0.868	0.868
SVM	0.869	0.867	0.867
Decision Tree	0.845	0.845	0.845

Case II

Algorithm	Precision	Recall	F measure
ANN	0.873	0.873	0.867
SVM	0.862	0.859	0.859
Decision Tree	0.854	0.854	0.854

Case III

Algorithm	Precision	Recall	F measure
ANN	0.873	0.873	0.873
SVM	0.862	0.86	0.86
Decision Tree	0.854	0.854	0.854

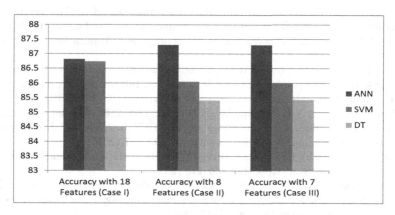

Fig. 4. Accuracy comparison of ANN SVM and DT

From the selected AI models, SVM requires tunning of kernel related and other control parameters. Various parameters of learning models are tuned to the best of our knowledge and efforts. However, with the use of different sets of parameters and change in the set feature, further tuning of learning model parameters may be required.

5   Conclusion

In this work, we addressed the music mood classification problem and presented an automated music mood classification technique. We presented the literature survey of the state-of-the-art in music mood recognition and presented a theoretical analysis of proposed technique with existing work. The significant feature of the proposed technique is to use crowd-sourced platforms to label the songs; that eliminates the subjectivity of one's perception of mood on a song. We have evaluated our system on a large and unbiased dataset of songs and the experimental results show that ANN performs better than the other two algorithms.

The work presented so far can be extended in various constructive ways. We list the possible extensions and applications below.

– The mood classes can have more categories for music like passion, anger, melancholy etc.
– Lyrics based features can be added to aid the classification further.
– The technique can be used to organize music on personal systems as per mood.

References

1. Balkwill, L.L., Thompson, W.F.: A cross-cultural investigation of the perception of emotion in music: psychophysical and cultural cues. Music Percept. Interdisc. J. **17**(1), 43–64 (1999)
2. Bartoszewski, M., Kwasnicka, H., Markowska-Kaczmar, U., Myszkowski, P.B.: Extraction of emotional content from music data. In: Proceedings of the 7th Computer Information Systems and Industrial Management Applications, pp. 293–299 (2008)
3. Chen, S.H., Lee, Y.S., Hsieh, W.C., Wang, J.C.: Music emotion recognition using deep Gaussian process. In: Asia-Pacific Signal and Information Processing Association Annual Summit and Conference (APSIPA), pp. 495–498 (2015)
4. Chen, Y.A., Yang, Y.H., Wang, J.C., Chen, H.: The AMG1608 dataset for music emotion recognition. In: Proceedings of the IEEE International Conference on Acoustics, Speech and Signal Processing, pp. 693–697 (2015)
5. Corrêa, D.C., Rodrigues, F.A.: A survey on symbolic data-based music genre classification. Expert Syst. Appl. **60**, 190–210 (2016)
6. Dag, A., Topuz, K., Oztekin, A., Bulur, S., Megahed, F.M.: A probabilistic data-driven framework for scoring the preoperative recipient-donor heart transplant survival. Decis. Support Syst. **86**, 1–12 (2016)
7. Downie, X., Laurier, C., Ehmann, M.: The 2007 MIREX audio mood classification task: lessons learned. In: Proceedings of the 9th International Conference on Music Information Retrieval, pp. 462–467 (2008)
8. Ekman, P.: An argument for basic emotions. Cogn. Emotion **6**(3–4), 169–200 (1992)
9. Feng, Y., Zhuang, Y., Pan, Y.: Popular music retrieval by detecting mood. In: Proceedings of the 26th Annual International ACM SIGIR Conference on Research and Development in Informaion Retrieval, pp. 375–376. ACM (2003)
10. Fritz, T., et al.: Universal recognition of three basic emotions in music. Curr. Biol. **19**(7), 573–576 (2009)
11. Gabrielsson, A.: Emotion perceived and emotion felt: same or different? Musicae Scientiae **5**(1-suppl), 123–147 (2001)
12. Goyal, S., Kim, E.: Application of fuzzy relational interval computing for emotional classification of music. In: Proceedings of the 8th IEEE Conference on Norbert Wiener in the 21st Century, pp. 1–8 (2014)

13. Han, J., Pei, J., Kamber, M.: Data Mining: Concepts and Techniques. Elsevier, Amsterdam (2011)
14. He, H., Jin, J., Xiong, Y., Chen, B., Sun, W., Zhao, L.: Language feature mining for music emotion classification via supervised learning from lyrics. In: Kang, L., Cai, Z., Yan, X., Liu, Y. (eds.) ISICA 2008. LNCS, vol. 5370, pp. 426–435. Springer, Heidelberg (2008). https://doi.org/10.1007/978-3-540-92137-0_47
15. Hu, X., Downie, J.S.: When lyrics outperform audio for music mood classification: a feature analysis. In: ISMIR, pp. 619–624
16. Hu, X., Yang, Y.H.: The mood of Chinese Pop music: representation and recognition. J. Assoc. Inf. Sci. Technol. **68**(8), 1899–1910 (2017)
17. Huang, M., Rong, W., Arjannikov, T., Jiang, N., Xiong, Z.: Bi-modal deep boltzmann machine based musical emotion classification. In: Villa, A.E.P., Masulli, P., Pons Rivero, A.J. (eds.) ICANN 2016. LNCS, vol. 9887, pp. 199–207. Springer, Cham (2016). https://doi.org/10.1007/978-3-319-44781-0_24
18. Lu, L., Liu, D., Zhang, H.J.: Automatic mood detection and tracking of music audio signals. IEEE Trans. Audio Speech Lang. Process. **14**(1), 5–18 (2006)
19. Mayer, R., Neumayer, R., Rauber, A.: Combination of audio and lyrics features for genre classification in digital audio collections. In: Proceedings of the 16th ACM International Conference on Multimedia, pp. 159–168 (2008)
20. Nest, E.: Analyzer documentation (1999). https://doi.org/10.1016/s0305-9006(99)00007-0, http://docs.echonest.com.s3-website-us-east-1.amazonaws.com/static/AnalyzeDocumentation.pdf
21. Pao, T.L., Cheng, Y.M., Yeh, J.H., Chen, Y.T., Pai, C.Y., Tsai, Y.W.: Comparison between weighted D-KNN and other classifiers for music emotion recognition. In: Proceedings of the 3rd International Conference on Innovative Computing Information and Control, pp. 530–530 (2008)
22. Platt, J.: Sequential minimal optimization: a fast algorithm for training support vector machines (1998)
23. Russell, J.A.: A circumplex model of affect. J. Pers. Soc. Psychol. **39**(6), 1161 (1980)
24. Saari, P., Eerola, T.: Semantic computing of moods based on tags in social media of music. IEEE Trans. Knowl. Data Eng. **26**(10), 2548–2560 (2014)
25. Scherer, K.R.: Emotion as a multicomponent process: a model and some cross-cultural data. Rev. Pers. Soc. Psychol. (1984)
26. Speck, J.A., Schmidt, E.M., Morton, B.G., Kim, Y.E.: A comparative study of collaborative vs. traditional musical mood annotation. In: ISMIR, pp. 549–554 (2011)
27. Thayer, R.E.: The Biopsychology of Mood and Arousal (1990)
28. Turnbull, D., Barrington, L., Torres, D., Lanckriet, G.: Towards musical query-by-semantic-description using the cal500 data set. In: Proceedings of the 30th Annual International ACM SIGIR Conference on Research and Development in Information Retrieval, pp. 439–446 (2007)
29. Xue, H., Xue, L., Su, F.: Multimodal music mood classification by fusion of audio and lyrics. In: He, X., Luo, S., Tao, D., Xu, C., Yang, J., Hasan, M.A. (eds.) MMM 2015. LNCS, vol. 8936, pp. 26–37. Springer, Cham (2015). https://doi.org/10.1007/978-3-319-14442-9_3
30. Yang, X., Dong, Y., Li, J.: Review of data features-based muc emotion recognition methods. Multimed. Syst. 1–25
31. Yang, Y.H., Su, Y.F., Lin, Y.C., Chen, H.H.: Music emotion recognition: the role of individuality. In: Proceedings of the International Workshop on Human-Centered Multimedia, pp. 13–22 (2007)
32. Zentner, M., Grandjean, D., Scherer, K.R.: Emotions evoked by the sound of music: characterization, classification, and measurement. Emotion **8**(4), 494 (2008)

Applying Big Data Intelligence for Real Time Machine Fault Prediction

Amrit Pal$^{(\boxtimes)}$ ⓘ and Manish Kumar

Indian Institute of Information Technology Allahabad, Allahabad 211015, UP, India
{rs167,manish}@iiita.ac.in

Abstract. Continuous use of mechanical systems requires precise maintenance. Automatic monitoring of such systems generates a large amount of data which require intelligent mining methods for processing and information extraction. The problem is to predict the faults generated with ball bearing which severely degrade operating conditions of machinery. We develop a distributed fault prediction model based on big data intelligence that extracts nine essential features from ball bearing dataset through distributed random forest. We also perform a rigorous simulation analysis of the proposed approach and the results ensure the accuracy/correctness of the method. Different types of fault classes are considered for prediction purpose and classification is done in a supervised distributed environment.

Keywords: Distributed environment · Random forest · Ball bearing
Fault prediction · Spark · Parallel processing · Decision tree

1 Introduction

Ball bearing is a critical element of rotating machinery which is widely used in the industry. A minor malfunction in the bearing can generate mechanical noise and sometimes leads to a severe breakdown in the machinery [1]. Therefore, bearing fault diagnosis is an urgent issue. Moreover, with the progressive growth of industrial revolution in the field of machinery, brings forth problematic challenges in mining the data on account of enormous amount of data generated.

Very often, massive data holds scarce information which is scattered all over the data. Hence, it is crucial to efficiently mine valuable information and to use solely this selected information for precision. Furthermore, the problem is to choose a decisive classifier that can work sufficiently well in a distributed environment for high dimensional dataset. A number of classifiers, such as Support Vector Machine (SVM), Artificial Neural Network (ANN), Naive Bayes, Logistic Regression, Decision Trees, etc. have been put forward in the recent years and showed their brilliance and relevance in the diagnostic field [2]. Unfortunately, several obstacles still exist when it comes to handling larger data with more different kinds of faults to be diagnosed. Random Forest algorithm [3] is better known for its ability to work with large data. But traditional Random Forest

ⓒ Springer Nature Switzerland AG 2018
A. Mondal et al. (Eds.): BDA 2018, LNCS 11297, pp. 376–391, 2018.
https://doi.org/10.1007/978-3-030-04780-1_26

algorithm is often not suitable for massive data and needs a distributed platform for parallel computation.

Studies on parallel and distributed platform have been quite popular in the recent years with big achievements [4]. Hadoop's [5] Map-Reduce programming model provides an efficient framework for processing gigantic datasets in an extremely parallel mining. However, the intermediate results acquired in each iteration are written to the Hadoop Distributed File System (HDFS) and therefore, the cost for I/O operation time increases and communication requires large resources. Another good distributed platform, appropriate for data mining, is Apache Spark [6]. Spark's core abstraction is Resilient Distributed Dataset (RDD). Unlike Hadoop, RDD [7] allows to store data in memory and to perform computation directly from the memory. Thus, it saves a large number of disk I/O time and speeds up the data processing. It is of great value and significance to solve big data issue in real time by using Spark.

With the focus on predicting faults in a distributed environment, we introduce a distributed ball bearing fault diagnosis mechanism using random forest classifier. The overall method is summarized as follows.

- It is implemented on Apache Spark for distributed mining. Data is partitioned into data blocks and each block is processed in parallel to train the fault prediction model independently. Thus, a decrement in time complexity is achieved.
- It is important to extract only the useful features from the raw data to improve the accuracy. Therefore, statistical processing of features is done prior to classification.
- For diagnosing the faults, random forest classifier is used as it is known for working adroitly in a distributed environment.

In the next Sect. 2, we discuss about the current study in the literature related to our work, Sect. 3 provides a brief recall about the random forest classification algorithm and Apache Spark for cluster computing, the overall proposed distributed fault prediction mechanism is elaborated in Sect. 4, finally, experiment results and conclusion are discussed in Sect. 5 and Sect. 6 respectively.

2 Related Work

Since majority of the machines uses bearing, it is evident that bearing fault diagnosis is an important issue that requires immediate attention. Techniques such as vibration and Acoustic Emission(AE) signal monitoring are popular for detecting bearing faults [8]. But considering its reliability for fault diagnosis, vibration analysis is the most useful approach [9–11]. The vibration analysis mainly includes time and frequency domain. Both the analysis is widely used for feature extraction in order to improve the performance of the diagnostic systems [12,13].

The most important task in fault diagnosis is classification. A number of Artificial Neural Networks such as Multilayer Perceptron, Radial Basis Function,

Probabilistic Neural Network, have been studied to explore the diagnosis of bearing fault [14]. The study reveals that the techniques are not suitable enough for multi-class classification problems and consume more training time. Peng et al. [15] have investigated the bearing fault using random forest and C4.5 algorithm. The experiment shows that random forest works better in classifying the ball bearing fault effectively. Yang et al. [16] have analyzed the possibilities of the random forest algorithm in machine fault diagnosis. And the experimental results indicate the reliability of the random forest based diagnosis method.

Random Forest is recognized to be one of the finest machine learning classification algorithm that can handle large data efficiently [17]. Focusing on the accuracy of the algorithm, different methods have been used to improve the random forest, such as, including only uncorrelated high performing tree and re-sampling of training data [18,19]. However, with the rampant growth of data day-to-day, it becomes more complex in its characteristic as well as its expansion in dimension, making it difficult for the traditional random forest algorithm to manage this data [20].

Management of larger data gives rise to more memory problem, leading to which numerous models of parallel/distributed computing were proposed. Dae et al. [21] have introduced a parallel version of C4.5 algorithm based on Map-Reduce. The results illustrate that it exhibits both time efficiency and scalability. Wang et al. [22] and Han et al. [23] have aimed at the improvement of the random forest algorithm using Map-Reduce. Both the experiments evidently show that random forest, working in parallel, confirms the effectiveness of classifying massive data. However, Hadoop Map-Reduce I/O operation time is costly and thus an in-memory distributed platform is required to improve the performance of the algorithm.

Bhagat et al. [24] have presented the data preprocessing technique-SMOTE for multi-class imbalance data using random forest. In order to handle a large dataset as needed, this technique was also adapted to big data using Map-Reduce. The proposed system was implemented on Apache Hadoop and Apache Spark platform. It has been observed that spark gives better results in terms of time complexity. Zhang et al. [25] have proposed a method for predicting air quality based on a Spark implementation of random forest algorithm. The results reveal that the proposed method can obtain better value both in precision and recall. The effectiveness and scalability have also been achieved through the method when dealing with big data.

As per the mechanical industry revolution is concerned, the amount of data generated by such industrial devices is very high. So, this tremendous increment of data has put a challenge in managing the data. Based on the existing research, we propose a distributed mining scheme in which direct streaming of the signal data is passed to the distributed storage and distributed processing is applied for detecting faults.

Table 1. Notations

Notation	Description	Notation	Description
D	Training dataset	B_i	Bootstrap sample
ϑ_k	A random sample of features	RF	Random Forest
F	Set of Features	f	Number of features ($|F|$)
X	Set of input vectors	x_i	i^{th} input vector for training
n	Total number of observations	y_i	Class label for i^{th} observation
Y	Feature matrix	$feat()$	Feature extraction process
s^d	One single observation of generated signal	S^d	Collection of d dimension signals
T_i	Generated tree from i^{th} sample	$t_{i,j}$	Tree generated from i^{th} RDD and j^{th} sample from that RDD
N	Number of decision trees in random forest		

3 Preliminaries

3.1 Notations

All the notations used in this entire work are mentioned in Table 1.

3.2 Random Forest

Random forest [3] is an ensemble of decision trees. It uses a bootstrap method for sampling the input data. Given a training set D, N decision trees are built which are further used for the prediction of class label testing data. Each of the decision trees votes for the final class; majority class will be the final classification result. The essential steps for creation of a random forest are shown in Algorithm 1. From a feature pool F, a random sample of features is selected ($\vartheta_k \in F$). A decision tree is generated using the available data and sampled feature set ϑ_k. To ensure independence among each selection of ϑ_k with a previously selected sample set ϑ_{k-1}, sampling with replacement is used. The generated tree can be interpreted as a classifier $c(x, \vartheta_k)$ with X as the input vector. For a collection of decision trees $c_1(X), c_2(X), ..., c_N(X)$ over a training dataset, selected randomly from complete input data, a marginal function can be used as a parameter to decide the class. This complete process can be mapped as a random forest classifier. There can be k different trees, an ensemble of the trees is calculated using Eq. 1 [26]:

$$C(X) = \sum_{i=1}^{N} \frac{1}{N} c_i(X) \qquad (1)$$

Algorithm 1. Random Forest

Input: Training dataset D $(n \times F)$, Number of Trees N
Output: Random Forest
1: Select a bootstrap sample B_i as the training sample
2: Select B_i as a single node
3: Calculate gain ratio for best split
4: Repeat 3 until no node to split
5: Repeat from 1 to 3, N times
6: Combine the decision trees $(T_1, T_2, ..., T_N)$
7: **return** Random Forest

3.3 Apache Spark

Apache Spark [6] is a rapid cluster computing technology for distributed computing. The main feature of spark is its in-memory cluster computing capability that increases the processing speed. It supports multiple languages, such as scala, java, and python. RDD (Resilient Distributed Datasets) [7] is the Spark's core abstraction. It is fault tolerant and an immutable collection of elements that can be operated in parallel. RDDs can be created in two ways: parallelizing an existing collection, or referencing a dataset in an external storage system, such as shared file system, HDFS or any Hadoop input format. Once an RDD is created, it can be operated in parallel. There are two types of RDD operations: transformation and action. Transformation creates a new RDD from an existing one and action runs a computation on the dataset and returns a value. Since all transformations in spark are lazy operations, they are computed only when an action requires a result to be returned.

4 Distributed Fault Prediction

It is very intuitive that a large dataset requires distributed processing for efficiently mining information from the data. This results in the division of entire mining problem into sub-problems while maintaining the independence among them. For this purpose, we adopt HDFS and RDD based Spark distributed processing system over a cluster of nodes. Data partitions are created according to the nodes available in the cluster. The data is loaded using spark RDD from HDFS blocks. Either a dataset or in any real time scenario, machine-generated signal (s^d) can be recorded using an appropriate hardware to form a complete collection of signals (S^d) which can be used for fault detection, where, d is the dimension of the signal. Relevant information from these d dimensional signals is extracted in form of features by applying a feature extraction process (Eq. 2). This process is applied to all the partitions in parallel and a feature matrix is generated which is used for fault diagnosis.

$$Y = feat(S^d) \tag{2}$$

where, Y is a feature matrix having f number of features and $feat$ is the extraction process that extracts f number of features from (S^d) as explained in Appendix A. For performing fault classification, we use random forest algorithm due to its ability to work in distributed environment. The algorithm is trained using in-memory computations of the distributed Spark processing system. Considering a cluster setup using Apache spark, feature matrix Y will be available in form of RDD to the workers in the cluster. Each worker generates the random forest RF_i from RDD_i as follows:

$$RDD_1 \rightarrow \{t_{11}(x, \vartheta_k), t_{12}(x, \vartheta_k), ...\} \rightarrow RF_1$$
$$RDD_2 \rightarrow \{t_{21}(x, \vartheta_k), t_{22}(x, \vartheta_k), ...\} \rightarrow RF_2$$
$$\vdots \qquad (3)$$
$$\vdots$$
$$RDD_z \rightarrow \{t_{z1}(x, \vartheta_k), t_{z2}(x, \vartheta_k), ...\} \rightarrow RF_z$$

The final random forest can be generated using the union of all the decision trees or the union of the random forests using Eq. 4.

$$RF = RF_1 \cup RF_2 \cup ... \cup RF_z \qquad (4)$$

Our proposed approach involves the following two steps: 1. Data acquisition and feature extraction 2. Perform classification using distributed random forest for fault prediction, which is depicted in the block diagram in Fig. 1.

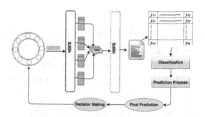

Fig. 1. Proposed fault prediction model

4.1   Data Acquisition and Feature Extraction

To test the utility of our prediction model, we use the data produced by the NSF I/UCR Center for Intelligent Maintenance Systems with the support from Rexnord Corporation in Milwaukee, WI [27]. The states of operation of the experimental apparatus are as follows: The rotating speed of the four bearings, installed on a shaft, is kept constant at 2000 RPM by an AC motor; using a spring mechanism, a radial load of 6000 lbs is applied on each bearing; and PCB 353B33 high sensitivity quartz ICP accelerometers are installed on the bearing housing. All failures are occurred after exceeding defined lifetime of the bearings, which is more than 100 million rotations. There are three categories of faults as shown in Fig. 2. To predict these faults, the required data is collected as follows:

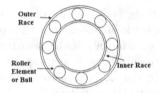

Fig. 2. Three categories of bearing faults presented as outer race, inner race, and ball (roller element)

1. The first dataset contains vibration readings of 4 bearings; each attached with two accelerometers. At the end of the test-to-failure experiment, it was found that bearing 1 and bearing 2 are in good condition, inner race defect is occurred in bearing 3 and roller element (ball) defect in bearing 4.
2. The second dataset also contains vibration readings of 4 bearings; each attached with one accelerometer. At the end of the test-to-failure experiment, outer race failure is occurred in bearing 1 and bearing 2; 3 and 4 are found in good conditions.
3. The third dataset again contains vibration readings of 4 bearings; each attached with one accelerometer. After finishing test-to-failure experiment, it is found that outer race failure occurred in bearing 3 while the other three bearings are found in good conditions.

Data Pre-processing. Ball bearings induce very low amplitude in vibration signals. So, it is arduous to identify bearings' localized faults in comparison to the faults like rotor unbalance produced by other machinery components. Also, the spectra of raw vibration signals contain very little diagnostic information regarding the bearing. Therefore, instead of processing the raw vibration data, the vibration signals are first processed by statistical parameters and then transformed into a perspicuous structure for further use. Statistical information about these features can be altered since fluctuation in vibration signals may occur arbitrarily.

For feature extraction, the raw vibration data is stored in Hadoop Distributed File System (HDFS) and is loaded into spark RDD. The RDD splits the data for distributed computing.

We extract 9 time domain features, which include parameters: RMS, mean, variance, skewness, kurtosis, maximum value, crest factor (CF), impulse factor (IF) and shape factor (SF); using formulations described in Appendix A.

The feature matrix after the feature extraction is represented as $Y = n \times f$. F represents the feature set: $F = F_1, F_2, ..., F_{10}$, where, the first 9 columns are the extracted features ("Kurtosis", "Skewness", "rms", "variance", "maximum value", "crest factor", "mean", "impulse factor", "shape factor") and the last column is the class label (0 for non-faulty, 1 for inner race fault, 2 for roller element fault and 3 for outer race fault). The feature matrix Y has n observations and their corresponding class levels, so the feature matrix can be represented as: $Y = ((x_1, y_1), (x_2, y_2), ..., (x_n, y_n))$, where, x_i is the i_{th} training observation and y_i is the output of x_i observation. The observation x_i can take any value for these features. This data is further used for classification of the bearing fault applying a random forest classifier. The working of the random forest classifier in the distributed environment is presented in the next section.

4.2   Distributed Random Forest (DRF) for Fault Prediction

Implementation of the distributed prediction environment requires independence among the sub-problems. Random forest is a collection of decision trees; the generation process of one tree is independent of the other. Considering this independence, the tree generation process can be done in parallel from available data. Thus, we apply a parallel random forest approach for bearing fault classification in a distributed environment. Where the individual tree is built in driver node using in-memory computation. The process starts with loading RDD_i, which contains a data chunk from generated feature matrix. There can be multiple RDDs, based on the size of the feature matrix. A *bootstrap* sampling is performed over selected RDD, which results in a sample B_i. The bootstrapped data is used to grow the decision tree T_i while selecting the feature subset ϑ_k from F. Multiple decision trees are built in parallel for different RDDs [25]. There are two phases of implementing random forest in distributed environment: The first phase is to create the random forest training model and the second is to predict the class label using the trained model.

Training Phase: The training data is loaded into spark RDD and split into multiple partitions. Each partition holds a block of the data and builds a subset of the forest. Each block generates a number of decision trees. These decision trees collectively form the random forest.

Prediction Phase: After completing the training phase, classification phase begins. The testing data is read as RDD and the prediction is evaluated in each block. Prediction is estimated by taking a majority vote of the decision trees built in the previous phase. The final predicted class is the majority voted class by the decision trees. The entire process is illustrated in Fig. 3.

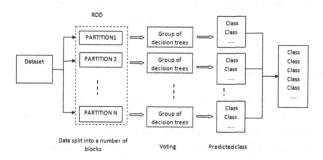

Fig. 3. A flowchart of how the classifying steps of the random forest are organized.

Algorithm 2. Training of Random Forest

Input: $Y =: (x_1, y_1), (x_2, y_2)....(x_n, y_n), F, N, Max\_Depth$
1: $RandomForest(Y, F, N, Max\_Depth)$:
2: $RF = \{\}$
3: for i in 1,2, ...,N do
4: $\quad B_i = $ A Boostrap sample from Y
5: $\quad \vartheta_k = $ subset of F
6: $\quad h(i) = InMemoryBuildTree(B_i, \vartheta_k, Max\_Depth)$
7: $\quad RF = RF$ union $h(i)$
8: end for
9: **return** RF

Algorithms 2 and 3 show the pseudo code of distributed random forest, which takes the following arguments as input: feature matrix Y, number of trees to be grown in the forest N, feature set F, depth of the tree Max_Depth. In the two pseudo codes: ϑ_k is the randomly selected feature subset from the feature set F, B_i is the bootstrap sample from Y for tree T_i. Let, BrL and BrR is the left and right branch of tree and $Node$ represents the nodes in the tree. $T_1, T_2, ..., T_N$ are the trees to be grown in the random forest.

In Algorithm 2, an empty set of random forest RF is first initialized, and gradually, it collects all the grown trees through the proposed method. $T_1, T_2, ..., T_N$ are built over different bootstrap samples containing subset of features and are grown on different nodes in the cluster.

For each tree creation, $InMemoryBuildTree$ is called on different bootstrap samples. The $InMemoryBuildTree$ algorithm starts with the root node and divides the data into the left subtree (BrL) and right subtree (BrR) recursively until the stopping condition is met; Max_Depth is used as the stopping criteria. Once, all the trees are grown, they are collected at master node. Master node uses maximum voting to predict the class of a test observation.

5   Experiment Results

To evaluate our model, we divided the total 30352 samples into 30 groups, each consisting of 287 samples from the first dataset, 131 samples from the second dataset and 593 samples from the third dataset. Thus, each group consists of 1011 samples.

5.1   Feature Analysis

For each group, we considered each of 9 essential features one by one and found out the average value of the features for each class of fault (non-faulty, inner race fault, outer race fault, and roller element fault). Figures 4 and 5 illustrate these analyses. Every point of discontinuity in the curve corresponds to the average value of the respective feature of the respective group. Essential features are extracted from the raw data (discussed in Sect. 4.1).

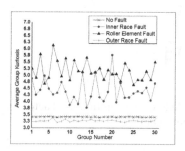

(a) Average Group Kurtosis vs Group Number

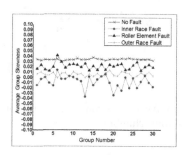

(b) Average Group Skewness vs Group Number

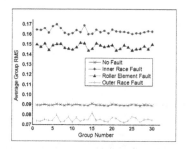

(c) Average Group RMS vs Group Number

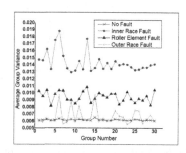

(d) Average Group Variance vs Group Number

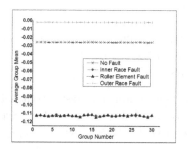

(e) Average Group Mean vs Group Number

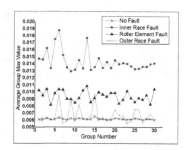

(f) Average Group Max Value vs Group Number

Fig. 4. Analysis of average value of features (kurtosis, skewness, RMS, variance, mean, and max value) corresponding to each group

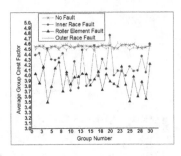
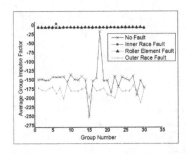

(a) Average Group Crest Factor vs Group Number

(b) Average Group Impulse Factor vs Group Number

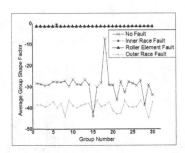

(c) Average Group Shape Factor vs Group Number

Fig. 5. Analysis of average value of crest factor, impulse factor, and shape factor, corresponding to each group

5.2 Feature Matrix Partitions

The feature matrices after the feature extraction are: The training matrix is of size 22764×10 and the testing matrix is of size 7588×10.

The datasets are tested in two different ways:

1. Training and testing on individual datasets: The training and testing datasets contain the mutually exclusive and exhaustive set of files for each dataset separately. Thus, we have six sets of files, two for each dataset (one testing and one training).
   The training of the classification model is done using Algorithm 2 with training dataset and testing of the model produces output in form of class labels.
2. Training and testing on combined dataset: The training and testing dataset contains the mutually exclusive and exhaustive set of files from all the datasets combined (75% from each dataset for training and 25% for testing). Thus, we have two sets of files (one for testing and one for training). Here also, the training is done through Algorithm 2 and the algorithm is tested on these sets of files and the output (accuracies) is recorded.

Algorithm 3. Build In-Memory Tree

Input: $Bi, \vartheta_k, Max\_Depth$
Output: Trained trees
1: $InMemoryBuildTree(Bi, \vartheta_k, Max\_Depth)$:
2: Create a root node $Node$
3: If all the samples are from the same class C, then terminate and label all samples as class C
4: else
5: Select ϑ^* from ϑ_k with highest information gain
6: $BrL, BrR = FindBestSplit(\vartheta^*)$
7: **if** $StoppingCriteria(BrL)$ do **then**
8:     $Left\_Prediction = FindPrediction(BrL)$
9: **else**
10:     $InMemoryBuildTree(Node- > Left, BrL)$
11: **end if**
12: **if** $StoppingCriteria(BrR)$ do **then**
13:     $Right\_Prediction = FindPrediction(BrR)$
14: **else**
15:     $InMemoryBuildTree(Node- > Right, BrR)$
16: **end if**
17: **return** tree

Table 2 shows the details of the data. Where, data is split into training and testing datasets: 75% for training and 25 % for testing. The training dataset is trained in parallel and each RDD generates a group of decision trees.

5.3   Correctness and Training Time Analysis

Classification of unlabelled test data is performed and the final result in form of class label is generated. The accuracy of the system is calculated and shown in Table 3 for different datasets. The proposed approach shows good accuracy for fault prediction considering each dataset separately. While a small deterioration of accuracy in case of combined dataset is perceived due to the increase in the number of outliers for aggregation of the datasets. Figure 6 shows the confusion matrix for the prediction done by the proposed model. The results show that about 99% time non faulty ball bearing is classified/labeled as non faulty, 99% time inner race fault is accurately classified, 99% time roller element fault is correctly classified, and 91% time outer race fault is correctly classified. We have also compared the DRF approach as shown in Table 4 with well known prediction algorithms such as bayesian classification, decision tree, and support vector machine. The DRF due to its distributed nature shows better training time performance because slave nodes parallel trains the forest of decision trees.

Table 2. Dataset details in terms of number of observations, and its division (75% in training and 25% in testing)

Dataset s. no	Training data size (75%)	Testing data size (25%)	Total size
1	1617	539	2156
2	738	246	984
3	3336	1112	4448
Total	5691	1897	7588

Table 3. Classification accuracy for different datasets considering combined dataset (all datasets at once), and each dataset separately

Dataset	Sample tested	Accuracy
Combined dataset	7588	86.12%
Dataset1	2156	99.25%
Dataset2	984	98.57%
Dataset3	4448	97.68%

Table 4. Comparison of DRF, bayesian classification, decision tree and SVM in context of accuracy and training time

Technique	Training time (Sec.)	Accuracy (%)
DRF	43	86.12
Bayesian Classification	206	83
Decision Tree	312	84
SVM	365	85

5.4 Scalability Analysis

Scalability analysis is done for performance evaluation through a scalability factor S_{Fact} using Eq. 5.

$$S_{Fact} = \frac{F_T}{C_T} \tag{5}$$

where, F_T is the time required in first case in which only two CPU cores are available. This time acts as a base for calculating the rate of gain of scalability while increasing the processing resources. C_T is the time required in all subsequent cases, when we continuously increase the number of available cores. An ideal scenario will be that the S_{Fact} should increase in the same ratio as the number of cores is increasing. The actual scenario is different from ideal as shown in the Fig. 7, it shows the analysis of the complete process with varying number of cores. Analysis shows that a good amount of scalability can be achieved using the proposed approach.

		Prediction outcome			
		Not Faulty	Inner Race	Roller Element	Outer Race
Actual value	Not Faulty	0.99418	0	0.00253	0.00331
	Inner Race	0.00742	0.99072	0.00185	0
	Roller Element	0.00743	0	0.99256	0
	Outer Race	0.08173	0	0	0.91826

Fig. 6. Confusion matrix for the generated fault predictions

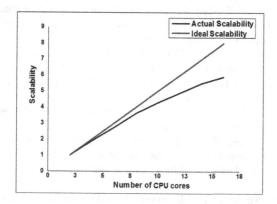

Fig. 7. Amount of scalability achieved using proposed distributed fault prediction mechanism

Overall analysis shows that applying the proposed large scale mining approach for finding the mechanical faults results in a good amount of scalability. The proposed model performs well and can be adopted for large collection of machinery.

6   Conclusion

Data generated by mechanical instruments/tools is scaling at a very high rate. Due to this scaling of data, more advanced featured techniques are desired to get relevant information from it. In this paper, we have applied a distributed mining technique to predict the faults in a ball bearing. A distributed fault prediction model is proposed which takes the advantages of large scale in-memory data mining technique. The model uses a distributed random forest algorithm for fault classification. Results and analyses show that our model performs sufficiently well and provides a good amount of accuracy as well as scalability. Thus, in conclusion, we can recommend that the proposed methodology is very effective and it can successfully diagnose the bearing condition. There are several future research opportunities in this context. One of the most important fundamental problem is to dissolve the primary problem into independent sub-problems. Applying more complex mining algorithms in a distributed manner can result in better performance and accuracy. While doing so, the problem of outliers exists as the signals are generated by different rotary components, which need to be dealt with.

References

1. Lacey, S.: The role of vibration monitoring in predictive maintenance. Asset Manag. Maintenance J. **24**(1), 42 (2011)
2. Widodo, A., Yang, B.S., Han, T.: Combination of independent component analysis and support vector machines for intelligent faults diagnosis of induction motors. Expert Syst. Appl. **32**(2), 299–312 (2007)
3. Breiman, L.: Random forests. Mach. Learn. **45**(1), 5–32 (2001)
4. Kuang, L., Hao, F., Yang, L.T., Lin, M., Luo, C., Min, G.: A tensor-based approach for big data representation and dimensionality reduction. IEEE Trans. Emerg. Top. Comput. **2**(3), 280–291 (2014)
5. Apache Hadoop website (2017). http://hadoop.apache.org/. Accessed 07 Dec 2017
6. Apache Spark website (2017). https://spark.apache.org/. Accessed 07 Dec 2017
7. Zaharia, M.: Resilient distributed datasets: a fault-tolerant abstraction for in-memory cluster computing. In: Proceedings of the 9th USENIX Conference on Networked Systems Design and Implementation, p. 2. USENIX Association, April 2012
8. Othman, M.S., Nuawi, M.Z., Mohamed, R.: Vibration and Acoustic Emission Signal Monitoring for Detection of Induction Motor Bearing Fault
9. Eren, L., Devaney, M.J.: Bearing damage detection via wavelet packet decomposition of the stator current. IEEE Trans. Instrum. Meas. **53**(2), 431–436 (2004)
10. Devaney, M.J., Eren, L.: Detecting motor bearing faults. IEEE Instrum. Meas. Mag. **7**(4), 30–50 (2004)
11. Mendel, E., et al.: Automatic bearing fault pattern recognition using vibration signal analysis. In: IEEE International Symposium on Industrial Electronics, ISIE 2008, pp. 955–960. IEEE, June 2008
12. Rashid, M.M., Amar, M., Gondal, I., Kamruzzaman, J.: A data mining approach for machine fault diagnosis based on associated frequency patterns. Appl. Intell. **45**(3), 638–651 (2016)
13. Thelaidjia, T., Moussaoui, A., Chenikher, S.: Feature extraction and optimized support vector machine for severity fault diagnosis in ball bearing. Eng. Solid Mech. **4**(4), 167–176 (2016)
14. Al-Raheem, K.F., Abdul-Karem, W.: Rolling bearing fault diagnostics using artificial neural networks based on Laplace wavelet analysis. Int. J. Eng. Sci. Technol. **2**(6) (2010)
15. Peng, H.W., Chiang, P.J.: Control of mechatronics systems: ball bearing fault diagnosis using machine learning techniques. In: 2011 8th Asian Control Conference (ASCC), pp. 175–180. IEEE, May 2011
16. Yang, B.S., Di, X., Han, T.: Random forests classifier for machine fault diagnosis. J. Mech. Sci. Technol. **22**(9), 1716–1725 (2008)

17. Abdulsalam, H., Skillicorn, D.B., Martin, P.: Classification using streaming random forests. IEEE Trans. Knowl. Data Eng. **23**(1), 22–36 (2011)
18. Bharathidason, S., Venkataeswaran, C.J.: Improving classification accuracy based on random forest model with uncorrelated high performing trees. Int. J. Comput. Appl **101**(13), 26–30 (2014)
19. Bernard, S., Adam, S., Heutte, L.: Dynamic random forests. Pattern Recogn. Lett. **33**(12), 1580–1586 (2012)
20. Murdopo, A.: Distributed decision tree learning for mining big data streams. Master of Science Thesis, European Master in Distributed Computing (2013)
21. Dai, W., Ji, W.: A mapreduce implementation of C4. 5 decision tree algorithm. Int. J. Database Theor. Appl. **7**(1), 49–60 (2014)
22. Wang, S., Zhai, J., Zhang, S., Zhu, H.: An ordinal random forest and its parallel implementation with mapreduce. In: 2015 IEEE International Conference on Systems, Man, and Cybernetics (SMC), pp. 2170–2173. IEEE, October 2015
23. Han, J., Liu, Y., Sun, X.: A scalable random forest algorithm based on mapreduce. In: 2013 4th IEEE International Conference on Software Engineering and Service Science (ICSESS), pp. 849–852. IEEE, May 2013
24. Bhagat, R.C., Patil, S.S.: Enhanced SMOTE algorithm for classification of imbalanced big-data using random forest. In: 2015 IEEE International Advance Computing Conference (IACC), pp. 403–408. IEEE, June 2015
25. Zhang, C., Yuan, D.: Fast fine-grained air quality index level prediction using random forest algorithm on cluster computing of spark. In: 2015 IEEE 12th International Conference on Ubiquitous Intelligence and Computing and 2015 IEEE 12th International Conference on Autonomic and Trusted Computing and 2015 IEEE 15th International Conference on Scalable Computing and Communications and Its Associated Workshops (UIC-ATC-ScalCom), pp. 929–934. IEEE, August 2015
26. Murphy, K.P.: Machine Learning: A Probabilistic Perspective. MIT press, Cambridge (2012)
27. Qiu, H., Lee, J., Lin, J., Yu, G.: Wavelet filter-based weak signature detection method and its application on rolling element bearing prognostics. J. Sound Vib. **289**(4), 1066–1090 (2006)

PRISMO: Priority Based Spam Detection Using Multi Optimization

Mohit Agrawal[1(✉)] and R. Leela Velusamy[2]

[1] Teradata India Private Limited, Hyderabad, India
mohit.agrawal@teradata.com
[2] National Institute of Technology, Tiruchirappalli, Tamil Nadu, India
leela@nitt.edu

Abstract. The rapid growth of social networking sites such as Twitter, Facebook, Google+, MySpace, Snapchat, Instagram, etc., along with its local invariants such as Weibo, Hyves, etc., has made them infiltrated with a large amount of spamming activities. Based on the features, an account or content can be classified as spam or benign. The presence of some irrelevant features decreases the performance of the classifier, understandability of dataset, and the time requirement for training and classification increases. Therefore, Feature subset selection is an essential phase in the process of machine learning mechanism. The objective of feature subset selection is to choose a subset of size 's' (s < n) from the total set of 'n' features that results in the least classification error. The feature subset selection problem can be represented as a problem of optimization in which the objective is to choose the near-optimal subset of features. Based on the literature survey, it is found that the classifier will offer its best performance if the data with high dimension is reduced such that it includes only appropriate features having lesser redundancy. The contribution of this paper comprises feature subset and its cost optimization simultaneously. The fundamental aspect PRISMO is to generate a primary feature subset through various optimization algorithms for the initialization stage. Further, the subset has been generated using the initial feature set based on their priority using basic rules of conjunction and disjunction. To evaluate the overall efficiency of PRISMO, various experiments were carried out using different dataset. The obtained result shows that the proposed model effectively reduces the cardinality of features without any bias to a specific dataset and any degradation to the classifier accurateness.

Keywords: Dimensionality reduction · Genetic Algorithm · Machine learning
Spamming · Particle Swarm Optimization · Web 2.0

1 Introduction

In a cybernetic era, the performance and security are the foremost concern for researcher all over. The dyadic tie among social actors (such as individual or an organization) can be termed as a social network [1]. The main objective of the social network is information sharing among different users. The escalating growth of social media such as Twitter, LinkedIn, Google+, Facebook, Instagram, Snapchat, MySpace, Weibo, Hyves, etc., has

© Springer Nature Switzerland AG 2018
A. Mondal et al. (Eds.): BDA 2018, LNCS 11297, pp. 392–401, 2018.
https://doi.org/10.1007/978-3-030-04780-1_27

been impelled with the user-friendly characteristic of Web 2.0 [2]. The ever-rising popularity of social networking sites made them a platform to perform unsolicited activities such as link farming, Sybil attack, Phishing, Spamming, etc.

Spamming is an e-crime that is performed with the use of the electronic messaging system to send unsolicited or irrelevant messages in form of advertisement, insults, hate speech, fake friend, dishonest reviews, etc. The reason for these activities may be personal or commercial. Spamming is economically feasible because of negligible operating cost. The spammer is the term used for individual who performs the spamming activity. Since a decade, classification of spam or spammers is carried through machine learning.

The main objective of machine learning is to achieve higher classification accuracy meanwhile reducing the overall time period required for classification. The data with high dimensionality degrades the performance and overall accuracy due to the presence of redundant features or noise in the dataset [3]. The identification of most optimal features from the given set of features is termed as Feature Subset Selection (FSS). In feature subset selection, the dataset 'D' having 'i' instances and 'n' dimensions (variable/sample) represented by x-$\{x_1, x_2...x_n\}$, and objective classification variables, C (Class variable), the mapping of feature from 'n' dimensions to 's' dimensions (output data) can be represented as:

$$F : R^{x*n} - > R^{x*s}$$

Where, R^{x*m} is an input dataset consisting having 'x' instances and 'n' features, and R^{x*s} is the resultant dataset with 'x' instances and 's' features. The major objective is to choose the subset of 's' features that optimally distinguishes the class variable.

Feature selection can further be classified into three categories, viz., wrapper, filter, and hybrid methodologies. Filter based approaches rank the features based on correlation, distance measures, etc. Filter mechanisms are computationally cheaper compared to wrapper method because the classifier is having no role for rank computation. Filter based methods avoid the correlation among features and depend on the characteristic of every single feature, meanwhile, wrapper-based methods select a subset of features based on accuracy estimation by classifiers. The wrapper methods provide better accuracy compared to filter-based approaches, but the computation cost is more in wrapper-based mechanisms. The hybrid model combines both individual and correlation measures among the measures with the use of classifier. In this work, the hybrid method has been considered for feature subset selection.

Exhaustive search can also be used for FSS, but a dataset with 'n' dimensions can have feature combination of 2^n, that makes it practically infeasible being a problem from NP-complete class [4]. Subset selection can be regarded as an optimization problem, where original feature set is further optimized to find a subset of relevant features eliminating the redundant features. In this paper, Genetic Algorithm (GA), Particle Swarm Optimization (PSO), and Simulated Annealing (SA) are combined to propose a hybrid model namely PRISMO for FSS. Further, classification is carried out using different classifiers namely Random Forest, Naïve Bayes, JRIP, and J48.

The rest of the paper is organized as follows: Sect. 2 gives an overview of existing mechanisms. The brief overview of the used mechanism is illustrated in Sect. 3. Section 4 elaborates the working of the proposed model. Section 5 illustrates the experimental results and observations with a comparison along with existing methodologies. Finally, Sect. 6 concludes the paper.

2 Related Work

Crawford et al. illustrates a comparative study among various spam detection approach using machine learning. The review centric features have been divided into distinct categories, such as combining a bag of words with term frequency features, syntactic and stylometric features, review characteristic feature, and linguistic inquiry and word count output (LIWC) [5].

The application differences among social media and e-mail spam have been highlighted by Castillo et al. [6]. The anti-spam method based on Identification, Rank, and Limit has been highlighted. Human, Bot, and Cyborg have been differentiated based on twitter behavior, content, and properties of an account [7]. Chu et al., achieved a true positive rate (TPR) of 96.0% with the use of Weka with 10-fold cross-validation.

Zhang et al. [8] proposed a method combining mutation operator along with Binary Particle Swarm Optimization (MBPSO) for feature selection followed by classification using decision tree with C4.5. 94.27%, 91.02%, and 97.51% are the achieved accuracy, sensitivity, and specificity respectively.

Huang et al. [9] proposed the mechanism that uses Genetic Algorithm as feature selection algorithm followed by Support Vector Machine (SVM). The optimization of parameters namely Penalty 'C' and Gamma ' ' along with feature selection has been done. Accuracy along with the weight of features has been considered as a fitness function to optimize the population.

Yardi et al. [10] observed that the amount the messages posted by spammers is high compared to legitimate users, and they are likely to follow a number of accounts compared to the number of followers they have resulting into the high follower-to-following ratio.

In [11], a hierarchical clustering algorithm has been used to divide spammers into advertising, self-promotion, abusive, and malicious. The accuracy of spammer detection is achieved to be 89.76% and the minimum accuracy of 78.88% has been achieved to identify spammers in each divided category.

Gupta et al. [12] observed that Bitly displaying a warning page on recognition malicious links is not an effective approach. Some short URL based feature has been identified and being coupled with the domain-specific features and achieved an accuracy of 86.41%.

Costa et al. [13], investigated on diverse types of tip spam. Different attributes namely Content attributes, User attributes, Place attributes, and Social attributes have been considered to identify a tip being irregular tip such as local marketing, pollution and bad- mounting. The classification even produced a remarkable result by considering only one feature, content attribute achieved the best accuracy of almost 73%.

Benvenuto et al. [14] used Content Attribute and User Behavior Attribute for spam detection and achieved a true positive rate (TPR) of 70.1% with basic classification. In [15], Lazy Associative Classifier (LAC) is further extended to let itself selecting a subset of cases to be labeled, performing active classification. The recall calculation for promoters, spammers, and legitimates is found to be 100%, 53.25%, and 99.22% respectively.

Lee et al. [16], proposed an optimal model for spam detection based on Random Forest (RF). Two parameters of RF had been optimized to maximize the detection rate. An optimal number of features has been selected using (i) one parameter optimization (ii) multiple parameter optimizations. The detection rates were found to be 95.1098% with 26 features and 95.0011% with 19 features using one parameter and multiple parameter optimizations respectively.

There are many naturally inspired metaheuristic algorithms such as [18–20], etc., that can be used as an FSS algorithm.

3 Methodology

The main underlying objective of PRISMO is to improve the overall accuracy of the classifier and cutting down the computation complexity. By using the method of feature selection for dimensionality reduction the better results can be obtained. Following techniques has been used in this paper; Genetic Algorithm, Simulated Annealing, and Particle Swarm Optimization for modeling a priority-based hybrid model for FSS.

3.1 Genetic Algorithm

GA is basically inspired by Darwin's theory of evolution [17]. It is one of the best alternatives to many conventional search optimization methodologies such as Depth First Search (DFS), linear programming, and so on. In GA, the set of intermediate solutions called candidate solutions are further optimized using the theory of Darwin with repetitive computations. The offspring with superior fitness will continue to exist for the upcoming generation. Based on mutation and crossover it generates the new population and the fitness of the candidate solution is calculated until the termination condition is met.

3.2 Particle Swarm Optimization

In PSO, the search space of a problem consists of many particles. The fitness value is evaluated by each particle for the current position. Each particle then moves through the search space with some random perturbations, which is updated using Eqs. 1 and 2. Eventually, the swarm altogether moves towards the optimum fitness value.

$$v = \omega * vel + c_1 * random() * (pBest - x) + c_2 * random() * (nBest - x) \quad (1)$$

$$x = x + vel \quad (2)$$

In Eq. 1, pBest and nBest is the best position a particle traversed and best position of neighbor so far respectively. The random() function returns a random value from

[0, 1], which gets updated every time. The impact of earlier velocity to the current one is controlled by inertia weight and represented by 'ω'. If $\omega > 1$, then particle favors past value and the current value for $\omega < 1$. Constant c_1 represents the cognitive component for measuring the degree of self-confidence. Similarly, c_2 represents social component to find capability of the particle to find a better solution [8].

3.3 Simulated Annealing

SA is a probabilistic technique for getting global optimum for a given problem. Problems, where determining a local optimum is more important as compared to global optimum in a fixed amount of time. SA may be desirable as brute-force search mechanism. For every step, X(n + 1) state of the current state X(n) is considered in SA. Movement of the system to X(n + 1) or staying in X(n) is determined probabilistically. Where X(n + 1) and X(n) represents new and old candidate solution respectively. Typically, this step is repeated until the system finds the optimal solution. The flow chart of SA.

4   Proposed Work

The proposed mechanism is basically divided into three phases, viz., Data Collection, Feature Subset Selection, and Classification.

4.1 Data Collection

In this paper, mainly two datasets are collected from (i) Twitter dataset with 54 million users, 1.9 billion links, and 1.8 billion tweets from the trending topic during 2009. The features related to content and social behavior is identified with the total of 62 features altogether [21]. (ii) Apontador, a Brazilian location-based social network containing 3538 tips with 1063 local marketing, 1716 pollution, and 759 bad-mounting tips. It contains 60 features containing 5 place-based, 12 social based, 11 users, based, and 32 content-based features [22]. The tip that contains local advertisement is termed as local marketing. Whereas, pollution contain tip based on irrelevant places. Finally, bad-mounting refers to aggressive comments related to places. The attributes considered for the dataset are a content attribute, user attribute, place attribute, and social attribute.

4.2 Feature Subset Selection

The FSS module consists of two layers namely feature Selection phase and feature formation phase. Feature Selection phase basically consists of multiple optimization algorithms. The data collected from data collection phase is input to Layer 1 of FSS. In Layer 1 (Feature Selection Phase), three optimization algorithms, viz., GA, PSO, and SA were considered for purpose of feature subset selection. The feature subset output of each individual algorithm is named as F1, F2, and F3 respectively. The feature subsets result in a distinct set of features with different cardinality. But, the minimum cardinality among F1, F2, and F3 is selected and named as n1. So that, there will be a maximum probability of selecting a cardinality of features using PRISMO that will be less than n1. The best n1 features are selected from all selected features from each feature selection algorithm. Above obtained feature sets were given input to Layer 2 (Feature Formation Phase), where a new set of the feature was created and named as F4 using rules of conjunction and disjunction. In this Layer, the most common features selected by all optimization techniques were considered in the resulting set.

Similarly, the set of features commonly selected by any two optimization techniques were considered as priority 2 features. The combination of priority 1 and priority 2 subsets dives the resultant subset with s features. Remaining features selected by only one optimization technique is considered at least key features and discarded from the resultant subset. Then, all feature subsets were given input to classification phase and were further tested with different classifiers.

4.3   Classification

The output of FSS phase is given as input to classification phase, which consists of multiple classifiers namely Random Forest, Naïve Bayes, JRIP, and J48. The output of individual algorithms of Layer 1 of the proposed model, viz., FS1, FS2, FS3 were also given as input to classification phase. Further, the feature subsets i.e. FS1, FS2, FS3, and FS4 were tested with different classifiers and a comparative study is carried out in the following Section (Fig. 1).

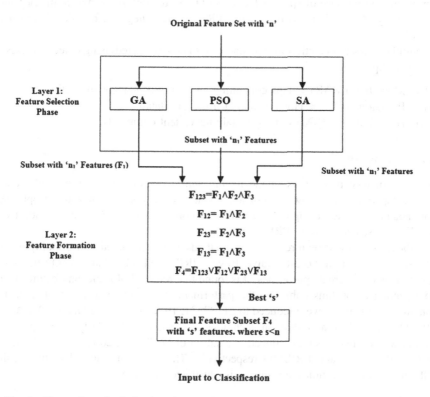

Fig. 1. Illustration of priority based spam detection using multi optimization algorithm.

5 Experiment and Result Analysis

The experimental analysis was carried out on two different datasets. First was global dataset from popular social networking site namely twitter with 62 features and second is a location-based social network of Brazil namely Apontador containing 60 features.

5.1 Experimental Setup

The implementation of proposed feature subset selection is done using MATLAB R2012a over Intel Core 2 Quad CPU running at 3.0 GHz. Both the dataset as described above has been used for feature subset selection phase. The obtained result has been verified using four different classifiers in WEKA (Waikato environment for knowledge analysis) tool namely Random Forest, Naïve Bayes, JRIP, and J48. 10-fold cross-validation has been used for training purpose. Basically, it divides the dataset into ten equal partitions. For every ten folds, nine partitions were used for training the model and one partition was used for validation.

5.2 Performance Metrics

For evaluation purpose of spam detection in twitter dataset, performance metrics such as Accuracy (ACC), True positive rate (TPR), and True negative rate (TNR) has been considered.

Similarly, spam detection in Apontador has been evaluated using detection rates of following tips:

  (i)   Marketing (MAR) = Tips containing an advertisement for marketing.
 (ii)   Pollution (POL) = Tips containing irrelevant details.
(iii)   Bad-Mounting (BM) = Tips containing violent comments.

5.3 Comparison

The performance parameters of the feature subset obtained using proposed method were compared with following four approaches i.e., Classification without optimization, Feature selection using GA, Feature selection using PSO, Feature selection using SA, Feature selection using PRISMO.

Above mentioned feature selection methods were compared using multiple classifiers, viz., Random Forest, Naïve Bayes, JRIP, and J48. Table 1 illustrates the comparison of different approaches. It is found that removal of irrelevant features using optimization algorithms enhances the performance of classifiers. The twitter dataset containing 62 features were reduced to 21, 23, 21, 19 features using GA, PSO, SA, and PRISMO as a feature reduction technique respectively. Similarly, the Apontador dataset with 60 features was reduced to a subset of 18, 24, 22, 20 features with the use of GA, PSO, SA, and PRISMO respectively. The detail of selected features using different feature selection mechanisms is illustrated in Table 1.

From Tables 2 and 3, it can be inferred that different feature subset selection mechanisms perform better when used with a random forest as a classifier. In Table 2, the Twitter dataset is analyzed with 60 unique features and it can be observed that feature reduction increases the performance along with reducing the dimensionality of the dataset. In the analysis, PRISMO is found to perform better compared to other techniques with the reduction of dimensionality to 19 features and accuracy of 88.26%. PRISMO achieves a TPR and TNR of 72.4% and 96.2% respectively which outperforms the existing approach with TPR and TNR of 70% and 96% respectively [14].

In Table 3, it can be observed that GA reduces the feature set to 18 features compared to PRISMO with 20 features. But the detection rate of marketing tip, pollution tip, and the bad-mounting tip is found to perform better compared to GA and other feature reduction techniques. The detection rate of PRISMO is 82.5%, 77.3%, and 65.2% for detecting tip spam such as marketing, pollution, and bad-mounting respectively compared to the detection rate of 76.4%, 68.8%, and 56.6% by existing method [13].

Table 1. Selected feature subset for optimization algorithms, viz., GA, PSO, SA, PRISMO.

Dataset	Total features	Feature selection algorithms			
		GA	PSO	SA	PRISMO
Twitter	62	2, 4, 6, 13, 14, 15, 16, 18, 23, 25, 26, 27, 33, 35, 41, 47, 50, 56, 60, 61, 62	58, 41, 9, 2, 7, 5, 36, 6, 45, 31, 33, 43, 62, 8, 23, 35, 21, 54, 4, 30, 60, 27, 61	41, 31, 21, 39, 58, 1, 60, 14, 15, 35, 28, 16, 2, 12, 33, 5, 8, 3, 26, 36, 57	2, 33, 35, 41, 60, 4, 6, 14, 15, 16, 23, 26, 62, 58, 5, 36, 31, 8, 21
Apontador	60	2, 5, 8, 10, 11, 16, 26, 27, 29, 33, 37, 41, 44, 48, 51, 57, 59, 60	29, 60, 47, 37, 53, 38, 46, 21, 59, 42, 17, 12, 58, 22, 9, 15, 11, 1, 25, 41, 19, 14, 8, 16	10, 33, 2, 16, 5, 37, 53, 42, 38, 46, 21, 47, 17, 59, 11, 26, 27, 8, 60, 6, 56, 25	2, 5, 8, 10, 11, 16, 26, 27, 29, 33, 60, 47, 37, 53, 38, 46, 21, 59, 42

Table 2. ACC, TPR, and TNR of different feature selection approach for the Twitter dataset.

FSS	Classifier												No. of features selected
	Random forest			Naive Bayes			JRIP			J48			
	ACC	TPR	TNR	ACC	TPR	TR	ACC	TPR	TNR	ACC	TPR	TNR	
NIL	88.16	71.8	96.3	85.91	70.7	93.5	87.6	71.3	95.8	83.5	70.7	90	62
GA	88.73	72.4	96.9	86.19	72.4	93.1	87.32	70.1	95.9	87.35	71	92.5	21
PSO	88.54	73.2	96.2	85.35	74.9	90.6	87.6	69.3	96.8	87.23	70.7	95.5	23
SA	87.98	70.4	96.8	83.09	56.3	96.5	86.47	65.4	97	83.94	68.2	91.8	21
PRISMO	88.26	72.4	96.2	85.25	71.8	92	87.04	69.9	95.6	86.66	71	94.5	19

Table 3. ACC, TPR, and TNR of different feature selection approaches for Apontador dataset.

FSS approaches	Classifiers												No. of features selected
	Random forest			Naïve Bayes			JRIP			J48			
	MAR	POL	BM	MAR	POL	BM	MAR	POL	BM	MAR	POL	BM	
Nil	**76.4**	**68.8**	**56.6**	44.6	20.5	90.4	73.3	6.0	86.7	81.7	71.4	66.3	60
GA	**80**	**78**	**63.5**	52.1	15.8	91.6	75.4	83.5	52.9	78.8	71.1	57.3	18
PSO	**76.3**	**74.7**	**64.6**	36.6	22.3	82.6	69.6	77.7	48.9	72.5	78	56.7	24
SA	**82.1**	**75.5**	**60.1**	56.7	12.8	89.3	72.9	86.8	48.3	76.7	72.5	70.2	22
PRISMO	**82.5**	**77.3**	**65.2**	47.5	12.5	90.4	71.7	83.9	46.6	77.5	75.8	57.3	20

Table 4. Time for building model using different FSS mechanisms along with random forest.

Dataset	Time using feature selection mechanism (ms)				
	Nil	GA	PSO	SA	PRISMO
Twitter	1240	640	1003	620	620
Apontador	2230	700	830	780	720

Table 4 highlights the time (ms) for building the model in WEKA. In previous results, it was clear that random forest is performing better compared to another classification mechanism. So, Random forest is considered for the analysis of time constraint. It shows that due to optimal feature selection the time taken to build a model is optimized by using different feature subset selection mechanism.

6 Conclusion and Future Work

In this paper, a priority-based approach for feature selection, PRISMO has been proposed for spam detection. The first layer of PRISMO uses multiple optimization algorithms, viz., GA, PSO, and SA, for feature reduction. The first layer of PRISMO selected the feature subset that produced a good distinction between spam and benign data. The information of the first layer is used by the second layer to determine the features with more priority using the rules of conjunction and disjunction. PRISMO was compared with some existing approaches for spam detection.

For evaluating the performance of feature selection algorithms, four classifiers were used in WEKA. The experiment was performed on Twitter and Apontador dataset. The experimental analysis shows that PRISMO performs better than existing solutions both in terms of accuracy and f-measure. Also, the resultant set of optimization algorithms were compared with PRISMO and it is found that PRISMO performs better. Further, the ROC curve illustrates that PRISMO has a higher true positive rate and low false positive rate compared to existing mechanisms. The disadvantage of PRISMO is over-dependence on constraints, viz., population, mutation rate, number of particles, the rate of crossover, etc. Therefore, in upcoming work minimizing the dominance of these parameters along with reducing the computational cost will be paid more attention.

References

1. Wasserman, S., et al.: Social Network Analysis, Methods and Applications, pp. 505–555. Cambridge University Press, Cambridge (1994)
2. Murugesan, S.: Understanding Web 2.0. IT Prof. **9**(4), 34–41 (2007)
3. Bruzzone, L., et al.: A novel approach to the selection of spatially invariant features for the classification of hyperspectral images with improved generalization capability. IEEE Trans. Geosci. Remote Sens. **47**(9), 3180–3191 (2009)
4. Davies, S., et al.: NP-completeness of searches for smallest possible feature sets. In: Association for the Advancement of Artificial Inteligence (AAAI) fall Symposium on Relevance, pp. 37–39 (1994)
5. Crawford, M., et al.: Survey of review spam detection using machine learning techniques. J. Big Data **2**(1), 1–24 (2015)
6. Castillo, C., et al.: Know your neighbours: web spam detection using the web topology. In: ACM SIGIR, pp. 423–430 (2007)
7. Chu, Z., et al.: Detecting automation of twitter accounts: are you a human, bot, or cyborg? IEEE Trans. Dependable Secure Comput. **9**(6), 811–824 (2012)
8. Zhang, Y., et al.: Binary PSO with mutation operator for feature selection using decision tree applied to spam detection. Knowl.-Based Syst. **64**, 22–31 (2014)
9. Huang, C.L., Wang, C.J.: A GA-based feature selection and parameters optimization for support vector machines. Expert Syst. Appl. **31**(2), 231–240 (2006)
10. Yardi, S., et al.: Detecting Spam in a Twitter Network. First Monday **15**(1) (2010)
11. Aggarwal, A., et al.: Detection of spam tipping behavior on foursquare. In: Proceedings of the 22nd International Conference on World Wide Web Companion. International World Wide Web Conferences Steering Committee (2013)
12. Gupta, N., et al.: bit.ly/malicious: deep dive into short URL based e-crime detection. In: 2014 APWG Symposium on Electronic Crime Research (eCrime). IEEE (2014)
13. Costa, H., et al.: Pollution, bad-mouthing, and local marketing: the underground of location-based social networks. Inf. Sci. **279**, 123–137 (2014)
14. Benevenuto, F., et al.: Detecting spammers on Twitter. In: Collaboration, Electronic Messaging, Anti-Abuse and Spam Conference (CEAS), vol. 6 (2010)
15. Benevenuto, F., et al.: Practical detection of spammers and content promoters in online video sharing systems. IEEE Trans. Syst. Man Cybern. B Cybern. **42**(3), 688–701 (2012)
16. Lee, S.M., et al.: Spam detection using feature selection and parameters optimization. In: 2010 International Conference on Complex, Intelligent and Software Intensive Systems (CISIS). IEEE (2010)
17. Goldberg, D.E., et al.: Genetic algorithms and machine learning. Mach. Learn. **3**(2), 95–99 (1988)
18. Zhang, Y., et al.: Multivariate approach for alzheimer's disease detection using stationary wavelet entropy and predator-prey particle swarm optimization. J. Alzheimers. Dis., 1–15 (2017)
19. Rajamohana, S.P., et al.: Hybrid optimization algorithm of improved binary particle swarm optimization (iBPSO) and cuckoo search for review spam detection. In: Proceedings of the 9th International Conference on Machine Learning and Computing, pp. 238–242. ACM (2017)
20. Sohrabi, et al.: A feature selection approach to detect spam in the Facebook social network. Arab. J. Sci. Eng. **43**(2), 949–958 (2018)
21. https://homepages.dcc.ufmg.br/ ∼ fabricio/spammerscollection.html
22. https://homepages.dcc.ufmg.br/ ∼ fabricio/spamcollectionApontadorElsevier.html

Malware Detection Using Machine Learning and Deep Learning

Hemant Rathore$^{(\boxtimes)}$, Swati Agarwal, Sanjay K. Sahay, and Mohit Sewak

Department of CS and IS, BITS Pilani, Goa Campus, Goa, India
{hemantr,swatia,ssahay,p20150023}@goa.bits-pilani.ac.in

Abstract. Research shows that over the last decade, malware have been growing exponentially, causing substantial financial losses to various organizations. Different anti-malware companies have been proposing solutions to defend attacks from these malware. The velocity, volume, and the complexity of malware are posing new challenges to the anti-malware community. Current state-of-the-art research shows that recently, researchers and anti-virus organizations started applying machine learning and deep learning methods for malware analysis and detection. We have used opcode frequency as a feature vector and applied unsupervised learning in addition to supervised learning for malware classification. The focus of this tutorial is to present our work on detecting malware with (1) various machine learning algorithms and (2) deep learning models. Our results show that the Random Forest outperforms Deep Neural Network with opcode frequency as a feature. Also in feature reduction, Deep Auto-Encoders are overkill for the dataset, and elementary function like Variance Threshold perform better than others. In addition to the proposed methodologies, we will also discuss the additional issues and the unique challenges in the domain, open research problems, limitations, and future directions.

Keywords: Auto-encoders · Cyber security · Deep learning
Machine learning · Malware detection

1 Introduction

In the digital age, malware have impacted a large number of computing devices. The term malware come from **mal**icious soft**ware** which are designed to meet the harmful intent of a malicious attacker. Malware can compromise computers/smart devices, steal confidential information, penetrate networks, and cripple critical infrastructures, etc. These programs include viruses, worms, trojans, spyware, bots, rootkits, ransomware, etc. According to Computer Economics[1], financial loss due to malware attack has grown quadruple from \$3.3 billion in 1997 to \$13.3 billion in 2006. Every few years the definition of *Year of Mega Breach* has to be recalibrated based on attacks performed in that particular

---

[1] https://www.computereconomics.com/article.cfm?id=1225.

© Springer Nature Switzerland AG 2018
A. Mondal et al. (Eds.): BDA 2018, LNCS 11297, pp. 402–411, 2018.
https://doi.org/10.1007/978-3-030-04780-1_28

year. Recently in 2016, *WannaCry ransomware attack*[2] crippled the computers of more than 150 countries, doing financial damage to different organizations. In 2016, Cybersecurity Ventures[3] estimated the total damage due to malware attacks was $3 trillion in 2015 and is expected to reach $6 trillion by 2021.

Antivirus software (such as Norton, McAfee, Avast, Kaspersky, AVG, Bitdefender, etc.) is a major line of defense for malware attacks. Traditionally, an antivirus software used the signature-based method for malware detection. Signature is a short sequence of bytes which can be used to identify known malware. But the signature-based detection system cannot provide security against zero-day attacks. Also, malware generation toolkits like Zeus [1] can generate thousands of variant of the same malware by using different obfuscation techniques. Signature generation is often a human-driven process which is bound to become infeasible with the current malware growth.

In the past few years, researchers and anti-malware communities have reported using machine learning and deep learning based methods for designing malware analysis and detection system. We surveyed these systems and divided the existing literature into two lines of research. (1) **feature extraction and feature reduction**: In malware analysis, features can be generated in two different ways: static analysis and dynamic analysis. In static analysis, features are extracted without executing the code whereas in dynamic analysis features are derived while running the executable. Ye et al. [17] used Windows API calls obtained from the static analysis as they can reflect true intent or behavior of an attacker. Their experiments show that few API calls like *OpenProcess, CloseHandle, CopyFileA* etc. always co-occur in malicious executables. Raff et al. [20] concluded that byte level n-gram could gather a lot of information about maliciousness from the code section as compared to portable executable header or import sections in a binary file. Strings also contain crucial semantic details, and they can often reflect the attacker's real intent and goals. Studies show that in a particular malware family, sample executables often share a similar group of opcodes [16]. Also, few opcodes are more dominant in malicious files as compare to benign executables which can act as a distinguisher. During malware analysis often features vector become extensively large, and it can have a negative impact during modeling. Literature shows various feature selection methods like document frequency [8], information gain [7], max-relevance algorithm [18] have been used in various malware detection systems. Azar [19] performed feature reduction using auto-encoders (in turn reducing the memory requirement) and applied various classification algorithms to achieve higher accuracy. David et al. [2] used a deep stack of de-noising auto-encoders implemented as deep belief network to generate the reduced feature set. (2) **Building Classification Models**: After feature extraction each file can be represented as a feature vector which can be used by the classification algorithm to build a model for malware detection. Firdausi et al. [3] used naive bayes, J48, decision tree, k-

---

[2] https://www.cbsnews.com/news/wannacry-ransomware-attacks-wannacry-virus-losses.

[3] https://cybersecurityventures.com/hackerpocalypse-cybercrime-report-2016.

nearest neighbor, multi-level perceptron and support vector machine on features extracted (using dynamic analysis) and achieved the highest accuracy of 96.8% with J48. Moskovitch [8] generated feature vectors with the byte n-gram method and applied feature selection based on document frequency and gain ratio. They reported highest accuracy by selecting top 300 5-gram terms with decision tree and artificial neural network. In 2013, Santos et al. [11] generated a combined feature vector from the static analysis (sequence of opcode frequency) and dynamic features (system call, exception, etc.) from a sample of 1000 malicious and 1000 benign files. Hardy et al. [4] in 2016 used Windows API calls as features with stacked autoencoder for malware detection and achieved an accuracy of 96.85%.

2 Experimental Setup

We formulate the problem of malware analysis and detection as a binary classification problem where malware and benign are the two classes. Figure 1 shows the proposed approach is a multi-step process consisting of various phases performing several tasks: collection of the dataset, disassembling of executable files, feature extraction, dimension reduction, building classification models, and empirical analysis of the results based on different metrics. We discuss each of these phases in the following subsections.

2.1 Dataset

To conduct our experiments, we gathered malware and benign executables from different sources. We downloaded malware samples from an open source repository known as Malicia Project[4]. In Malicia Project, Nappa et al. [9] have collected 11,688 malware samples on Windows platform belonging to a total of 55 different malware families The data collection is performed over a span of 11 months (07/03/2012 to 25/03/2013) from more than 500 drive-by download servers also known as exploit servers. Typically these servers are deployed for a lifetime of 16 h while some servers even operated for months to spread the malware files. Many malware executables in the dataset will connect to the internet without user consent to perform some cybercrime operation. Most of the malicious executable will also repack themselves on an average of 5.4 times in a day to evade the antivirus signature-based detection system. Thus opcode frequency as a feature can be an excellent measure to detect these malware.

To collect benign executable samples for our dataset, we gathered default files installed in different Windows operating system. VirusTotal[5] is an antivirus aggregator that can be used to check whether an executable is malicious or benign. We declare a sample as non-malicious/benign when all the anti-virus from virustotal.com declares it as harmless. We combine the malware and benign executable files downloaded from different sources (Malicia and Windows) and use it as our experimental dataset. Thus the dataset contains 11,688 malware and 2,819 benign executable files.

---

[4] http://malicia-project.com/.
[5] https://www.virustotal.com/.

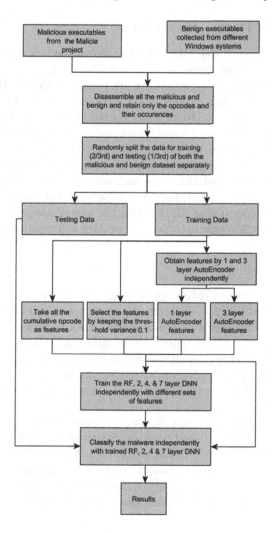

Fig. 1. Flowchart for the classification of malware with different sets of features. (Source: Sewak et al. [12])

2.2 Disassembling of Malicious and Benign Executables

As discussed in Sect. 2.1 our data set consist of 14,507 executable files. To generate the features, we disassemble them by converting an executable file (.exe) to assembly code (.asm). We used object dump utility which is a part of GNU Binutils package[6]. During disassembling few executable files were found to be corrupted or encrypted thus those files were removed from the dataset. Finally, we used 2,819 benign and 11,308 malware executables to generate the feature vector and to build the classification model.

---

[6] https://www.gnu.org/software/binutils/.

2.3 Creation of Feature Vector Space

In any machine learning algorithm, the feature vector is a critical component. We generate our feature vector by the static analysis of executable files. In static analysis, discriminatory attributes are collected without the execution of code. Literature shows that various static attribute such as Windows API calls [15,17], strings [15], opcode [10,14], and control flow graph [15], etc. are used to separate the malicious and benign executables. We used opcode frequency as a discriminatory feature. Firstly an exhaustive feature list called as *master opcode list* of 1,600 unique opcodes was created. We future generate a feature vector where rows represent the file name, and columns represent the opcode frequency. Each entry in the vector space represents the number of occurrence of a particular opcode in that file. Finally, the vector space of 2819 × 1600 for benign and 11308 × 1600 for malware executables was generated.

2.4 Other Issues

Since there is a significant difference between the number of malware (11,308) and benign executables (2,819) in our dataset, thus it will lead to class imbalance problem. Various methods are available to solve class imbalance problem like random sampling (oversampling/undersampling), cluster-based sampling, ADASYN [5], etc. We used Adaptive Synthetic sampling approach for imbalanced learning (ADASYN) which is an oversampling method for minority class tuples. It synthetically generates data points of minority class based on the k-nearest neighbor algorithm.

As discussed in Sect. 2.3, our dataset contains a large number of features and executable files thus we used cross-validation to generalize our model to an independent dataset. We used 3-fold cross validation in all our experiments. In rotation estimation (a.k.a. cross-validation) data is split into three equal parts where two blocks are used to training the model, and remaining one block is used for testing. The above exercise is done three times to accommodate all possible combinations.

3 Modelling Malware Detection

As discussed in Sect. 2, malware detection is a binary classification problem. After disassembling the executable samples (malware/benign), successfully generating the feature vectors and using ADASYN, the next steps are dimensionality reduction and then finally building the classification models.

3.1 Dimensionality Reduction

In statistics and machine learning, dimensionality reduction is a process of reducing the number of features under consideration. Our feature vector suffers from the curse of dimensionality since the total number of the unique opcode is 1,600.

When we further analyzed our feature set, we found that for few opcodes the corresponding frequency is zero since the particular opcodes are deprecated. Also for few opcodes, the count was relatively less because they are platform specific and the platform is deprecated. A model created on a dataset suffering from the curse of dimensionality will take a longer time to train and is inefficient in space complexity as well. To choose an optimal number of features we are using different variants of dimensionality reduction methods.

1. *None:* In this method all the opcodes are taken into account for building a classification model without using any feature reduction. We use this as a baseline for different feature reduction methods.
2. *Variance Threshold:* It is a method used to remove the features with low variations. We have removed the attributes with a variance of less than 0.1 assuming they have less prediction power.
3. *Auto-Encoders:* In deep learning auto-encoders are unsupervised learning methods which require only feature vector (opcode frequency), and not class labels for dimensionality reduction.
   (a) A single layer auto-encoder (Non Deep Auto-Encoder), also referred to as AE-1L which contain one encoder layer and a decoder layer.
   (b) A 3-layer stacked auto-encoder(Deep Auto-Encoder), also referred to as AE-3L which contain three encoders followed by three decoders.

For our experiments, all the auto-encoders use Exponential Linear Unit (ELU) function at all the layers except in the last layer which uses linear activation function. In AE-1L, the input directly connects to bottleneck layer which in turn link to the output layer. In both the auto-encoder (AE-1L and AE-3L) models, the bottleneck layer consists of 32 ELU nodes. Thus the architecture of AE-1L is (Input-32-Output) where bottleneck layer will behave as both encoder and decoder. In case of AE-3L where encoder consists of two additional hidden layers connected in sequential order containing 128 and 64 nodes respectively. Similarly, AE-3L decoder comprised of two hidden layers of similar width but connected in reverse order. Thus architecture of AE-3L will be (Input-128-64-32-64-128-Output). For training of both the auto-encoders (AE-1L and AE-3L), the mean square error is used as a loss function over a batch size of 64 samples. Instead of using standard stochastic gradient we have used Adam optimizer [6] to train a batch over 120 epochs. The Fig. 2 shows the training and validation loss for AE-1L during a complete cycle. The plot shows mean squared error loss (y-axis) for training and validation which are converging around 120 epoch (x-axis).

3.2   Building the Learning Model

In this paper, we used both machine learning and deep learning based approaches to build the classification models. Based on learning methods we divided our work into two case studies: (1) model based on the Random Forest (RF). In the previous studies [10, 14] conducted on the Malicia dataset [9], we found that

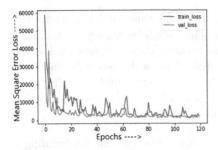
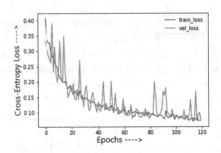

Fig. 2. Plot for AE-1L shows mean squared error loss (y-axis) for training and validation across 120 epochs (x-axis) (Source: Sewak et al. [13])

Fig. 3. Plot for DNN-2L shows cross entropy loss (y-axis) for training and validation across 120 epochs (x-axis) (Source: Sewak et al. [13])

tree-based classifier performs better as compared to other classifiers while among tree based classifier RF outperforms others. Thus we finally choose RF from the set of standard classifiers. (2) models based on deep learning.

1. Deep Neural Network using two hidden layers (DNN-2L)
2. Deep Neural Network using four hidden layers (DNN-4L)
3. Deep Neural Network using seven hidden layers (DNN-7L)

We designed multiple models of different depths to learn features at the different level of abstraction. In DNNs all the hidden layers contain ELU activation function except the last. Since malware detection is a binary classification problem, the last layer comprises of softmax activation (sigmoid) function. All the DNNs contain Adam optimizer [6] instead of gradient decent since in general, they have faster convergence rate. Also, we used cross entropy loss function and to avoid overfitting problems we used a dropout rate of 0.1. In DNN-2L, the two hidden layers contain 1024 and 32 nodes respectively. DNN-4L contain four layers with 2^{12-2i} nodes in each layer. Thus DNN-4L hidden layers contains (1024, 256, 64, 16) nodes. The DNN-7L has seven layers with 2^{11-i} nodes in i^{th} hidden layer. Thus DNN-7L hidden layer contain (1024, 512, 256, 128, 64, 32, 16) nodes. Figure 3 shows the training and validation loss for DNN-2L for a complete cycle of 120 epochs. In this figure, both training and validation loss are gradually decreasing as the model parameters are getting trained in each epoch and finally converged around 120 epoch. Also, something training loss is more than validation loss which is counterintuitive but is it because of the drop-out rate (0.1) during the training cycle.

4 Results

In this section, we will discuss the experimental results obtained after feature reduction (refer Sect. 2.3) with classification models (refer Sect. 3) using various evaluation metrics (accuracy, recall, selectivity, and precision).

Table 1. Results with Features Reduction, Classification Models, Accuracy, Recall/True Positive Rate (TPR), Selectivity/True Negative Rate (TNR), Precision/Positive Predictive Value (PPV) (Source: Sewak et al. [12])

Features	Classifiers	Accuracy	TPR	TNR	PPV
None	RF	99.74	99.48	100.0	100.0
VT	RF	99.78	99.59	99.97	99.97
AE-1L	RF	99.41	98.86	99.97	99.97
AE-3L	RF	99.36	98.72	100.0	100.0
None	DNN-2L	97.79	96.33	99.26	99.24
VT	DNN-2L	98.84	98.32	99.37	99.37
AE-1L	DNN-2L	96.95	94.57	99.37	99.34
AE-3L	DNN-2L	96.25	93.75	98.79	98.74
None	DNN-4L	97.42	95.38	99.48	99.46
VT	DNN-4L	98.69	97.96	99.42	99.42
AE-1L	DNN-4L	98.99	98.29	99.70	99.70
AE-3L	DNN-4L	97.16	98.61	95.68	95.85
None	DNN-7L	96.15	99.05	93.20	93.66
VT	DNN-7L	96.20	98.89	93.48	93.89
AE-1L	DNN-7L	98.99	98.61	99.81	99.81
AE-3L	DNN-7L	93.60	87.97	99.31	99.23

Table 1 reveals that for different feature reduction methods we found that VT (combined with RF) based attribute reduction achieved the highest accuracy of 99.78% which is marginally higher than no reduction (None and RF) 99.74% in the feature set. AE-1L performed better than deeper Auto-Encoder (AE-3L) and obtained the highest accuracy (99.41%) with RF. AE-3L based reduction performed lowest in all the methods. Highest True Positive Rate (TPR) of 99.59% was archived by VT (and RF) followed by None, and highest True Negative Rate (TNR) of 100% was achieved by no feature reduction (None and RF).

Table 1 shows that among different classification models, RF outperformed the deep learning models and achieved the highest accuracy of 99.7% (RF and VT). RF again produced the second highest accuracy with no feature reduction. Between different deep learning models, DNN-3L and DNN-7L both combined with AE-1L attained an accuracy of 98.99%. Highest TPR and TNR were produced by RF with VT and None as feature reduction respectively.

5 Conclusion

In the last few years malware have become a significant threat. Classical defense mechanism (like signature-based malware detection) used by anti-virus will fail to cope up new age malware challenges. In this paper, we have modeled malware analysis and detection as machine learning and deep learning problem. We have used best practices in building these models (like cross-validation, fixing class imbalance problem, etc.). We expertly handled the curse of dimensionality by using various feature reduction methods (None, AE-1L and AE-3L). Finally, we compared the models build using RF and DNN (DNN-2L, DNN-4L, and DNN-7L).

Based on our results random forest outperforms all the three deep neural network models in malware detection. We achieved the highest accuracy of 99.78% with random forest and variance threshold which is an improvement of 1.26% on previously reported the best accuracy. Also in feature reduction, variance threshold outplayed auto-encoders in improving the model performance. Another significant contribution of our investigation is a comparison of different combinations of auto-encoder (of depth 1 and 3) and deep neural network (of depth 2, 4 and 7) for malware detection. To our surprise, the best result did not come from any of the deep learning models which indicates that deep leaning may be overkill for Malicia dataset and the trained models are moving towards overfitting.

The same models can be used to detect more complex malware (polymorphic and metamorphic) in the future. Further, it will be interesting to see the effectiveness of other deep learning techniques like recurrent neural network, long short-term memory, etc. for malware detection.

References

1. What is zeus? (2011). https://www.sophos.com/en-us/medialibrary/pdfs/technical%20papers/sophos%20what%20is%20zeus%20tp.pdf
2. David, O.E., Netanyahu, N.S.: Deepsign: deep learning for automatic malware signature generation and classification. In: 2015 International Joint Conference on Neural Networks (IJCNN), pp. 1–8. IEEE (2015)
3. Firdausi, I., Erwin, A., Nugroho, A.S., et al.: Analysis of machine learning techniques used in behavior-based malware detection. In: 2010 Second International Conference on Advances in Computing, Control and Telecommunication Technologies (ACT), pp. 201–203. IEEE (2010)
4. Hardy, W., Chen, L., Hou, S., Ye, Y., Li, X.: Dl4md: a deep learning framework for intelligent malware detection. In: Proceedings of the International Conference on Data Mining (DMIN), p. 61. The Steering Committee of The World Congress in Computer Science, Computer Engineering and Applied Computing (2016)
5. He, H., Bai, Y., Garcia, E.A., Li, S.: Adasyn: adaptive synthetic sampling approach for imbalanced learning. In: IEEE International Joint Conference on Neural Networks IJCNN 2008. (IEEE World Congress on Computational Intelligence), pp. 1322–1328. IEEE (2008)
6. Kingma, D.P., Ba, J.: Adam: a method for stochastic optimization. arXiv preprint arXiv:1412.6980 (2014)
7. Masud, M.M., et al.: Cloud-based malware detection for evolving data streams. ACM Trans. Manage. Inf. Syst. (TMIS) **2**(3), 16 (2011)
8. Moskovitch, R., et al.: Unknown malcode detection using OPCODE representation. In: Ortiz-Arroyo, D., Larsen, H.L., Zeng, D.D., Hicks, D., Wagner, G. (eds.) EuroIsI 2008. LNCS, vol. 5376, pp. 204–215. Springer, Heidelberg (2008). https://doi.org/10.1007/978-3-540-89900-6_21

9. Nappa, A., Rafique, M.Z., Caballero, J.: Driving in the cloud: an analysis of drive-by download operations and abuse reporting. In: Rieck, K., Stewin, P., Seifert, J.-P. (eds.) DIMVA 2013. LNCS, vol. 7967, pp. 1–20. Springer, Heidelberg (2013). https://doi.org/10.1007/978-3-642-39235-1_1

10. Sahay, S.K., Sharma, A.: Grouping the executables to detect malwares with high accuracy. Procedia Comput. Sci. **78**, 667–674 (2016)

11. Santos, I., Brezo, F., Ugarte-Pedrero, X., Bringas, P.G.: Opcode sequences as representation of executables for data-mining-based unknown malware detection. IET Inf. Sci. **231**, 64–82 (2013)

12. Sewak, M., Sahay, S.K., Rathore, H.: Comparison of deep learning and the classical machine learning algorithm for the malware detection. In: 2018 19th IEEE/ACIS International Conference on Software Engineering, Artificial Intelligence, Networking and Parallel/Distributed Computing (SNPD), pp. 293–296. IEEE (2018)

13. Sewak, M., Sahay, S.K., Rathore, H.: An investigation of a deep learning based malware detection system. In: Proceedings of the 13th International Conference on Availability, Reliability and Security, p. 26. ACM (2018)

14. Sharma, A., Sahay, S.K.: An effective approach for classification of advanced malware with high accuracy. arXiv preprint arXiv:1606.06897 (2016)

15. Ye, Y., Li, T., Adjeroh, D., Iyengar, S.S.: A survey on malware detection using data mining techniques. ACM Comput. Surv. (CSUR) **50**(3), 41 (2017)

16. Ye, Y., Li, T., Chen, Y., Jiang, Q.: Automatic malware categorization using cluster ensemble. In: Proceedings of the 16th ACM SIGKDD International Conference on Knowledge Discovery and Data Mining, pp. 95–104. ACM (2010)

17. Ye, Y., Wang, D., Li, T., Ye, D.: IMDS: intelligent malware detection system. In: Proceedings of the 13th ACM SIGKDD International Conference on Knowledge Discovery and Data Mining, pp. 1043–1047. ACM (2007)

18. Ye, Y., Wang, D., Li, T., Ye, D., Jiang, Q.: An intelligent pe-malware detection system based on association mining. J. Comput. Virol. **4**(4), 323–334 (2008)

19. Yousefi-Azar, M., Varadharajan, V., Hamey, L., Tupakula, U.: Autoencoder-based feature learning for cyber security applications. In: 2017 International Joint Conference on Neural Networks (IJCNN), pp. 3854–3861. IEEE (2017)

20. Zak, R., Raff, E., Nicholas, C.: What can n-grams learn for malware detection? In: 2017 12th International Conference on Malicious and Unwanted Software (MALWARE), pp. 109–118. IEEE (2017)

Spatial Co-location Pattern Mining

Venkata M. V. Gunturi[✉]

IIT-Ropar, Rupnagar, India
`gunturi@iitrpr.ac.in`

Abstract. Given a spatial dataset containing instances of a set of spatial Boolean feature-types, the problem of spatial co-location pattern mining aims to determine a subset of feature-types which are frequently co-located in space. Spatial Co-location patterns have a wide range of applications in the domains such as ecology, public health and public safety. For instance, in an ecological dataset containing event instances corresponding to different bird species and vegetation types, spatial co-location patterns may revel that a particular species of birds prefer a particular kind of trees for their nests. Similarly, in a crime dataset, spatial co-location may revel a pattern that drunk-driving cases are co-located with bar locations. This article presents a gentle introduction to spatial co-location pattern mining. It introduces a well studied interest measure called *participation index* for co-location mining and, then discusses an algorithm to determine patterns having high participation index in a spatial dataset.

1 Introduction

Widespread use of spatial computing technologies [10,19–21,24] has lead to increasing interest in mining interesting and non-trivial patterns from spatial data. Over the years several works have made progress towards this end (e.g., [5,7,9,11–13,15,22,23]) by exploring different aspects of the problem of finding patterns from data which is embedded in geographic space.

One of influential results in the area of pattern mining for geographic space includes the work done in the area of *Spatial Co-Location* pattern mining. Given a spatial dataset containing feature-instances of spatial boolean feature-types, the problem of spatial co-location pattern mining aims to determine a subset of feature-types whose instances occur together frequently. Mining spatial co-location patterns has a wide range of applications in domains such as ecology, criminology and public health. Table 1 illustrates some sample patterns.

This article presents a gentle introduction to the problem of spatial co-location pattern mining. Section 2 introduces some basic concepts and presents the problem in a formal manner. In Sect. 3, we present a traditional algorithm for mining co-location patterns. Section 4 formally establishes the anti-monotonicity property of the participation index, a popular interest measure used in co-location pattern mining. Finally in Sect. 5, we summarize some of the recent results in the area and conclude in Sect. 6.

© Springer Nature Switzerland AG 2018
A. Mondal et al. (Eds.): BDA 2018, LNCS 11297, pp. 412–421, 2018.
https://doi.org/10.1007/978-3-030-04780-1_29

Table 1. Sample spatial co-location patterns in various spatial datasets.

Sno	Feature-type A	Feature-type B	Interpretation
1	Nile Crocodile location	Egyptian Plover bird location	Nile crocodile and the plover birds have a symbiotic relationship
2	Bar locations	Reports of Drunk-driving	This pattern may be observed near the bar closing times [15]. Perhaps the bar customers drive back after bar closing
3	Locations of patients with respiratory problems	High particulate matter locations	Perhaps the air pollution contributes to the respiratory problems

2 Basic Concepts and Problem Definition

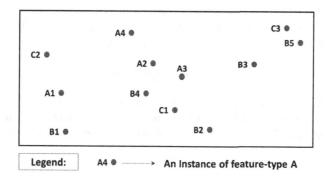

Fig. 1. A sample spatial dataset.

Definition 1. *Spatial Boolean Feature (f^i): comprises of those features which can be precisely defined in boolean form (i.e., present or absent) for any given location (x, y). Examples include, ecological features such as presence of bird nest, a particular type of tree, a particular animal species, etc. Spatial boolean features can also be used to define a particular type of spatial event. Examples of this category include, incidence locations of events such as forest fires, a particular disease, etc. In case one is interested in studying spatial aspect of diseases, the event location is the spatial location (home or office) of the patient. Note that in a dataset, each of the feature types f^i's would be associated a set of locations $\mathcal{L}_{f^i}$ where the particular feature type f^i occurs.*

Definition 2. *Spatial Co-Location Pattern (C_i): Given a set of Spatial Boolean feature-types $\mathcal{F} = \{f^1, f^2, f^3, \ldots, f^i\}$, a Spatial co-location C_i is defined as a non-empty subset of given feature-types $\mathcal{F}$, i.e., $C_i = \{f^1, f^2, \ldots, f^\gamma\}$ and $C_i \subseteq \mathcal{F}$.*

Definition 3. *Row instance of Spatial Co-location pattern C_i: Given a spatial[1] Co-location pattern $C_i = \{f^1, f^2, \ldots, f^\gamma\}$, a row instance of C_i is defined a particular collection of the instances (occurrences) of the feature types in C_i. In other words, a row instance of C_i is a collection of some specific features-instances of the feature-types in C_i. The key aspect to note is that, for any particular collection of feature-instances $\mathcal{G}(C_i) = \{G_{f^1}, G_{f^2}, \ldots, G_{f^\gamma}\}$ to become a valid row instance of C_i, each of the feature-instances in $\mathcal{G}(C_i)$ should be a neighbor of all the other feature-instances in $\mathcal{G}(C_i)$. In other words, $\forall$ p and q where $G_{f^p} \in \mathcal{G}(C_i)$ and $G_{f^q} \in \mathcal{G}(C_i)$, the feature-instance G_{f^p} and G_{f^q} are spatial neighbors (as per a given relation $\mathcal{R}$). Note that the notion of spatial neighbor is formally defined along with the input of the problem. Typically, they are defined using Euclidean, Geodetic or Network distance thresholds.*

Definition 4. *Table instance of Co-location pattern C_i: A collection of all the valid row instances of a Co-location pattern C_i are called as a table instance of C_i.*

Definition 5. *Participation ratio $Pr(f^\alpha, C_i)$ of a feature type f^α in a Co-location pattern C_i is the fraction of instances of the feature-type f^α which participate in any row instance of the spatial Co-location pattern C_i. More formally, $Pr(f^\alpha, C_i)$ is represented by the following equation:*

$$Pr(f^\alpha, C_i) = \frac{\#\ distinct\ instances\ of\ f^\alpha\ participating\ in\ C_i}{\#\ instances\ of\ f^\alpha}$$

Definition 6. *Participation index of a Co-location $C_i = \{f^1, f^2, \ldots, f^\gamma\}$ ($C_i \subseteq \mathcal{F}$) is defined as $\min_{f^j \in C_i}\{Pr(f^j, C_i)\}$. In other words, participation index of a Co-location pattern C_i is the minimum of all participation ratios' of its component feature-types.*

2.1   Problem Definition

The problem of spatial co-location pattern mining is defined as follows:
Input:

1. A set of N Spatial Boolean feature-type $\mathcal{F} = \{f^1, f^2, f^3, \ldots, f^N\}$.
2. A set of M feature-instances $P = \{p_1, p_2, \ldots, p_M\}$. Each $p_i \in P$ is a tuple <instance-id, spatial feature-type f^θ, location l>. Here, the feature-type $f^\theta \in \mathcal{F}$.
3. A neighbor relation $\mathcal{R}$. Given any two feature-instances p and q, the neighbor relation outputs "1" or "0" depending on whether p and q are deemed to be spatial neighbors or not. $\mathcal{R}$ is symmetric and reflexive.
4. Minimum prevalence threshold θ of the participation index.

Output:

1. Co-locations patterns whose participation index value is greater than θ.

---

[1] Whenever the context is clear, we drop the keyword "Spatial" from "Spatial Co-location" to maintain clarity of text.

3   Co-location Miner Algorithm

This section describes an elemental algorithm for determining spatial co-locations which internally uses spatial join operations. In its most basic form spatial co-location miner is quite similar to the traditional Apriori algorithm [2,3]. It generates candidate patterns in increasing size and at each stage, it uses patterns of size $k - 1$ to generate patterns of size k. Algorithm 1 illustrates a pseudo-code of the algorithm.

Algorithm 1. Spatial Co-Location Miner

1: $P \leftarrow$ Prevalent size-1 co-location set along with their table instances.
2: Generate size-2 co-location rules
3: **for** size k co-location pattern $kin\{3, 4, \dots, N\}$ **do**
4:    Generate candidate prevalent patterns of size k from the candidate patterns of size $k - 1$.
5:    Generate table instances and prune based on neighborhood.
6:    Prune based on participation index threshold.
7: **end for**
8: Output the maximal sized co-location patterns.

In the first step (refer Algorithm 1), the algorithm generates spatial co-location patterns of size-1. These would be trivially be all the feature-types (singleton) present in the input dataset. For instance, consider the sample spatial dataset illustrated in Fig. 1. For this input dataset, the size-1 co-location patterns would simply be all the individual feature-types, A, B and C. Figure 2 illustrates the table instances of these trivial co-location patterns.

Candidate Spatial Co-location pattern		
A	**B**	**C**
A1	B1	C1
A2	B2	C2
A3	B3	C3
A4	B4	
	B5	

(Instances)

Fig. 2. Table instances of size-1 co-locations patterns.

Following this, in the second step (refer Algorithm 1), the algorithm generates size-2 candidate patterns. Table instances of these size-2 candidate patterns are generated by computing the spatial inner join of all the instances of all spatial features. The output of the spatial inner join procedure would be the pairs of feature-instances which satisfy the neighbor relation $\mathcal{R}$ (given in the input). In other words, it returns all the pairs of feature-instances which are deemed to be "neighbors" according to the neighbor relation $\mathcal{R}$. For this step, one may use sweeping based spatial join algorithms such as the one proposed in [4]. Figure 3 illustrates a sample result of the spatial join algorithm on the dataset shown in Fig. 1. Figure 4 shows the result in a tabular form along with the table instances of all the size-2 patterns ($< A, B >$, $< B, C >$, $< A, C >$). At this stage, the algorithm computes the participation indices of all each of candidate size-2 patterns. Figure 4 reports these indices at the bottom. Following this, any size-2 pattern whose participation index is less than the given threshold θ is pruned out.

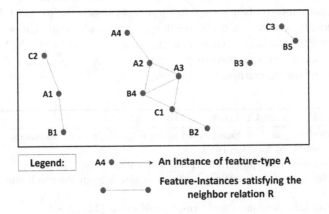

Fig. 3. Size-2 co-locations pattern instances.

Following the generation of size-2 candidate patterns, the algorithm enters a loop (line 3 Algorithm 1). In each iteration of the loop, at step-4, the algorithm generates candidate patterns of size k ($k = 3, 4, \ldots, N$) using the patterns of size $k - 1$. This procedure is explained next through our example size-2 patterns shown in Figs. 3 and 4.

Candidate Spatial Co-location pattern		
A-B	**B-C**	**A-C**
A1-B1	B4-C1	A1-C2
A2-B4	B2-C1	A3-C1
A3-B4	B5-C3	
0.4	**0.50**	**0.60**

Participation Index

Fig. 4. Table instances of size-2 co-locations patterns.

Consider our size-2 patterns illustrated in Fig. 4. As can be noted, we have three size-2 patterns, viz., $< A, B >$, $< B, C >$, $< A, C >$. The algorithm takes pairs of size-2 patterns which have a common feature-type to create size-3 patterns. In our example, it would take $< A, B >$ and $< B, C >$ (or $\{< A, B > < A, C >\}$ or $\{< B, C > < A, C >\}$). Note that in any particular iteration of the loop, step-4 would take two size $k - 1$ patterns p and q having $k - 2$ common features. Following this the algorithm creates a size k pattern comprising of the following: (a) $k - 2$ features common to both p and q, (b) $k - 1$th feature of p, (c) $k - 1$th feature of q. While creating a pattern of size k, it verifies whether all of its subsets of size $k - 1$ have participation index greater than θ (minimum prevalence threshold) or not. For instance, in our example, while creating the pattern $< A, B, C >$, it checks if all size-2 subsets of $< A, B, C >$, i.e., $< A, B >$, $< B, C >$, $< A, C >$, have participation index greater θ. In case any of the subsets does pass the threshold, then the pattern $< A, B, C >$ would have been pruned. In our example, θ is set to 0.20.

After generating size k patterns from size $k-1$ patterns, the algorithm computes (at step-5) the table instances of all size k patterns. In our example, we have only one size k pattern, viz., $< A, B, C >$. For creating the table instances of size k patterns, the algorithms *joins* the table instances of size $k-1$ patterns. For our pattern $< A, B, C >$, these would be the table instances of the $< A, B >$ and $< B, C >$ (any two table instances can be joined). The row instances of $< A, B >$ and $< B, C >$ are processed in the followed way. Assume a row instance $< A_x, B_y >$ of $< A, B >$ and a row instance $< B_y, C_z >$ of $< B, C >$. These row instances would be joined on the value on the common attribute B to form a row instance of $< A, B, C >$. However, before accepting $< A_x, B_y, C_z >$ as a valid row instance of $< A, B, C >$, the algorithm tests if A_x and C_z adhere to the definition of spatial neighbor $\mathcal{R}$. If not, then the row instance $< A_x, B_y, C_z >$ is discarded. Figure 5 illustrates the size-3 pattern and its table instance for our example. In our example, the row instances $< A3, B4 >$ and $< B4, C1 >$ joined to form the row instance $< A3, B4, C1 >$ of $< A.B, C >$. Note that the join of row instances $< A2, B4 >$ and $< B4, C1 >$ would not be accepted as the $A2$ and $C1$ are not spatial neighbors (refer Fig. 3).

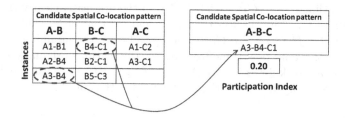

Fig. 5. Size-3 co-locations pattern instances.

Following the computation of the table instances of size k patterns, the algorithm determines its participation index. Any pattern with a participation index less that θ is pruned out. The loop on line 3 of Algorithm 1 iterates until the size of the pattern reaches the maximum possible of N (i.e., the number of feature-types in the dataset). In our example shown in Fig. 1, the maximum possible size of co-location pattern can be 3 $(< A, B, C)$.

4   Theoretical Analysis

Lemma 1. *Participation index of a Co-location pattern $C_i = \{f^1, f^2, \ldots, f^\gamma\}$ is anti-monotonic in number of features.*

Proof. Consider a spatial Co-location pattern $C_i = \{f^1, f^2, \ldots, f^\gamma\}$. Participation ratio $Pr(f^k, C_i)$ of a feature type f^k in the Co-location pattern C_i is given by the following equation:

$$Pr(f^k, C_i) = \frac{\#\ distinct\ instances\ of\ f^k\ participating\ in\ C_i}{\#\ instances\ of\ f^k}$$

Participation index of C_i is defined as $\min_{f^j \in C_i}\{Pr(f^j, C_i)\}$. In other words, participation index of the Co-location pattern C_i is the lowest participation ratio among all its component feature-types. For establishing anti-monotonicity, we need to prove that participation index of another pattern $C_j = \{f^1, f^2, \ldots, f^\gamma, f^\beta\}$ which contains one additional feature-type f^β (in addition to $f^1, f^2, \ldots, f^\gamma$) is less than (or equal to) the participation index of C_i.

Without loss of generality, consider any feature type f^α in C_i. Now, a new feature-type f^β is being included in C_i (to create C_j). Consequently, the participation ratio of f^α would either decrease further or remain the same (at best). This happens because numerator in the participation ratio is likely to decrease or (at best) remain the same. Basically, now for a feature-instance of f^α to be included in a row instance of C_j it would have to be a spatial neighbor of one additional feature-type f^β (in addition to features $f^1, f^2, \ldots, f^\gamma$). So the number of distinct feature-instances of f^α in the numerator can either remain same or decrease. In other words, $Pr(f^\alpha, C_j) \leq Pr(f^\alpha, C_i)$. Using a similar argument, one can also show $\forall f^j \in C_i \; Pr(f^j, C_j) \leq Pr(f^j, C_i)$.

Now, consider the newly added feature-type f^β. We have the following two possibilities: (i) $Pr(f^\beta, C_j) < Pr(f^j, C_j) \forall f^j \in C_i$ or, (ii) there exists some $f^j \in C_i$ such that $Pr(f^\beta, C_j) > Pr(f^j, C_j)$. In the first case, clearly the participation index of C_j would be less than (or equal to) that of C_i. This is because $Pr(f^j, C_j) \leq Pr(f^j, C_i) \; \forall f^j \in C_i$. In the second case, the minimum of the participation ratios would now come from one of the features originally present in C_i. However, as we know $Pr(f^j, C_j) \leq Pr(f^j, C_i) \; \forall f^j \in C_i$, the participation index of C_j would again be less than (or equal to) that of C_i.

5   Other Works in Spatial Co-Location Pattern Mining

Spatial Co-locations were originally proposed in [22, 25, 26]. Over the years, several researchers have extended the traditional definition of co-location patterns along multiple aspects. Following is brief summary of some of those works.

[8, 9, 14, 15] brought in the notion of time into co-location patterns. In these works, feature-instances have both spatial and temporal co-ordinates. [14, 15] defined the co-location patterns as partially ordered sets. More specifically, the event instances participating in the co-location pattern were now related using a spatio-temporal neighbor relation. On the other hand, [8, 9] explored co-location patterns in the domain of moving objects. Here, the input to the problem was a series of snapshots of locations of feature-types over a time window. Each snapshot was a collection of feature-instances at a particular time point. Given such a dataset, the goal was to determine co-location patterns which appear persistently over a set of time frames (but not necessarily in consecutive snapshots). In other words, for a set of feature-types to be considered as a valid pattern, they must form a valid co-location pattern (as per a certain threshold on participation index) at-least a certain number (given as threshold) of snapshots in a given time-window.

[5,6] explored the statistical significance aspect of co-location pattern discovery. The notion of statistical significance is of particular importance in co-location pattern mining. For instance consider the following scenario. We have two features α and β in the dataset which are independent of each other but abundant in quantity. In such a case, feature instances of α and β are likely to fall next to each other and the co-location pattern mining algorithm would return $\{\alpha, \beta\}$ as a valid pattern. Such spurious patterns can also arise if the feature-types α and β are auto-correlated (i.e., feature instances have a natural tendency to cluster). In such a case, if clusters of two independent features happen to overlap with each other, the participation index of the pattern $\{\alpha, \beta\}$ would increase dramatically. To this end, [5,6] proposed a novel Null model design where they take the underlying data-distribution characteristics of features to create a random dataset. Following this they compare (using p-values) the participation index of the pattern obtained in the real dataset against the random dataset to check if the obtained pattern was statistically significant. Basically, their approach creates several random datasets (having the same data distribution as the original real dataset), and counts the number random datasets which had a higher participation index for the pattern C_i generated by the algorithm on the real dataset. The higher the number, the greater is the chance that C_i was a random pattern.

[1] explored use of cross-k function [16] as an interest measure (instead of participation index) in co-location pattern mining. They extended the traditional co-location pattern mining by attempting to determine the *sub-region* of the given study region where the given two feature-types are associated to the maximum extent. The primary challenge addressed in this paper was the non-monotonic nature the cross-k function, whereby a larger area may have a smaller value of the interest measure but its subset (a smaller area inside the large area) may have a larger value of the interest measure.

[17,18] proposed algorithms which exploited high performance computing (GPUs) for determining spatial co-location patterns.

6 Conclusion

Spatial Co-location pattern mining delves into the problem of determining frequently co-occurring spatial feature-types in spatial dataset. These patterns are fundamentally different from the traditional association rules defined for transactions databases. Spatial co-locations have use-cases in several application domains such as ecology, public health and public safety. Several works explored this problem from multiple aspects. Participation index was one of the popular interest measures used in co-location pattern mining.

References

1. Agarwal, P., Verma, R., Gunturi, V.M.V.: Discovering spatial regions of high correlation. In: 2016 IEEE 16th International Conference on Data Mining Workshops (ICDMW), pp. 1082–1089 (2016)
2. Agrawal, R., Imieliński, T., Swami, A.: Mining association rules between sets of items in large databases. In: Proceedings of the 1993 ACM SIGMOD International Conference on Management of Data, SIGMOD 1993, pp. 207–216 (1993)
3. Agrawal, R., Srikant, R.: Fast algorithms for mining association rules in large databases. In: Proceedings of the 20th International Conference on Very Large Data Bases, VLDB 1994, pp. 487–499 (1994)
4. Arge, L., Procopiuc, O., Ramaswamy, S., Suel, T., Vitter, J.S.: Scalable sweeping-based spatial join. In: Proceedings of the 24th International Conference on Very Large Data Bases, VLDB 1998, pp. 570–581 (1998)
5. Barua, S., Sander, J.: Mining statistically significant co-location and segregation patterns. IEEE Trans. Knowl. Data Eng. 26(5), 1185–1199 (2014)
6. Barua, S., Sander, J.: SSCP: mining statistically significant co-location patterns. In: Pfoser, D., et al. (eds.) SSTD 2011. LNCS, vol. 6849, pp. 2–20. Springer, Heidelberg (2011). https://doi.org/10.1007/978-3-642-22922-0_2
7. Cao, H., Mamoulis, N., Cheung, D.W.: Mining frequent spatio-temporal sequential patterns. In: Fifth IEEE International Conference on Data Mining (ICDM 2005), pp. 82–89 (2005)
8. Celik, M., Kang, J.M., Shekhar, S.: Zonal co-location pattern discovery with dynamic parameters. In: Seventh IEEE International Conference on Data Mining (ICDM), pp. 433–438 (2007)
9. Celik, M., Shekhar, S., Rogers, J.P., Shine, J.A.: Mixed-drove spatiotemporal co-occurrence pattern mining. IEEE Trans. Knowl. Data Eng. 20(10), 1322–1335 (2008)
10. Güting, R.H.: An introduction to spatial database systems. VLDB J. 3(4), 357–399 (1994)
11. Koperski, K., Han, J.: Discovery of spatial association rules in geographic information databases. In: Egenhofer, M.J., Herring, J.R. (eds.) SSD 1995. LNCS, vol. 951, pp. 47–66. Springer, Heidelberg (1995). https://doi.org/10.1007/3-540-60159-7_4
12. Li, X., Han, J., Lee, J.-G., Gonzalez, H.: Traffic density-based discovery of hot routes in road networks. In: Papadias, D., Zhang, D., Kollios, G. (eds.) SSTD 2007. LNCS, vol. 4605, pp. 441–459. Springer, Heidelberg (2007). https://doi.org/10.1007/978-3-540-73540-3_25
13. Liu, Z., Huang, Y.: Mining co-locations under uncertainty. In: Nascimento, M.A., Sellis, T., Cheng, R., Sander, J., Zheng, Y., Kriegel, H.-P., Renz, M., Sengstock, C. (eds.) SSTD 2013. LNCS, vol. 8098, pp. 429–446. Springer, Heidelberg (2013). https://doi.org/10.1007/978-3-642-40235-7_25
14. Mohan, P., Shekhar, S., Shine, J.A., Rogers, J.P.: Cascading spatio-temporal pattern discovery: a summary of results. In: Proceedings of the SIAM International Conference on Data Mining (SDM), pp. 327–338 (2010)
15. Mohan, P., Shekhar, S., Shine, J.A., Rogers, J.P.: Cascading spatio-temporal pattern discovery. IEEE Trans. Knowl. Data Eng. 24(11), 1977–1992 (2012)
16. Ripley, B.D.: The second-order analysis of stationary point processes. J. Appl. Probab. 13(2), 255–266 (1976)

17. Sainju, A.M., Aghajarian, D., Jiang, Z., Prasad, S.K.: Parallel grid-based coloca-tion mining algorithms on GPUs for big spatial event data. IEEE Transactions on Big Data (2018). https://doi.org/10.1109/TBDATA.2018.2871062
18. Sainju, A.M., Jiang, Z.: Grid-based colocation mining algorithms on GPU for big spatial event data: a summary of results. In: Gertz, M., et al. (eds.) SSTD 2017. LNCS, vol. 10411, pp. 263–280. Springer, Cham (2017). https://doi.org/10.1007/978-3-319-64367-0_14
19. Shekhar, S., Chawla, S.: Spatial Databases: A Tour. Prentice Hall (2003). (ISBN 013-017480-7)
20. Shekhar, S., Chawla, S., Ravada, S., Fetterer, A., Liu, X., Lu, C.T.: Spatial databases - accomplishments and research needs. IEEE Trans. Knowl. Data Eng. **11**(1), 45–55 (1999)
21. Shekhar, S., Feiner, S.K., Aref, W.G.: Spatial computing. Commun. ACM **59**(1), 72–81 (2015)
22. Shekhar, S., Huang, Y.: Discovering spatial co-location patterns: a summary of results. In: Jensen, C.S., Schneider, M., Seeger, B., Tsotras, V.J. (eds.) SSTD 2001. LNCS, vol. 2121, pp. 236–256. Springer, Heidelberg (2001). https://doi.org/10.1007/3-540-47724-1_13
23. Wang, S., Huang, Y., Wang, X.S.: Regional co-locations of arbitrary shapes. In: Nascimento, M.A., et al. (eds.) SSTD 2013. LNCS, vol. 8098, pp. 19–37. Springer, Heidelberg (2013). https://doi.org/10.1007/978-3-642-40235-7_2
24. Worboys, M., Duckham, M.: GIS: A computing perspective. CRC (2004). ISBN: 0415283752
25. Yoo, J.S., Shekhar, S., Celik, M.: A join-less approach for co-location pattern min-ing: a summary of results. In: Fifth IEEE International Conference on Data Mining (ICDM) (2005)
26. Yoo, J.S., Shekhar, S., Smith, J., Kumquat, J.P.: A partial join approach for min-ing co-location patterns. In: Proceedings of the 12th Annual ACM International Workshop on Geographic Information Systems, GIS 2004, pp. 241–249 (2004)

Author Index

Agarwal, Sonali   89
Agarwal, Swati   402
Agrawal, Mohit   392
Agrawal, Shriyansh   155
Andrews, Shiju   228
Asthana, Ayushi   115

Bansal, Aayushi   204
Bansal, Suyash   204
Batra, Shivani   204
Bhalla, Subhash   217

Casturi, Rao   73
Chakravarthy, Sharma   33
Chaudhary, Sanjay   250
Chouhan, Satyendra Singh   363

Dewan, Prateek   170

Ganesh, Sakthi   307
Gujar, Sujit   55
Gunturi, Venkata M. V.   412
Gupta, Mohit   115
Gupta, Sonu   170

Himaja, D.   348
Hota, Chittaranjan   266

Ishikawa, Yoshiharu   3

Jain, Mohit   266
Johnson, Jimson   228
Joshi, Nishant   115

Komar, Kanthi Sannappa   33
Kothari, Ravi   335
Kumar, Manish   376
Kumaraguru, Ponnurangam   170

Ma, Qinying   295
Maruthi Padmaja, T.   348

Mehedy, Lenin   15
Mehndiratta, Pulkit   115
Mishra, Sanket   266
Mohan, S. Lalit   155
Mohania, Mukesh   15
Mondal, Anirban   126

Nath, Keshab   188
Niyogi, Rajdeep   363

Pal, Amrit   376
Pandey, Vaibhav   282
Paruchuri, Praveen   55
Patel, Ashish Kumar   363
Patidar, Rashmi Girirajkumar   217
Priya, Rashmi   322
Punn, Narinder Singh   89

Radha Krishna, P.   348
Ramesh, Dharavath   322
Rathore, Hemant   402
Reddy, Y. Raghu   155
Roy, Swarup   188

Sachdeva, Shelly   170, 204
Sahay, Sanjay K.   402
Saini, Poonam   282
Sakkeer, M. A.   228
Santra, Abhishek   33
Sewak, Mohit   402
Shah, Purnima   250
Sharma, Nonita   100
Shrestha, Shashank   217
Singh, Atul   126
Siva Naga Sasank, B.   266
Sridhar, K. T.   228
Srinivasa, Srinath   139
Srivastava, Siddharth   335
Subbanarasimha, Raksha Pavagada   139
Sugiura, Kento   3
Sunderraman, Rajshekhar   73

Takao, Daiki   3
Talukder, Asoke K.   307
Tripathi, Anurag   335

Velusamy, R. Leela   392
Verma, Dinesh   15
Vo, Hoang Tam   15

Wang, Jialin   295

Yadav, Sourabh   100

Zhang, Yanchun   295

Printed in the United States
By Bookmasters